Fundamentals of Digital
Image Processing

Fundamentals of Digital Image Processing

A Practical Approach
with Examples in Matlab

Chris Solomon

School of Physical Sciences, University of Kent, Canterbury, UK

Toby Breckon

School of Engineering, Cranfield University, Bedfordshire, UK

WILEY-BLACKWELL

A John Wiley & Sons, Ltd., Publication

This edition first published 2011, © 2011 by John Wiley & Sons, Ltd

Wiley-Blackwell is an imprint of John Wiley & Sons, formed by the merger of Wiley's global Scientific, Technical and Medical business with Blackwell Publishing.

Registered office: John Wiley & Sons Ltd, The Atrium, Southern Gate, Chichester, West Sussex, PO19 8SQ, UK

Editorial Offices:
9600 Garsington Road, Oxford, OX4 2DQ, UK

111 River Street, Hoboken, NJ 07030-5774, USA

For details of our global editorial offices, for customer services and for information about how to apply for permission to reuse the copyright material in this book please see our website at www.wiley.com/wiley-blackwell

Library of Congress Cataloguing-in-Publication Data

Solomon, Chris and Breckon, Toby
   Fundamentals of digital image processing : a practical approach with examples in Matlab / Chris Solomon and Toby Breckon
       p. cm.
   Includes index.
   Summary: "Fundamentals of Digital Image Processing is an introductory text on the science of image processing and employs the Matlab programming language to illustrate some of the elementary, key concepts in modern image processing and pattern recognition drawing on specific examples from within science, medicine and electronics"— Provided by publisher.
   ISBN 978-0-470-84472-4 (hardback) – ISBN 978-0-470-84473-1 (pbk.)
  1. Image processing–Digital techniques.   2. Matlab.   I. Breckon, Toby.   II. Title.
   TA1637.S65154 2010
   621.36'7—dc22

                                          2010025730

This book is published in the following electronic formats: eBook 9780470689783; Wiley Online Library 9780470689776

A catalogue record for this book is available from the British Library.

Set in 10/12.5 pt Minion by Thomson Digital, Noida, India

3   2013

Contents

Preface

Scope of this book

This is an introductory text on the science *(and art)* of image processing. The book also employs the Matlab programming language and toolboxes to illuminate and consolidate some of the elementary but key concepts in modern image processing and pattern recognition.

The authors are firm believers in the old adage, "***Hear*** *and forget. . .,* ***See*** *and remember. . .,* ***Do*** *and know*". For most of us, it is through good examples and gently guided experimentation that we really learn. Accordingly, the book has a large number of carefully chosen examples, graded exercises and computer experiments designed to help the reader get a real grasp of the material. All the program code (.m files) used in the book, corresponding to the examples and exercises, are made available to the reader/course instructor and may be downloaded from the book's dedicated web site – www.fundipbook.com.

Who is this book for?

For undergraduate and graduate students in the technical disciplines, for technical professionals seeking a direct introduction to the field of image processing and for instructors looking to provide a hands-on, structured course. This book intentionally starts with simple material but we also hope that relative experts will nonetheless find some interesting and useful material in the latter parts.

Aims

What then are the specific aims of this book ? Two of the principal aims are –

- To introduce the reader to some of the key concepts and techniques of modern image processing.

- To provide a framework within which these concepts and techniques can be understood by a series of examples, exercises and computer experiments.

These are, perhaps, aims which one might reasonably expect from *any* book on a technical subject. However, we have one further aim namely to provide the reader with the fastest, most direct route to acquiring a real hands-on understanding of image processing. We hope this book will give you a real fast-start in the field.

Assumptions

We make no assumptions about the reader's mathematical background beyond that expected at the undergraduate level in the technical sciences – ie reasonable competence in calculus, matrix algebra and basic statistics.

Why write this book?

There are already a number of excellent and comprehensive texts on image processing and pattern recognition and we refer the interested reader to a number in the appendices of this book. There are also some exhaustive and well-written books on the Matlab language. What the authors felt was lacking was *an image processing book which combines a simple exposition of principles with a means to quickly test, verify and experiment with them in an instructive and interactive way.*

In our experience, formed over a number of years, Matlab and the associated image processing toolbox are extremely well-suited to help achieve this aim. It is simple but powerful and its key feature in this context is that it enables one to *concentrate on the image processing concepts and techniques* (i.e. the real business at hand) while keeping concerns about programming syntax and data management to a minimum.

What is Matlab?

Matlab is a programming language with an associated set of specialist software toolboxes. It is an industry standard in scientific computing and used worldwide in the scientific, technical, industrial and educational sectors. Matlab is a commercial product and information on licences and their cost can be obtained direct by enquiry at the web-site www.mathworks.com. Many Universities all over the world provide site licenses for their students.

What knowledge of Matlab is required for this book?

Matlab is very much part of this book and we use it extensively to demonstrate how certain processing tasks and approaches can be quickly implemented and tried out in practice. Throughout the book, we offer comments on the Matlab language and the best way to achieve certain image processing tasks in that language. Thus the learning of concepts in image processing and their implementation within Matlab go hand-in-hand in this text.

Is the book any use then if I don't know Matlab?

Yes. This is fundamentally a book about image processing which aims to make the subject accessible and practical. It is not a book about the Matlab programming language. Although some prior knowledge of Matlab is an advantage and will make the practical implementation easier, we have endeavoured to maintain a self-contained discussion of the concepts which will stand up apart from the computer-based material.

If you have not encountered Matlab before and you wish to get the maximum from this book, please refer to the Matlab and Image Processing primer on the book website (http://www.fundipbook.com). This aims to give you the essentials on Matlab with a strong emphasis on the basic properties and manipulation of images.

Thus, you do not have to be knowledgeable in Matlab to profit from this book.

Practical issues

To carry out the vast majority of the examples and exercises in the book, the reader will need access to a current licence for **Matlab** and the **Image Processing Toolbox** only.

Features of this book and future support

This book is accompanied by a dedicated website (http://www.fundipbook.com). The site is intended to act as a point of contact with the authors, as a repository for the code examples (Matlab .m files) used in the book and to host additional supporting materials for the reader and instructor.

About the authors

Chris Solomon gained a B.Sc in theoretical physics from Durham University and a Ph.D in Medical imaging from the Royal Marsden Hospital, University of London. Since 1994, he has been on the Faculty at the School of Physical Sciences where he is currently a Reader in Forensic Imaging. He has broad research interests focussing on evolutionary and genetic algorithms, image processing and statistical learning methods with a special interest in the human face. Chris is also Technical Director of Visionmetric Ltd, a company he founded in 1999 and which is now the UK's leading provider of facial composite software and training in facial identification to police forces. He has received a number of UK and European awards for technology innovation and commercialisation of academic research.

Toby Breckon holds a Ph.D in Informatics and B.Sc in Artificial Intelligence and Computer Science from the University of Edinburgh. Since 2006 he has been a lecturer in image processing and computer vision in the School of Engineering at Cranfield University. His key research interests in this domain relate to 3D sensing, real-time vision, sensor fusion, visual surveillance and robotic deployment. He is additionally a visiting member of faculty at *Ecole Supérieure des Technologies Industrielles Avancées* (France) and has held visiting faculty positions in China and Japan. In 2008 he led the development of image-based automatic threat detection for the winning Stellar Team system in the UK MoD Grand Challenge. He is a Chartered Engineer (CEng) and an Accredited Imaging Scientist (AIS) as an Associate of the Royal Photographic Society (ARPS).

Thanks

The authors would like to thank the following people and organisations for their various support and assistance in the production of this book: the authors families and friends for their support and (frequent) understanding, Professor Chris Dainty (National University of Ireland), Dr. Stuart Gibson (University of Kent), Dr. Timothy Lukins (University of Edinburgh), The University of Kent, Cranfield University, VisionMetric Ltd and Wiley-Blackwell Publishers.

For further examples and exercises see http://www.fundipbook.com

Using the book website

There is an associated website which forms a vital supplement to this text. It is:

www.fundipbook.com

The material on the site is mostly organised by chapter number and this contains –

EXERCISES: intended to consolidate and highlight concepts discussed in the text. Some of these exercises are numerical/conceptual, others are based on Matlab.

SUPPLEMENTARY MATERIAL: Proofs, derivations and other supplementary material referred to in the text are available from this section and are intended to consolidate, highlight and extend concepts discussed in the text.

Matlab CODE: The Matlab code to all the examples in the book as well as the code used to create many of the figures are available in the Matlab code section.

IMAGE DATABASE: The Matlab software allows direct access and use to a number of images as an integral part of the software. Many of these are used in the examples presented in the text.

We also offer a modest repository of images captured and compiled by the authors which the reader may freely download and work with. Please note that some of the example Matlab code contained on the website and presented in the text makes use of these images. **You will therefore need to download these images to run some of the Matlab code shown.**

We strongly encourage you to make use of the website and the materials on it. It is a vital link to making your exploration of the subject both practical and more in-depth. Used properly, it will help you to get much more from this book.

1

Representation

In this chapter we discuss the representation of images, covering basic notation and information about images together with a discussion of standard image types and image formats. We end with a practical section, introducing Matlab's facilities for reading, writing, querying, converting and displaying images of different image types and formats.

1.1 What is an image?

A digital image can be considered as a discrete representation of data possessing both spatial (layout) and intensity (colour) information. As we shall see in Chapter 5, we can also consider treating an image as a multidimensional signal.

1.1.1 Image layout

The two-dimensional (2-D) discrete, digital image $I(m, n)$ represents the response of some sensor (or simply a value of some interest) at a series of fixed positions ($m = 1, 2, \ldots, M$; $n = 1, 2, \ldots, N$) in 2-D Cartesian coordinates and is derived from the 2-D continuous spatial signal $I(x, y)$ through a sampling process frequently referred to as discretization. Discretization occurs naturally with certain types of imaging sensor (such as CCD cameras) and basically effects a local averaging of the continuous signal over some small (typically square) region in the receiving domain.

The indices m and n respectively designate the rows and columns of the image. The individual picture elements or pixels of the image are thus referred to by their 2-D (m, n) index. Following the Matlab® convention, $I(m, n)$ denotes the response of the pixel located at the mth row and nth column starting from a top-left image origin (see Figure 1.1). In other imaging systems, a column–row convention may be used and the image origin in use may also vary.

Although the images we consider in this book will be discrete, it is often theoretically convenient to treat an image as a continuous spatial signal: $I(x, y)$. In particular, this sometimes allows us to make more natural use of the powerful techniques of integral and differential calculus to understand properties of images and to effectively manipulate and

Fundamentals of Digital Image Processing – A Practical Approach with Examples in Matlab
Chris Solomon and Toby Breckon
© 2011 John Wiley & Sons, Ltd

Figure 1.1   The 2-D Cartesian coordinate space of an M x N digital image

process them. Mathematical analysis of discrete images generally leads to a linear algebraic formulation which is better in some instances.

The individual pixel values in most images do actually correspond to some physical response in real 2-D space (e.g. the optical intensity received at the image plane of a camera or the ultrasound intensity at a transceiver). However, we are also free to consider images in abstract spaces where the coordinates correspond to something other than physical space and we may also extend the notion of an image to three or more dimensions. For example, medical imaging applications sometimes consider full three-dimensional (3-D) recon-struction of internal organs and a time sequence of such images (such as a beating heart) can be treated (if we wish) as a single four-dimensional (4-D) image in which three coordinates are spatial and the other corresponds to time. When we consider 3-D imaging we are often discussing spatial volumes represented by the image. In this instance, such 3-D pixels are denoted as voxels (volumetric pixels) representing the smallest spatial location in the 3-D volume as opposed to the conventional 2-D image.

Throughout this book we will usually consider 2-D digital images, but much of our discussion will be relevant to images in higher dimensions.

1.1.2   Image colour

An image contains one or more colour channels that define the intensity or colour at a particular pixel location $I(m, n)$.

In the simplest case, each pixel location only contains a single numerical value representing the signal level at that point in the image. The conversion from this set of numbers to an actual (displayed) image is achieved through a colour map. A colour map assigns a specific shade of colour to each numerical level in the image to give a visual representation of the data. The most common colour map is the greyscale, which assigns all shades of grey from black (zero) to white (maximum) according to the signal level. The

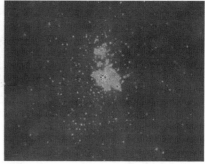

Figure 1.2 Example of grayscale (left) and false colour (right) image display *(See colour plate section for colour version)*

greyscale is particularly well suited to intensity images, namely images which express only the intensity of the signal as a single value at each point in the region.

In certain instances, it can be better to display intensity images using a false-colour map. One of the main motives behind the use of false-colour display rests on the fact that the human visual system is only sensitive to approximately 40 shades of grey in the range from black to white, whereas our sensitivity to colour is much finer. False colour can also serve to accentuate or delineate certain features or structures, making them easier to identify for the human observer. This approach is often taken in medical and astronomical images.

Figure 1.2 shows an astronomical intensity image displayed using both greyscale and a particular false-colour map. In this example the *jet* colour map (as defined in Matlab) has been used to highlight the structure and finer detail of the image to the human viewer using a linear colour scale ranging from dark blue (low intensity values) to dark red (high intensity values). The definition of colour maps, i.e. assigning colours to numerical values, can be done in any way which the user finds meaningful or useful. Although the mapping between the numerical intensity value and the colour or greyscale shade is typically linear, there are situations in which a nonlinear mapping between them is more appropriate. Such nonlinear mappings are discussed in Chapter 4.

In addition to greyscale images where we have a single numerical value at each pixel location, we also have true colour images where the full spectrum of colours can be represented as a triplet vector, typically the (R,G,B) components at each pixel location. Here, the colour is represented as a linear combination of the basis colours or values and the image may be considered as consisting of three 2-D planes. Other representations of colour are also possible and used quite widely, such as the (H,S,V) (hue, saturation and value (or intensity)). In this representation, the intensity V of the colour is decoupled from the chromatic information, which is contained within the H and S components (see Section 1.4.2).

1.2 Resolution and quantization

The size of the 2-D pixel grid together with the data size stored for each individual image pixel determines the spatial resolution and colour quantization of the image.

The representational power (or size) of an image is defined by its resolution. The resolution of an image source (e.g. a camera) can be specified in terms of three quantities:

- *Spatial resolution* The column (C) by row (R) dimensions of the image define the number of pixels used to cover the visual space captured by the image. This relates to the sampling of the image signal and is sometimes referred to as the pixel or digital resolution of the image. It is commonly quoted as $C \times R$ (e.g. 640×480, 800×600, 1024×768, etc.)

- *Temporal resolution* For a continuous capture system such as video, this is the number of images captured in a given time period. It is commonly quoted in frames per second (fps), where each individual image is referred to as a video frame (e.g. commonly broadcast TV operates at 25 fps; 25–30 fps is suitable for most visual surveillance; higher frame-rate cameras are available for specialist science/engineering capture).

- *Bit resolution* This defines the number of possible intensity/colour values that a pixel may have and relates to the *quantization* of the image information. For instance a binary image has just two colours (black or white), a grey-scale image commonly has 256 different grey levels ranging from black to white whilst for a colour image it depends on the colour range in use. The bit resolution is commonly quoted as the number of binary bits required for storage at a given quantization level, e.g. binary is 2 bit, grey-scale is 8 bit and colour (most commonly) is 24 bit. The range of values a pixel may take is often referred to as the *dynamic range* of an image.

It is important to recognize that the bit resolution of an image does not necessarily correspond to the resolution of the originating imaging system. A common feature of many cameras is automatic gain, in which the minimum and maximum responses over the image field are sensed and this range is automatically divided into a convenient number of bits (i.e. digitized into N levels). In such a case, the bit resolution of the image is typically less than that which is, in principle, achievable by the device.

By contrast, the blind, unadjusted conversion of an analog signal into a given number of bits, for instance $2^{16} = 65\ 536$ discrete levels, does not, of course, imply that the true resolution of the imaging device as a whole is actually 16 bits. This is because the overall level of *noise* (i.e. random fluctuation) in the sensor and in the subsequent processing chain may be of a magnitude which easily exceeds a single digital level. The sensitivity of an imaging system is thus fundamentally determined by the noise, and this makes noise a key factor in determining the number of quantization levels used for digitization. There is no point in digitizing an image to a high number of bits if the level of noise present in the image sensor does not warrant it.

1.2.1   Bit-plane splicing

The visual significance of individual pixel bits in an image can be assessed in a subjective but useful manner by the technique of bit-plane splicing.

To illustrate the concept, imagine an 8-bit image which allows integer values from 0 to 255. This can be conceptually divided into eight separate image planes, each corresponding

BIT PLANE IMAGES   **SETTING BIT PLANES TO 0**

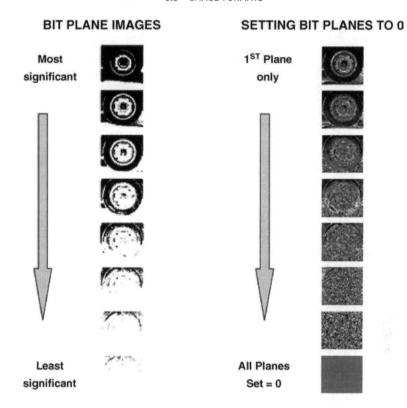

Figure 1.3   An example of bit-plane slicing a grey-scale image

to the values of a given bit across all of the image pixels. The first bit plane comprises the first and most significant bit of information (intensity $= 128$), the second, the second most significant bit (intensity $= 64$) and so on. Displaying each of the bit planes in succession, we may discern whether there is any visible structure in them.

In Figure 1.3, we show the bit planes of an 8-bit grey-scale image of a car tyre descending from the most significant bit to the least significant bit. It is apparent that the two or three least significant bits do not encode much useful visual information (it is, in fact, mostly noise). The sequence of images on the right in Figure 1.3 shows the effect on the original image of successively setting the bit planes to zero (from the first and most significant to the least significant). In a similar fashion, we see that these last bits do not appear to encode any visible structure. In this specific case, therefore, we may expect that retaining only the five most significant bits will produce an image which is practically visually identical to the original. Such analysis could lead us to a more efficient method of encoding the image using fewer bits – a method of *image compression*. We will discuss this next as part of our examination of image storage formats.

1.3   Image formats

From a mathematical viewpoint, any meaningful 2-D array of numbers can be considered as an image. In the real world, we need to effectively display images, store them (preferably

Table 1.1 Common image formats and their associated properties

Acronym	Name	Properties
GIF	Graphics interchange format	Limited to only 256 colours (8-bit); lossless compression
JPEG	Joint Photographic Experts Group	In most common use today; lossy compression; lossless variants exist
BMP	Bit map picture	Basic image format; limited (generally) lossless compression; lossy variants exist
PNG	Portable network graphics	New lossless compression format; designed to replace GIF
TIF/TIFF	Tagged image (file) format	Highly flexible, detailed and adaptable format; compressed/uncompressed variants exist

compactly), transmit them over networks and recognize bodies of numerical data as corresponding to images. This has led to the development of standard digital image formats. In simple terms, the image formats comprise a file header (containing information on how exactly the image data is stored) and the actual numeric pixel values themselves. There are a large number of recognized image formats now existing, dating back over more than 30 years of digital image storage. Some of the most common 2-D image formats are listed in Table 1.1. The concepts of lossy and lossless compression are detailed in Section 1.3.2.

As suggested by the properties listed in Table 1.1, different image formats are generally suitable for different applications. GIF images are a very basic image storage format limited to only 256 grey levels or colours, with the latter defined via a colour map in the file header as discussed previously. By contrast, the commonplace JPEG format is capable of storing up to a 24-bit RGB colour image, and up to 36 bits for medical/scientific imaging applications, and is most widely used for consumer-level imaging such as digital cameras. Other common formats encountered include the basic bitmap format (BMP), originating in the development of the Microsoft Windows operating system, and the new PNG format, designed as a more powerful replacement for GIF. TIFF, tagged image file format, represents an overarching and adaptable file format capable of storing a wide range of different image data forms. In general, photographic-type images are better suited towards JPEG or TIF storage, whilst images of limited colour/detail (e.g. logos, line drawings, text) are best suited to GIF or PNG (as per TIFF), as a lossless, full-colour format, is adaptable to the majority of image storage requirements.

1.3.1 Image data types

The choice of image format used can be largely determined by not just the image contents, but also the actual image data type that is required for storage. In addition to the bit resolution of a given image discussed earlier, a number of distinct image types also exist:

- *Binary images* are 2-D arrays that assign one numerical value from the set $\{0, 1\}$ to each pixel in the image. These are sometimes referred to as logical images: black corresponds to zero (an 'off' or 'background' pixel) and white corresponds to one (an 'on' or 'foreground' pixel). As no other values are permissible, these images can be represented as a simple bit-stream, but in practice they are represented as 8-bit integer images in the common image formats. A fax (or facsimile) image is an example of a binary image.

- *Intensity or grey-scale images* are 2-D arrays that assign one numerical value to each pixel which is representative of the intensity at this point. As discussed previously, the pixel value range is bounded by the bit resolution of the image and such images are stored as N-bit integer images with a given format.

- *RGB or true-colour* images are 3-D arrays that assign three numerical values to each pixel, each value corresponding to the red, green and blue (RGB) image channel component respectively. Conceptually, we may consider them as three distinct, 2-D planes so that they are of dimension C by R by 3, where R is the number of image rows and C the number of image columns. Commonly, such images are stored as sequential integers in successive channel order (e.g. $R_0G_0B_0$, $R_1G_1B_1$, ...) which are then accessed (as in Matlab) by $I(C, R, \text{channel})$ coordinates within the 3-D array. Other colour representations which we will discuss later are similarly stored using the 3-D array concept, which can also be extended (starting numerically from 1 with Matlab arrays) to four or more dimensions to accommodate additional image information, such as an alpha (transparency) channel (as in the case of PNG format images).

- *Floating-point* images differ from the other image types we have discussed. By definition, they do not store integer colour values. Instead, they store a floating-point number which, within a given range defined by the floating-point precision of the image bit-resolution, represents the intensity. They may (commonly) represent a measurement value other than simple intensity or colour as part of a scientific or medical image. Floating point images are commonly stored in the TIFF image format or a more specialized, domain-specific format (e.g. medical DICOM). Although the use of floating-point images is increasing through the use of high dynamic range and stereo photography, file formats supporting their storage currently remain limited.

Figure 1.4 shows an example of the different image data types we discuss with an example of a suitable image format used for storage. Although the majority of images we will encounter in this text will be of integer data types, Matlab, as a general matrix-based data analysis tool, can of course be used to process floating-point image data.

1.3.2   Image compression

The other main consideration in choosing an image storage format is compression. Whilst compressing an image can mean it takes up less disk storage and can be transferred over a network in less time, several compression techniques in use exploit what is known as *lossy compression*. Lossy compression operates by removing redundant information from the image. As the example of bit-plane slicing in Section 1.2.1 (Figure 1.3) shows, it is possible to remove some information from an image without any apparent change in its visual

Image: 24-bit RGB colour
Pixel Data Type: 3 x integer (0→255)
Image Format: JPEG

Image: 8-bit grayscale
Pixel Data Type: integer (0→255)
Image Format: GIF

Image: binary
Pixel Data Type: integer (0 or 1)
Image Format: PNG

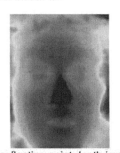

Image: floating point depth image Pixel
Data Type: floating point values Image
Format: TIFF (Copyright: Tim Lukins, UoE)

Figure 1.4   Examples of different image types and their associated storage formats

appearance. Essentially, if such information is visually redundant then its transmission is unnecessary for appreciation of the image. The form of the information that can be removed is essentially twofold. It may be in terms of fine image detail (as in the bit-slicing example) or it may be through a reduction in the number of colours/grey levels in a way that is not detectable by the human eye.

Some of the image formats we have presented, store the data in such a compressed form (Table 1.1). Storage of an image in one of the compressed formats employs various algorithmic procedures to reduce the raw image data to an equivalent image which appears identical (or at least nearly) but requires less storage. It is important to distinguish between compression which allows the original image to be reconstructed perfectly from the reduced data without any loss of image information (*lossless compression*) and so-called lossy compression techniques which reduce the storage volume (sometimes dramatically) at the expense of some loss of detail in the original image as shown in Figure 1.5, the lossless and lossy compression techniques used in common image formats can significantly reduce the amount of image information that needs to be stored, but in the case of lossy compression this can lead to a significant reduction in image quality.

Lossy compression is also commonly used in video storage due to the even larger volume of source data associated with a large sequence of image frames. This loss of information, itself a form of noise introduced into the image as *compression artefacts*, can limit the effectiveness of later image enhancement and analysis.

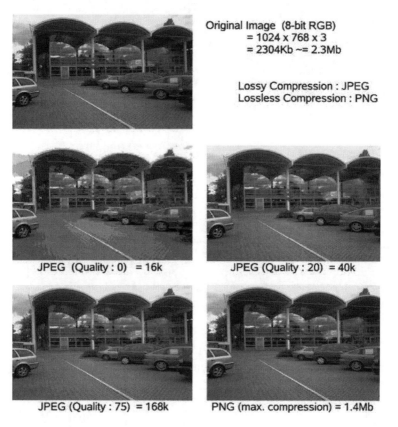

Original Image (8-bit RGB)
= 1024 x 768 x 3
= 2304Kb ~= 2.3Mb

Lossy Compression : JPEG
Lossless Compression : PNG

JPEG (Quality : 0) = 16k   JPEG (Quality : 20) = 40k

JPEG (Quality : 75) = 168k   PNG (max. compression) = 1.4Mb

Figure 1.5   Example image compressed using lossless and varying levels of lossy compression *(See colour plate section for colour version)*

In terms of practical image processing in Matlab, it should be noted that an image written to file from Matlab in a lossy compression format (e.g. JPEG) will not be stored as the exact Matlab image representation it started as. Image pixel values will be altered in the image output process as a result of the lossy compression. This is not the case if a lossless compression technique is employed.

An interesting Matlab exercise is posed for the reader in Exercise 1.4 to illustrate this difference between storage in JPEG and PNG file formats.

1.4 Colour spaces

As was briefly mentioned in our earlier discussion of image types, the representation of colours in an image is achieved using a combination of one or more colour channels that are combined to form the colour used in the image. The representation we use to store the colours, specifying the number and nature of the colour channels, is generally known as the colour space.

Considered as a mathematical entity, an image is really only a spatially organized set of numbers with each pixel location addressed as $I(C, R)$. Grey-scale (intensity) or binary images are 2-D arrays that assign one numerical value to each pixel which is representative of

Figure 1.6 Colour RGB image separated into its red (R), green (G) and blue (B) colour channels *(See colour plate section for colour version)*

the intensity at that point. They use a single-channel colour space that is either limited to a 2-bit (binary) or intensity (grey-scale) colour space. By contrast, RGB or true-colour images are 3-D arrays that assign three numerical values to each pixel, each value corresponding to the red, green and blue component respectively.

1.4.1 RGB

RGB (or true colour) images are 3-D arrays that we may consider conceptually as three distinct 2-D planes, one corresponding to each of the three red (R), green (G) and blue (B) colour channels. RGB is the most common colour space used for digital image representation as it conveniently corresponds to the three primary colours which are mixed for display on a monitor or similar device.

We can easily separate and view the red, green and blue components of a true-colour image, as shown in Figure 1.6. It is important to note that the colours typically present in a real image are nearly always a blend of colour components from all three channels. A common misconception is that, for example, items that are perceived as blue will only appear in the blue channel and so forth. Whilst items perceived as blue will certainly appear brightest in the blue channel (i.e. they will contain more blue light than the other colours) they will also have milder components of red and green.

If we consider all the colours that can be represented within the RGB representation, then we appreciate that the RGB colour space is essentially a 3-D colour space (cube) with axes R, G and B (Figure 1.7). Each axis has the same range $0 \rightarrow 1$ (this is scaled to 0–255 for the common 1 byte per colour channel, 24-bit image representation). The colour black occupies the origin of the cube (position $(0, 0, 0)$), corresponding to the absence of all three colours; white occupies the opposite corner (position $(1, 1, 1)$), indicating the maximum amount of all three colours. All other colours in the spectrum lie within this cube.

The RGB colour space is based upon the portion of the electromagnetic spectrum visible to humans (i.e. the continuous range of wavelengths in the approximate range

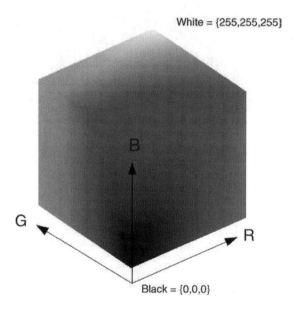

Figure 1.7   An illustration of RGB colour space as a 3-D cube *(See colour plate section for colour version)*

400–700 nm). The human eye has three different types of colour receptor over which it has limited (and nonuniform) absorbency for each of the red, green and blue wavelengths. This is why, as we will see later, the colour to grey-scale transform uses a nonlinear combination of the RGB channels.

In digital image processing we use a simplified RGB colour model (based on the CIE colour standard of 1931) that is optimized and standardized towards graphical displays. However, the primary problem with RGB is that it is perceptually nonlinear. By this we mean that moving in a given direction in the RGB colour cube (Figure 1.7) does not necessarily produce a colour that is perceptually consistent with the change in each of the channels. For example, starting at white and subtracting the blue component produces yellow; similarly, starting at red and adding the blue component produces pink. For this reason, RGB space is inherently difficult for humans to work with and reason about because it is not related to the natural way we perceive colours. As an alternative we may use perceptual colour representations such as HSV.

1.4.1.1   RGB to grey-scale image conversion

We can convert from an RGB colour space to a grey-scale image using a simple transform. Grey-scale conversion is the initial step in many image analysis algorithms, as it essentially simplifies (i.e. reduces) the amount of information in the image. Although a grey-scale image contains less information than a colour image, the majority of important, feature-related information is maintained, such as edges, regions, blobs, junctions and so on. Feature detection and processing algorithms then typically operate on the converted grey-scale version of the image. As we can see from Figure 1.8, it is still possible to distinguish between the red and green apples in grey-scale.

An RGB colour image, I_{colour}, is converted to grey scale, $I_{\text{grey-scale}}$, using the following transformation:

$$I_{\text{grey-scale}}(n, m) = \alpha I_{\text{colour}}(n, m, r) + \beta I_{\text{colour}}(n, m, g) + \gamma I_{\text{colour}}(n, m, b) \qquad (1.1)$$

Figure 1.8 An example of RGB colour image (left) to grey-scale image (right) conversion *(See colour plate section for colour version)*

where (n, m) indexes an individual pixel within the grey-scale image and (n, m, c) the individual channel at pixel location (n, m) in the colour image for channel c in the red r, blue b and green g image channels. As is apparent from Equation (1.1), the grey-scale image is essentially a weighted sum of the red, green and blue colour channels. The weighting coefficients $(\alpha, \beta$ and $\gamma)$ are set in proportion to the perceptual response of the human eye to each of the red, green and blue colour channels and a standardized weighting ensures uniformity (NTSC television standard, $\alpha = 0.2989, \beta = 0.5870$ and $\gamma = 0.1140$). The human eye is naturally more sensitive to red and green light; hence, these colours are given higher weightings to ensure that the relative intensity balance in the resulting grey-scale image is similar to that of the RGB colour image. An example of performing a grey-scale conversion in Matlab is given in Example 1.6.

RGB to grey-scale conversion is a noninvertible image transform: the true colour information that is lost in the conversion cannot be readily recovered.

1.4.2 Perceptual colour space

Perceptual colour space is an alternative way of representing true colour images in a manner that is more natural to the human perception and understanding of colour than the RGB representation. Many alternative colour representations exist, but here we concentrate on the Hue, Saturation and Value (HSV) colour space popular in image analysis applications.

Changes within this colour space follow a perceptually acceptable colour gradient. From an image analysis perspective, it allows the separation of colour from lighting to a greater degree. An RGB image can be transformed into an HSV colour space representation as shown in Figure 1.9.

Each of these three parameters can be interpreted as follows:

- H (hue) is the dominant wavelength of the colour, e.g. red, blue, green

- S (saturation) is the 'purity' of colour (in the sense of the amount of white light mixed with it)

- V (value) is the brightness of the colour (also known as luminance).

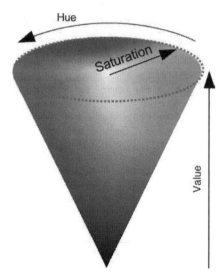

Figure 1.9 HSV colour space as a 3-D cone *(See colour plate section for colour version)*

The HSV representation of a 2-D image is also as a 3-D array comprising three channels (h, s, v) and each pixel location within the image, $I(n, m)$, contains an (h, s, v) triplet that can be transformed back into RGB for true-colour display. In the Matlab HSV implementation each of h, s and v are bounded within the range $0 \rightarrow 1$. For example, a blue hue (top of cone, Figure 1.9) may have a value of $h = 0.9$, a saturation of $s = 0.5$ and a value $v = 1$ making it a vibrant, bright sky-blue.

By examining the individual colour channels of images in the HSV space, we can see that image objects are more consistently contained in the resulting hue field than in the channels of the RGB representation, despite the presence of varying lighting conditions over the scene (Figure 1.10). As a result, HSV space is commonly used for colour-based

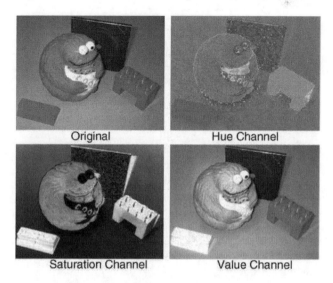

Figure 1.10 Image transformed and displayed in HSV colour space *(See colour plate section for colour version)*

image segmentation using a technique known as colour slicing. A portion of the hue colour wheel (a slice of the cone, Figure 1.9) is isolated as the colour range of interest, allowing objects within that colour range to be identified within the image. This ease of colour selection in HSV colour space also results in its widespread use as the preferred method of colour selection in computer graphical interfaces and as a method of adding false colour to images (Section 1.1.2).

Details of RGB to HSV image conversion in Matlab are given in Exercise 1.6.

1.5   Images in Matlab

Having introduced the basics of image representation, we now turn to the practical aspect of this book to investigate the initial stages of image manipulation using Matlab. These are presented as a number of worked examples and further exercises for the reader.

1.5.1   Reading, writing and querying images

Reading and writing images is accomplished very simply via the **imread** and **imwrite** functions. These functions support all of the most common image formats and create/export the appropriate 2-D/3-D image arrays within the Matlab environment. The function **imfinfo** can be used to query an image and establish all its important properties, including its type, format, size and bit depth.

Example 1.1

Matlab code	What is happening?
imfinfo('cameraman.tif')	%Query the cameraman image that %is available with Matlab %imfinfo provides information %ColorType is gray-scale, width is 256 . . . etc.
I1=imread('cameraman.tif');	%Read in the TIF format cameraman image
imwrite(I1,'cameraman.jpg','jpg');	%Write the resulting array I1 to %disk as a JPEG image
imfinfo('cameraman.jpg')	%Query the resulting disk image %Note changes in storage size, etc.

Comments

• Matlab functions: **imread**, **imwrite** and **iminfo**.

• Note the change in file size when the image is stored as a JPEG image. This is due to the (lossy) compression used by the JPEG image format.

1.5.2 Basic display of images

Matlab provides two basic functions for image display: *imshow* and *imagesc*. Whilst *imshow* requires that the 2-D array specified for display conforms to an image data type (e.g. intensity/colour images with value range 0–1 or 0–255), *imagesc* accepts input arrays of any Matlab storage type (uint 8, uint 16 or double) and any numerical range. This latter function then scales the input range of the data and displays it using the current/default colour map. We can additionally control this display feature using the *colormap* function.

Example 1.2

Matlab code	What is happening?
A=imread('cameraman.tif');	%Read in intensity image
imshow(A);	%First display image using imshow
imagesc(A);	%Next display image using imagesc
axis image;	%Correct aspect ratio of displayed image
axis off;	%Turn off the axis labelling
colormap(gray);	%Display intensity image in grey scale

Comments

- Matlab functions: *imshow*, *imagesc* and *colormap*.

- Note additional steps required when using *imagesc* to display conventional images.

In order to show the difference between the two functions we now attempt the display of unconstrained image data.

Example 1.3

Matlab code	What is happening?
B=rand(256).*1000;	%Generate random image array in range 0–1000
imshow(B);	%Poor contrast results using imshow because data %exceeds expected range
imagesc(B); axis image; axis off; colormap(gray); colorbar;	%imagesc automatically scales colourmap to data %range
imshow(B,[0 1000]);	%But if we specify range of data explicitly then %imshow also displays correct image contrast

Comments

- Note the automatic display scaling of *imagesc*.

If we wish to display multiple images together, this is best achieved by the **subplot** function. This function creates a mosaic of axes into which multiple images or plots can be displayed.

Example 1.4

Matlab code	What is happening?
B=imread('cell.tif');	%Read in 8-bit intensity image of cell
C=imread('spine.tif');	%Read in 8-bit intensity image of spine
D=imread('onion.png');	%Read in 8-bit colour image
subplot(3,1,1); imagesc(B); axis image;	%Creates a 3 × 1 mosaic of plots
axis off; colormap(gray);	%and display first image
subplot(3,1,2); imagesc(C); axis image;	%Display second image
axis off; colormap(jet);	%Set colourmap to jet (false colour)
subplot(3,1,3); imshow(D);	%Display third (colour) image

Comments

- Note the specification of different colour maps using **imagesc** and the combined display using both **imagesc** and **imshow**.

1.5.3 Accessing pixel values

Matlab also contains a built-in interactive image viewer which can be launched using the **imview** function. Its purpose is slightly different from the other two: it is a graphical, image viewer which is intended primarily for the inspection of images and sub-regions within them.

Example 1.5

Matlab code	What is happening?
B=imread('cell.tif');	%Read in 8-bit intensity image of cell
imview(B);	%Examine grey-scale image in interactive viewer
D=imread('onion.png');	%Read in 8-bit colour image.
imview(B);	%Examine RGB image in interactive viewer
B(25,50)	%Print pixel value at location (25,50)
B(25,50)=255;	%Set pixel value at (25,50) to white
imshow(B);	%View resulting changes in image
D(25,50,:)	%Print RGB pixel value at location (25,50)
D(25,50, 1)	%Print only the red value at (25,50)

```
D(25,50,:)=(255, 255, 255);        %Set pixel value to RGB white
imshow(D);                         %View resulting changes in image
```

Comments

- Matlab functions: *imview*.

- Note how we can access individual pixel values within the image and change their value.

1.5.4 Converting image types

Matlab also contains built in functions for converting different image types. Here, we examine conversion to grey scale and the display of individual RGB colour channels from an image.

Example 1.6

Matlab code   **What is happening?**
```
D=imread('onion.png');                   %Read in 8-bit RGB colour image

Dgray=rgb2gray(D);                       %Convert it to a grey-scale image

subplot(2,1,1); imshow(D); axis image;   %Display both side by side
subplot(2,1,2); imshow(Dgray);
```

Comments

- Matlab functions: *rgb2gray*.

- Note how the resulting grayscale image array is 2-D while the originating colour image array was 3-D.

Example 1.7

Matlab code   **What is happening?**
```
D=imread('onion.png');                       %Read in 8-bit RGB colour image.

Dred=D(:,:,1);                               %Extract red channel (first channel)
Dgreen=D(:,:,2);                             %Extract green channel (second channel)
Dblue=D(:,:,3);                              %Extract blue channel (third channel)

subplot(2,2,1); imshow(D); axis image;       %Display all in 2 × 2 plot

subplot(2,2,2); imshow(Dred); title('red');  %Display and label
subplot(2,2,3); imshow(Dgreen); title('green');
subplot(2,2,4); imshow(Dblue); title('blue');
```

Comments

- Note how we can access individual channels of an RGB image and extract them as separate images in their own right.

Exercises

The following exercises are designed to reinforce and develop the concepts and Matlab examples introduced in this chapter
   Matlab functions: ***imabsdiff, rgb2hsv.***

Exercise 1.1   Using the examples presented for displaying an image in Matlab together with those for accessing pixel locations, investigate adding and subtracting a scalar value from an individual location, i.e. $I(i, j) = I(i, j) + 25$ or $I(i, j) = I(i, j) - 25$. Start by using the grey-scale 'cell.tif' example image and pixel location $(100, 20)$. What is the effect on the grey-scale colour of adding and subtracting?
   Expand your technique to RGB colour images by adding and subtracting to all three of the colour channels in a suitable example image. Also try just adding to one of the individual colour channels whilst leaving the others unchanged. What is the effect on the pixel colour of each of these operations?

Exercise 1.2   Based on your answer to Exercise 1.1, use the ***for*** construct in Matlab (see *help for* at the Matlab command prompt) to loop over all the pixels in the image and brighten or darken the image.
   You will need to ensure that your program does not try to create a pixel value that is larger or smaller than the pixel can hold. For instance, an 8-bit image can only hold the values 0–255 at each pixel location and similarly for each colour channel for a 24-bit RGB colour image.

Exercise 1.3   Using the grey-scale 'cell.tif' example image, investigate using different false colour maps to display the image. The Matlab function ***colormap*** can take a range of values to specify different false colour maps: enter *help graph3d* and look under the *Color maps* heading to get a full list. What different aspects and details of the image can be seen using these false colourings in place of the conventional grey-scale display?
   False colour maps can also be specified numerically as parameters to the ***colormap*** command: enter *help colormap* for further details.

Exercise 1.4   Load an example image into Matlab and using the functions introduced in Example 1.1 save it once as a JPEG format file (e.g. sample.jpg) and once as a PNG format image (e.g. sample.png). Next, reload the images from both of these saved files as new images in Matlab, 'Ijpg' and 'Ipng'.
   We may expect these two images to be exactly the same, as they started out as the same image and were just saved in different image file formats. If we compare them by subtracting one from the other and taking the absolute difference at each pixel location we can check whether this assumption is correct.
   Use the ***imabsdiff*** Matlab command to create a difference image between 'Ijpg' and 'Ipng'. Display the resulting image using ***imagesc***.
   The difference between these two images is not all zeros as one may expect, but a noise pattern related to the difference in the images introduced by saving in a lossy compression format (i.e. JPEG) and a lossless compression format (i.e. PNG). The difference we see is due

to the image information removed in the JPEG version of the file which is not apparent to us when we look at the image. Interestingly, if we view the difference image with *imshow* all we see is a black image because the differences are so small they have very low (i.e. dark) pixel values. The automatic scaling and false colour mapping of *imagesc* allows us to visualize these low pixel values.

Exercise 1.5 Implement a program to perform the bit-slicing technique described in Section 1.2.1 and extract/display the resulting plane images (Figure 1.3) as separate Matlab images.

You may wish to consider displaying a mosaic of several different bit-planes from an image using the *subplot* function.

Exercise 1.6 Using the Matlab *rgb2hsv* function, write a program to display the individual hue, saturation and value channels of a given RGB colour image. You may wish to refer to Example 1.6 on the display of individual red, green and blue channels.

For further examples and exercises see http://www.fundipbook.com

2

Formation

All digital images have to originate from somewhere. In this chapter we consider the issue of image formation both from a mathematical and an engineering perspective. The origin and characteristics of an image can have a large bearing on how we can effectively process it.

2.1 How is an image formed?

The image formation process can be summarized as a small number of key elements. In general, a digital image s can be formalized as a mathematical model comprising a functional representation of the scene (the object function o) and that of the capture process (the point-spread function (PSF) p). Additionally, the image will contain additive noise n. These are essentially combined as follows to form an image:

$$\text{Image} = \text{PSF} * \text{object function} + \text{noise}$$
$$s = p * o + n \tag{2.1}$$

In this process we have several key elements:

- *PSF* this describes the way information on the object function is spread as a result of recording the data. It is a characteristic of the imaging instrument (i.e. camera) and is a deterministic function (that operates in the presence of noise).

- *Object function* This describes the object (or scene) that is being imaged (its surface or internal structure, for example) and the way light is reflected from that structure to the imaging instrument.

- *Noise* This is a nondeterministic function which can, at best, only be described in terms of some statistical noise distribution (e.g. Gaussian). Noise is a stochastic function which is a consequence of all the unwanted external disturbances that occur during the recording of the image data.

- *Convolution operator* $*$ A mathematical operation which 'smears' (i.e. convolves) one function with another.

Fundamentals of Digital Image Processing – A Practical Approach with Examples in Matlab
Chris Solomon and Toby Breckon
© 2011 John Wiley & Sons, Ltd

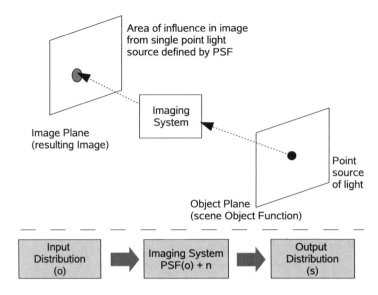

Figure 2.1 An overview of the image formation 'system' and the effect of the PSF

Here, the function of the light reflected from the object/scene (*object function*) is transformed into the image data representation by convolution with the *PSF*. This function characterizes the image formation (or capture) process. The process is affected by *noise*.

The PSF is a characteristic of the imaging instrument (i.e. camera). It represents the response of the system to a point source in the object plane, as shown in Figure 2.1, where we can also consider an imaging system as an input distribution (scene light) to output distribution (image pixels) mapping function consisting both of the PSF itself and additive noise (Figure 2.1 (lower)).

From this overview we will consider both the mathematical representation of the image formation process (Section 2.2), useful as a basis for our later consideration of advanced image-processing representations (Chapter 5), and from an engineering perspective in terms of physical camera imaging (Section 2.3).

2.2   The mathematics of image formation

The formation process of an image can be represented mathematically. In our later consideration of various processing approaches (see Chapters 3–6) this allows us to reason mathematically using knowledge of the conditions under which the image originates.

2.2.1   Introduction

In a general mathematical sense, we may view image formation as a process which transforms an *input distribution* into an *output distribution*. Thus, a simple lens may be viewed as a 'system' that transforms a spatial distribution of light in one domain (the object

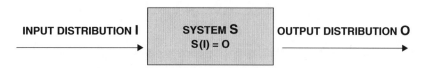

Figure 2.2   Systems approach to imaging. The imaging process is viewed as an operator S which acts on the input distribution I to produce the output O

plane) to a distribution in another (the image plane). Similarly, a medical ultrasound imaging system transforms a set of spatially distributed acoustic reflection values into a corresponding set of intensity signals which are visually displayed as a grey-scale intensity image. Whatever the specific physical nature of the input and output distributions, the concept of a *mapping* between an input distribution (the thing you want to investigate, see or visualize) and an output distribution (what you actually measure or produce with your system) is valid. The systems theory approach to imaging is a simple and convenient way of conceptualizing the imaging process. Any imaging device is a system, or a 'black box', whose properties are defined by the way in which an input distribution is mapped to an output distribution. Figure 2.2 summarizes this concept.

The process by which an imaging system transforms the input into an output can be viewed from an alternative perspective, namely that of the Fourier or *frequency domain*. From this perspective, images consist of a superposition of harmonic functions of different frequencies. Imaging systems then act upon the spatial frequency content of the input to produce an output with a modified spatial frequency content. Frequency-domain methods are powerful and important in image processing, and we will offer a discussion of such methods later in Chapter 5. First however, we are going to devote some time to understanding the basic mathematics of image formation.

2.2.2   Linear imaging systems

Linear systems and operations are extremely important in image processing because the majority of real-world imaging systems may be well approximated as *linear systems*. Moreover, there is a thorough and well-understood body of theory relating to linear systems. Nonlinear theory is still much less developed and understood, and deviations from strict linearity are often best handled in practice by approximation techniques which exploit the better understood linear methods.

An imaging system described by operator S is *linear* if for any two input distributions X and Y and any two scalars a and b we have

$$S\{aX + bY\} = aS\{X\} + bS\{Y\} \qquad (2.2)$$

In other words, applying the linear operator to a weighted sum of two inputs yields the same result as first applying the operator to the inputs independently and then combining the weighted outputs. To make this concept concrete, consider the two simple input

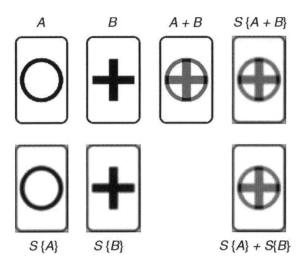

Figure 2.3 Demonstrating the action of a linear operator S. Applying the operator (a blur in this case) to the inputs and then linearly combining the results produces the same result as linearly combining them and then applying the operator

distributions depicted in Figure 2.3 consisting of a cross and a circle. These radiate some arbitrary flux (optical photons, X-rays, ultrasonic waves or whatever) which are imperfectly imaged by our linear system. The first row of Figure 2.3 shows the two input distributions, their sum and then the result of applying the linear operator (a blur) to the sum. The second row shows the result of first applying the operator to the individual distributions and then the result of summing them. In each case the final result is the same. The operator applied in Figure 2.3 is a *convolution* with Gaussian blur, a topic we will expand upon shortly.

2.2.3 Linear superposition integral

Consider Figure 2.4, in which we have some general 2-D input function $f(x', y')$ in an input domain (x', y') and the 2-D response $g(x, y)$ of our imaging system to this input in the output domain (x, y). In the most general case, we should allow for the possibility that each and every point in the input domain may contribute in some way to the output. If the system is linear, however, the contributions to the final output must combine linearly. For this reason, basic linear image formation is described by an integral operator which is called the *linear superposition integral*:

$$g(x,y) = \int\int f(x',y')h(x,y;x',y') \, dx' \, dy' \qquad (2.3)$$

This integral actually expresses something quite simple. To understand this, consider some arbitrary but specific point in the output domain with coordinates (x, y). We wish to know

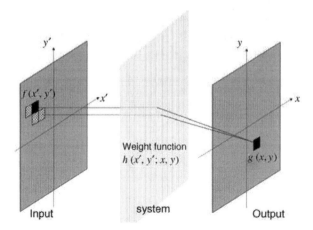

Figure 2.4 The linear superposition principle. Each point in the output is given by a weighted sum (integral) of all input points. The weight function is called the point-spread function and specifies the contribution of each and every input point to each and every output point

$g(x, y)$, the output response at that point. The linear superposition integral can be understood by breaking it down into three steps:

(1)   Take the value of the input function f at some point in the input domain (x', y') and multiply it by some weight h, with h determining the amount by which the input flux at this particular point contributes to the output point.

(2)   Repeat this for each and every valid point in the input domain multiplying by the appropriate weight each time.

(3)   Sum (i.e. integrate) all such contributions to give the response $g(x, y)$.

Clearly, it is the weighting function h which determines the basic behaviour of the imaging system. This function tells us the specific contribution made by each infinitesimal point in the input domain to each infinitesimal point in the resulting output domain. In the most general case, it is a function of *four* variables, since we must allow for the contribution of a given point (x', y') in the input space to a particular point (x, y) in the output space to depend on the exact location of *both points*. In the context of imaging, the weighting function h is referred to as the *point-spread function* (PSF). The reason for this name will be demonstrated shortly. First, however, we introduce an important concept in imaging: the (Dirac) delta or impulse function.

2.2.4   *The Dirac delta or impulse function*

In image processing, the delta or impulse function is used to represent mathematically a bright intensity source which occupies a very small (infinitesimal) region in space. It can be modelled in a number of ways,[1] but arguably the simplest is to consider it as the limiting form of a scaled rectangle function as the width of the rectangle tends to zero. The 1-D rectangle function is defined as

$$\text{rect}\left(\frac{x}{a}\right) = 1 \quad |x| < a/2$$

$$= 0 \quad \text{otherwise}$$

(2.4)

Accordingly, the 1-D and 2-D delta function can be defined as

$$\delta(x) = \lim_{a \to 0} \frac{1}{a} \text{rect}\left(\frac{x}{a}\right) \qquad \text{in 1-D}$$

$$\delta(x, y) = \lim_{a \to 0} \frac{1}{a^2} \text{rect}\left(\frac{x}{a}\right) \text{rect}\left(\frac{y}{a}\right) \qquad \text{in 2-D}$$

(2.5)

In Figure 2.5 we show the behaviour of the scaled rectangle function as $a \to 0$. We see that:

As $a \to 0$ the support (the nonzero region) of the function tends to a vanishingly small region either side of $x = 0$.

As $a \to 0$ the height of the function tends to infinity but the *total area under the function* remains equal to one.

$\delta(x)$ is thus a function which is zero everywhere except at $x = 0$ precisely. At this point, the function tends to a value of infinity but retains a finite (unit area) under the function.

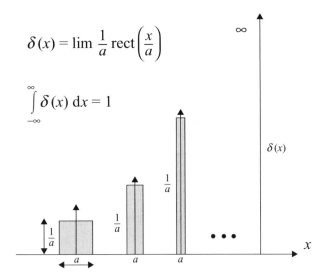

Figure 2.5   The Dirac delta or impulse function can be modelled as the limiting form of a scaled rectangle function as its width tends to zero. Note that the area under the delta function is equal to unity

_____

[1] An alternative is the limiting form of a Gaussian function as the standard deviation $\sigma \to 0$.

Thus:

$$\delta(x) = \infty \quad x = 0$$
$$= 0 \quad x \neq 0$$

$$(2.6)$$

$$\int_{-\infty}^{\infty} \delta(x)\,dx = 1$$

$$(2.7)$$

and it follows that a shifted delta function corresponding to an infinitesimal point located at $x = x_0$ is defined in exactly the same way, as

$$\delta(x-x_0) = \infty \quad x = x_0$$
$$= 0 \quad x \neq x_0$$

$$(2.8)$$

These results extend naturally to two dimensions and more:

$$\delta(x,y) = \infty \quad x = 0, y = 0$$
$$= 0 \quad \text{otherwise}$$

$$(2.9)$$

$$\int_{-\infty}^{\infty}\int_{-\infty}^{\infty} \delta(x,y)\,dx\,dy = 1$$

$$(2.10)$$

However, the most important property of the delta function is *defined by* its action under an integral sign, i.e. its so-called *sifting property*. The *sifting theorem* states that

$$\int_{-\infty}^{\infty} f(x)\delta(x-x_0)\,dx = f(x_0) \qquad \text{1-D case}$$

$$(2.11)$$

$$\int_{-\infty}^{\infty}\int_{-\infty}^{\infty} f(x,y)\delta(x-x_0,y-y_0)\,dx\,dy = f(x_0,y_0) \quad \text{2-D case}$$

This means that, whenever the delta function appears inside an integral, the result is equal to the remaining part of the integrand evaluated at those precise coordinates for which the delta function is nonzero.

In summary, the delta function is formally defined by its three properties of *singularity* (Equation (2.6)), *unit area* (Equation (2.10)) and the *sifting property* (Equation (2.11)).

Delta functions are widely used in optics and image processing as idealized representations of point and line sources (or apertures):

$$f(x,y) = \delta(x-x_0, y-y_0) \qquad \text{(point source located at } x_0, y_0)$$
$$f(x,y) = \delta(x-x_0) \qquad \qquad \text{(vertical line source located on the line } x = x_0)$$
$$f(x,y) = \delta(y-y_0) \qquad \qquad \text{(horizontal line source located on the line } y = y_0)$$
$$f(x,y) = \delta(ax+by+c) \qquad \text{(source located on the straight line } ax+by+c)$$

$$(2.12)$$

2.2.5   The point-spread function

The point-spread function of a system is defined as the response of the system to an input distribution consisting of a very small intensity point. In the limit that the point becomes infinitesimal in size, we can represent the input function as a Dirac delta function $f(x', y') = \delta(x' - x_0, y' - y_0)$. For a linear system, we can substitute this into the linear superposition integral, Equation (2.3), to give

$$g(x, y) = \int\int \delta(x' - x_0, y' - y_0) h(x, y; x', y') \, dx' \, dy' \qquad (2.13)$$

By applying the sifting theorem of Equation (2.11) this gives a response $g(x, y)$ to the point-like input as

$$g(x, y) = h(x, y; x_0, y_0) \qquad (2.14)$$

In other words, the response to the point-like input (sometimes termed the *impulse response*) is *precisely equal* to h. However, in the context of imaging, the weighting function in the superposition integral $h(x, y; x', y')$ is termed the point spread function or PSF because it measures the spread in output intensity response due to an infinitesimal (i.e. infinitely sharp) input point.

Why, then, is the PSF an important measure? Consider for a moment that any input distribution may be considered to consist of a very large ($\rightarrow \infty$) collection of very small ($\rightarrow$ infinitesimal) points of varying intensity. The PSF tells us what each of these points will look like in the output; so, through the linearity of the system, the output is given by the sum of the PSF responses. It is thus apparent that the *PSF of such a linear imaging system (in the absence of noise) completely describes its imaging properties* and, therefore, is of primary importance in specifying the behaviour of the system.

It also follows from what we have said so far that an ideal imaging system would possess a PSF that exactly equalled the Dirac delta function for all combinations of the input and output coordinates (i.e. all valid combinations of the coordinates (x, y, x', y')). Such a system would map each infinitesimal point in the input distribution to a corresponding infinitesimal point in the output distribution. There would be no blurring or loss of detail and the image would be 'perfect'. Actually, the fundamental physics of electromagnetic diffraction dictates that such an ideal system can never be achieved in practice, but it nonetheless remains a useful theoretical concept. By extension then, a good or 'sharp' imaging system will generally possess a *narrow* PSF, whereas a poor imaging system will have a broad PSF, which has the effect of considerably *overlapping* the output responses to neighbouring points in the input. This idea is illustrated in Figure 2.6, in which a group of points is successively imaged by systems with increasingly poor (i.e. broader) PSFs.

Figure 2.6   The effect of the system PSF. As the PSF becomes increasingly broad, points in the original input distribution become broader and overlap

2.2.6 Linear shift-invariant systems and the convolution integral

So far in our presentation, we have assumed that the system PSF may depend on the *absolute* locations of the two points in the input and output domains. In other words, by writing the PSF as a function of four variables in Equation (2.3), we have explicitly allowed for the possibility that the response at, let us say, point x_1, y_1 in the output domain due to a point x_1', y_1' in the input domain *may be quite different* from that at some other output point x_2, y_2. If this were the case in practice for all pairs of input and output points, then specification of the system PSF would be a very cumbersome business. As an example, consider discretizing both the input and output domains into even a modest number of pixels such as 128×128. Since every pixel in the input can in principle contribute to every pixel in the output in a different way, we would then require $128^4 \approx 268.4$ million values to fully specify the system behaviour.

Fortunately, most imaging systems (at least to a very good approximation) possess a property called *shift invariance* or *isoplanatism*. A shift-invariant system is characterized by a PSF that depends *only on the difference* between the coordinates in the input and output domains, not on their absolute values. Thus, a shift-invariant PSF in a 2-D imaging system has a functional form which makes it dependent not on four variables, but on two variables only:

$$h(x, y; x', y') = h(x'', y'') = h(x-x', y-y') \tag{2.15}$$

Shift invariance has a simple physical interpretation. Consider a source point in the input domain at coordinates x', y'. If we move this source point to different locations in the input domain, changing the values of the coordinates x', y', the corresponding response (i.e. its shape/functional form) in the output domain will remain the same but will simply be correspondingly translated in the output domain. In other words, *shift the input and the corresponding output just shifts too*. Figure 2.7[2] is a simulation of a photon-limited system in which photons in the input domain are randomly emitted and imaged by the linear shift-invariant system. This shows clearly how the image is built as the weighted superposition of shift-invariant responses to an increasing number of point-like inputs.

When we have a shift-invariant system, the linear superposition integral Equation (2.3) reduces to a simpler and very important form – a convolution integral:

$$g(x, y) = \int_{-\infty}^{\infty} \int_{-\infty}^{\infty} f(x', y')h(x-x', y-y') \, dx' \, dy' \tag{2.16}$$

Note that the infinite limits are used to indicate that the integration is to take place over all values for which the product of the functions is nonzero. The output $g(x, y)$ is said to be given by the convolution of the input $f(x, y)$ with the PSF $h(x, y)$. Note that the corresponding form for convolution of 1-D functions follows naturally:

$$g(x) = \int_{-\infty}^{\infty} f(x')h(x-x') \, dx' \tag{2.17}$$

---

[2] The Matlab code which produced Figure 2.7 is available at http://www.fundipbook.com/materials/.

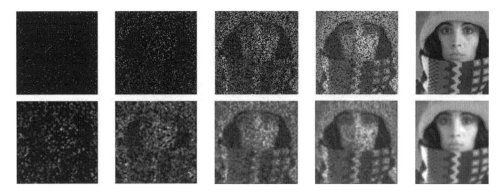

Figure 2.7 Each image in the top row shows arbitrary points in the input domain (increasing in number from left to right) whilst the images in the bottom row show the response of a LSI system to the corresponding input. The final image at the bottom right consists of a weighted superposition of responses to the input points, each of which has a similar mathematical shape or form

Convolution integrals are so important and common that they are often written in an abbreviated form:

$$g(x, y) = f(x, y) * * h(x, y) \quad \text{(2-D)}$$
$$g(x) = f(x) * h(x) \quad \text{(1-D)} \tag{2.18}$$

where the asterisks denote the operation of convolving the input function f with the system PSF h. In general, the function h in the convolution integrals above is called the kernel.

2.2.7 Convolution: its importance and meaning

It would be hard to overstate the importance of convolution in imaging. There are two main reasons:

(1) A very large number of image formation processes (and, in fact, measurement procedures in general) are well described by the process of convolution. In fact, if a system is both linear and shift invariant, then image formation is *necessarily* described by convolution.

(2) The *convolution theorem* (which we will discuss shortly) enables us to visualize and understand the convolution process in the spatial frequency domain. This equivalent frequency-domain formulation provides a very powerful mathematical framework to deepen understanding of both image formation and processing. This framework will be developed in Chapter 5 and also forms a major part of Chapter 6.

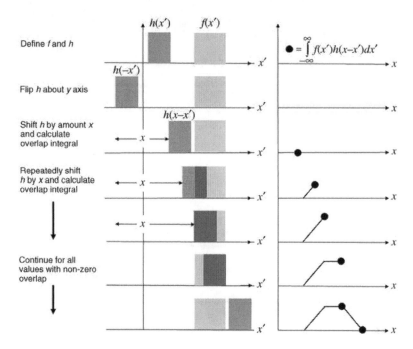

Figure 2.8   The calculation of a 1-D convolution integral

First, let us try to understand the mechanism of the convolution process. A convolution integral is a type of 'overlap' integral, and Figure 2.8 gives a graphical representation of how convolutions occur and are calculated. Basically, a flipped version of one of the functions is systematically displaced and the integral of their product (the overlap area) is calculated at each shift position. All shift coordinates for which the overlap area is nonzero define the range of the convolution.

We might well still ask, but why then is the physical process of forming an image so often described by convolution? This is perhaps best understood through a simple example (see Figure 2.9a–c). Gamma rays are too energetic to be focused, so that the very earliest gamma ray detectors used in diagnostic radionuclide imaging consisted of a gamma-ray-sensitive scintillation crystal encapsulated by a thick lead collimator with a *finite-size aperture* to allow normally incident gamma photons to strike and be absorbed in the crystal. The scan was acquired by raster-scanning with uniform speed over the patient; in this way, the derived signal is thus proportional to the gamma activity emanating from that region of the body lying beneath the aperture. Note that a finite-size aperture is absolutely necessary because reducing the aperture size to an infinitesimal size would permit a vanishingly small fraction of emitted photons to actually reach the detector, which is clearly useless for diagnostic purposes.

Let us first remind ourselves that, in general, the linear superposition integral allows for the possibility that *every point* in the input domain $f(x')$ may in principle contribute to the detector response at a chosen point x and does this through the PSF $h(x, x')$; thus, we have quite generally that $g(x) = \int f(x')h(x, x')\,dx'$. Consider then Figure 2.9a, where for

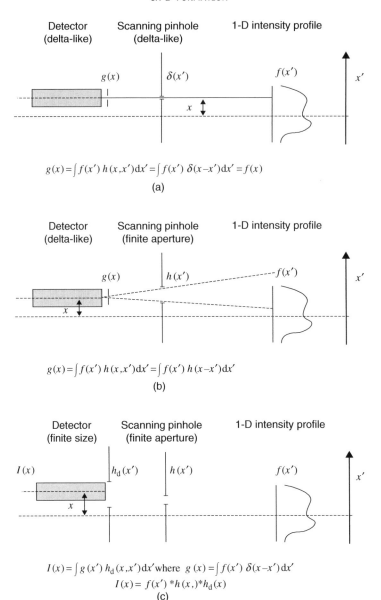

Figure 2.9 (a) A delta like detector views a source intensity profile through a delta-like scanning pinhole. The resulting image is a convolution of the source with the pinhole PSF. In principle, this replicates the source profile so that $g(x) = f(x)$ but this idealized response can never describe a real system. (b) Here, a delta like detector views a source intensity profile through a finite-size scanning pinhole. The resulting image is a convolution of the source with the pinhole PSF. All real systems exhibit such finite aperture effects. (c) A finite-size detector viewing a source intensity profile through a finite scanning pinhole. The resulting image is a convolution of the source intensity with both the pinhole and detector PSFs. In general, if N LSI elements are involved in the image formation process, the resulting image is described by convolving the source with each corresponding PSF

simplicity, we depict a 1-D detector and a 1-D intensity source. The idealized pinhole aperture may be approximated by a delta function $\delta(x')$. It is readily apparent that, if the detector were kept stationary but the pinhole aperture scanned along the x' axis, we would only get a nonzero detector response $g(x)$ at the detector when the shift coordinate $x = x'$. This situation is thus described by the linear superposition $g(x) = \int f(x')\delta(x-x')\,dx'$.[3]

If we now consider a *finite* pinhole aperture, as in Figure 2.9b, it is clear that many points in the input domain will now contribute to the detector response $g(x)$ but some will not. In fact, it is clear that only those points in the input domain for which the difference $x-x'$ lies below a certain threshold can contribute. Note that the PSF of this scanning aperture is determined by pure geometry and the detector response is thus independent of the absolute position x' of the aperture – the dependence on x' comes only as a result of how the input $f(x')$ changes.

Finally, consider Figure 2.9c. Here, we depict a finite detector aperture as well as a finite scanning aperture, and this results in a more complicated situation. However, this situation is simply described by an additional further convolution such that the actual recorded intensity at the finite detector is now given by $I(x) = \int g(x')h_{\mathrm{d}}(x-x')\,dx'$, where $g(x')$ is the intensity at some point on the detector and $h_{\mathrm{d}}(x')$ is the detector response (PSF). The way to understand this situation is depicted in Figure 2.10. Essentially, we break it down into two parts, first calculating the resulting intensity

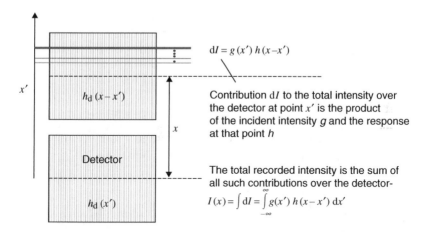

Figure 2.10 The response of a finite detector with response function $h_{\mathrm{d}}(x)$ to an incident intensity field $g(x)$ is given by their convolution. We consider the contribution to the total intensity recorded by the detector at some ordinate x' when the detector is displaced by some amount x. This is $dI = g(x')h_{\mathrm{d}}(x-x')$. By integrating all such contributions over the detector surface we obtain the convolution integral

---

[3] Strictly, this supposes that the effective sensitive area of the detector is infinitesimal and thus also well approximated by a delta response – this will occur either because it is infinitesimally small or because the scanning aperture lies very close to it.

at a fixed point on the detector and then considering how all such contributions make up the final recorded intensity. Referring to Figure 2.10, the detector has a response $h_d(x')$ when referred to the origin. We are interested in the total intensity recorded by this detector $I(x)$ when it is displaced some distance x from the origin. The contribution to the recorded intensity dI at some precise point with ordinate x' is given by the product of the incident intensity at that point $g(x')$ times the shifted response $h_d(x-x')$ – thus $dI = g(x')h_d(x-x')$. The total recorded intensity at the detector is thus given by all such contributions dI for which the ordinate x' lies within the dimensions of the detector dimension (i.e. for which $dI = g(x')h_d(x-x') \neq 0$. Thus, we obtain $I(x) = \int dI = \int_{-\infty}^{\infty} g(x')h_d(x-x')\,dx'$.

2.2.8 Multiple convolution: N imaging elements in a linear shift-invariant system

In Figure 2.9c we effectively have an imaging system in which two aperture functions describe the imaging properties: $h(x)$ and $h_d(x)$. Under these circumstances, we have seen that the recorded intensity $I(x)$ is given by successive convolutions of the input $f(x)$ with the PSF of the scanning aperture $h(x)$ and the PSF of the detector $h_d(x)$. Symbolically, we denote this by $I(x) = f(x) * h(x) * h_d(x)$. Provided the assumption of linearity and shift invariance remains valid, this result extends naturally to an arbitrary system of N imaging elements (e.g. a sequence of lenses in an optical imaging system). Thus, in general, *any processing sequence in which N linear and shift-invariant system elements act upon the input is described by a sequence of N convolutions of the input with the respective PSFs of the elements*. This is summarized in Figure 2.11.

2.2.9 Digital convolution

In digital image processing, signals are discrete not continuous. Under these conditions, convolution of two functions is achieved by discretizing the convolution integral. In the simple 1-D case, this becomes

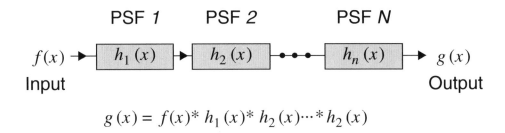

$$g(x) = f(x) * h_1(x) * h_2(x) \cdots * h_2(x)$$

Figure 2.11 The output of an LSI system characterized by N components each having a PSF $h_i(x)$ is the repeated convolution of the input with each PSF

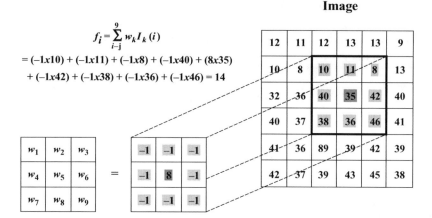

Figure 2.12 Discrete convolution. The centre pixel of the kernel and the target pixel in the image are indicated by the dark grey shading. The kernel is 'placed' on the image so that the centre and target pixels match. The filtered value of the target pixel is then given by a linear combination of the neighbourhood pixels, the specific weights being determined by the kernel values. In this specific case the target pixel of original value 35 has a filtered value of 14

$$g_j = \sum_i f_i h_{j-i} \tag{2.19}$$

where the indices j and i correspond to discrete values of x and x' respectively.

In two dimensions, the discrete convolution integral may be written as

$$g_{kl} = \sum_j \sum_i f_{ij} h_{k-i,l-j} \tag{2.20}$$

where the indices k and l correspond to x and y and i and j correspond to x' and y'. This equation somewhat belies the simplicity of the mechanism involved. The filter kernel h is only nonzero when both the shift coordinates $k-i$ and $l-j$ are small enough to lie within the spatial extent of h (i.e. are nonzero). For this reason, although the image f_{ij} has a large spatial extent, the *filter kernel* $h_{k-i,l-j}$, when it corresponds to a PSF, is typically of much smaller spatial extent. The convolved or filtered function is thus given by a weighted combination of all those pixels that lie beneath the filter kernel (see Figure 2.12). A simple example of 1-D discrete convolution in Matlab® is given in Example 2.1 and some simple examples of 2-D convolution using Matlab are given in Example 2.2. Note that 2-D convolution of two functions that are both of large spatial extent is much more computationally efficient when carried out in the Fourier domain. We will discuss this when we introduce the convolution theorem later in Chapter 5. Figures 2.13 and 2.14 result from the code in Examples 2.1 and 2.2.

Example 2.1

Matlab code	What is happening?
f=ones(64,1); f=f./sum(f);	%Define rectangle signal *f* and normalize
g=conv(f,f); g=g./sum(g);	%Convolve *f* with itself to give *g* and normalize
h=conv(g,g); h=h./sum(h);	%Convolve *g* with itself to give *h* and normalize
j=conv(h,h); j=j./sum(j);	%Convolve *h* with itself to give *j* and normalize

```
subplot(2,2,1),plot(f,'k-'); axis square;
   axis off;
subplot(2,2,2),plot(g,'k-'); axis square;
   axis off;
subplot(2,2,3),plot(h,'k-'); axis square;
   axis off;
subplot(2,2,4),plot(j,'k-'); axis square;
   axis off;
```

Comments

Matlab functions: ***conv***.

This example illustrates the repeated convolution of a 1-D uniform signal with itself. The resulting signal quickly approaches the form of the normal distribution, illustrating the central limit theorem of statistics.

Example 2.2

Matlab code	What is happening?
A=imread('trui.png');	%Read in image
PSF= fspecial('gaussian',[5 5],2);	%Define Gaussian convolution kernel
h=fspecial('motion',10,45);	%Define motion filter
B=conv2(PSF,A);	%Convolve image with convolution kernel
C=imfilter(A,h,'replicate');	%Convolve motion PSF using alternative function
D=conv2(A,A);	%Self-convolution – motion blurred with original
subplot(2,2,1),imshow(A);	%Display original image
subplot(2,2,2),imshow(B,[]);	%Display filtered image
subplot(2,2,3),imshow(C,[]);	%Display filtered image
subplot(2,2,4),imshow(D,[]);	%Display convolution image with
	itself (autocorrln)

Comments

Matlab functions: ***conv2***, ***imfilter***.

This example illustrates 2-D convolution. The first two examples show convolution of the image with a Gaussian kernel using two related Matlab functions. The final image shows the *autocorrelation* of the image given by the convolution of the function with itself.

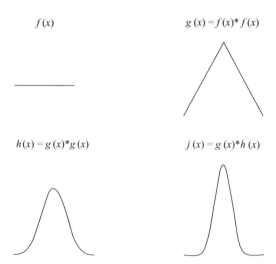

$$f(x) \qquad\qquad\qquad g(x) = f(x)*f(x)$$

$$h(x) = g(x)*g(x) \qquad\qquad j(x) = g(x)*h(x)$$

Figure 2.13   Repeated convolution of a 1-D rectangle signal. The output rapidly approaches the form of the normal distribution, illustrating the central limit theorem of statistics

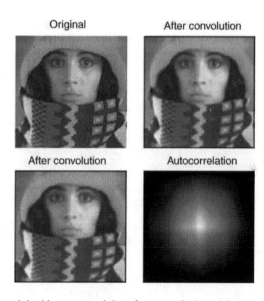

Original   After convolution

After convolution   Autocorrelation

Figure 2.14   Top left: original image; top right: after convolution with Gaussian kernel; bottom left: after convolution with Gaussian kernel; bottom right: convolution of the image with itself

2.3   The engineering of image formation

From our mathematical consideration of the image formation, we now briefly consider the engineering aspects of this process. It is not possible to consider every eventuality in this instance, but we instead limit ourselves to the most common and, hence, *practical* imaging scenarios.

2.3.1   The camera

In general, let us consider our imaging device to be a camera. A camera image of a conventional scene is essentially a projection of the 3-D world (i.e. the scene) to a 2-D representation (i.e. the image). The manner in which this 3-D to 2-D projection occurs is central to our ability to perform effective analysis on digital images.

If we assume an object/scene is illuminated by a light source (possibly multiple sources), then some of the light will be reflected towards the camera and captured as a digital image. Assuming we have a conventional camera, the light reflected into the lens is imaged onto the camera image plane based on the camera projection model. The camera projection model transforms 3-D world coordinates (X, Y, Z) to 2-D image coordinates (x, y) on the image plane. The spatial quantization of the image plane projection into a discretized grid of pixels in turn transforms the 2-D image coordinate on the image plane to a pixel position (c, r). The majority of scene images we deal with will be captured using a perspective camera projection.

Perspective projection: In general the perspective projection (Figure 2.15, right) can be stated as follows:

$$x = f\frac{X}{Z} \qquad y = f\frac{Y}{Z} \tag{2.21}$$

A position (X, Y, Z) in the scene (depth Z) is imaged at position (x, y) on the image plane determined by the focal length of the camera f (lens to image plane distance). The perspective projection has the following properties (illustrated in Figure 2.16):

- *Foreshortening* The size of objects imaged is dependent on their distance from the viewer. Thus, objects farther away from the viewer appear smaller in the image.

- *Convergence* Lines that are parallel in the scene appear to converge in the resulting image. This is known as perspective distortion, with the point(s) at which parallel lines meet termed the vanishing point(s) of the scene.

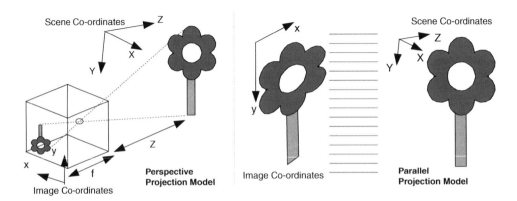

Figure 2.15   Camera projection models: perspective (left) and parallel/affine (right)

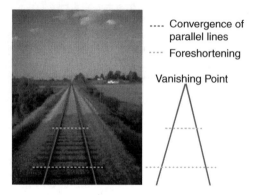

Figure 2.16 Effects of the perspective projection model on scene geometry in the image

To be more precise this is the pin-hole perspective projection model, as we assume all the light rays pass through a single point and, hence, the scene is inverted (horizontally and vertically) on the image plane (Figure 2.15, left). Realistically, we deploy a lens, which further complicates matters by introducing lens distortion (affecting f), but this optics-related topic is beyond the scope of this text. With a lens, the scene is still imaged upside down (and back to front), but this is generally dealt with by the camera and should be ignored for all applied purposes. It should be noted that the perspective projection is not easily invertible from an image alone – i.e. given 2-D image position (x, y) and camera focal length f it is not possible to recover (X, Y, Z). However, given knowledge of (X, Y) in the scene (measured relative to the camera position), the recovery of position depth Z may become possible (subject to noise and distortion). Accurate (X, Y) positions, relative to the exact camera position, are, however, difficult to determine and rarely recorded at time of image capture.

Orthographic projection: An alternative camera projection model to the perspective projection is the orthographic (or parallel) projection. This is used by some specialist imaging instruments; for instance, a flat-bed scanner produces an orthographic projection of a scanned document, a medical scanner produces an orthographic projection of the human body. The orthographic projection is simply denoted as

$$x = X \qquad y = Y$$

The orthographic projection is an affine transformation such that relative geometric relationships are maintained, as the scene to image plane projection is parallel. As such, the image features are not reversed (Figure 2.15, right), but notably *all* information relating to scene depth is lost. In an orthographic projection, size is independent of the distance from the viewer and parallel lines are preserved. The projection may have a scaling factor m such that $x = mX$ and $y = mY$, but this relates to regular image zooming or reduction and maintains these properties. Images captured over a close range (i.e. short focal length, macro-scale photography) or of a planar target can often be assumed to have an orthographic camera projection because of the negligible effects of the perspective projection over this range.

In addition to the projection model, the camera has a number of other characteristics that determine the final image formed from the scene. Notably any camera is of fixed spatial resolution (number of pixels making up the image) and has a fixed depth per pixel (number of bits used to represent each pixel) (see Section 2.1). Both of these are discussed in the next section regarding the image digitization process. In addition, every camera has unique lens distortion characteristics related to differences in manufacture (referred to as the intrinsic camera parameters) that together with similar subtle differences in the image sensor itself make it impossible to capture two identical digital images from a given camera source – *thus no two images are the same.*

2.3.2 The digitization process

From our discussion of how our image is formed by the camera as a 3-D to 2-D projective transform (Section 2.3.1), we now move on to consider the discretization of the image into a finite-resolution image of individual pixels. The key concept in digitization is that of quantization: the mapping of the continuous signal from the scene to a discrete number of spatially organized points (pixels) each with a finite representational capacity (pixel depth).

2.3.2.1 Quantization

Quantization in digital imaging happens in two ways: spatial quantization and colour quantization.

Spatial quantization corresponds to sampling the brightness of the image at a number of points. Usually a $C \times R$ rectangular grid is used but variations from rectangular do exist in specialist sensors. The quantization gives rise to a matrix of numbers which form an approximation to the analogue signal entering the camera. Each element of this matrix is referred to as a pixel – an individual picture element. The fundamental question is: How well does this quantized matrix approximate the original analogue signal ? We note that

(1) If n^2 samples are taken at regular intervals (uniform sampling, $n \times n$ image) within a bounding square, then the approximation improves as n increases. As spatial resolution increases so does image quality in terms of the reduction in approximation error between the original and the digitized image.

(2) In general, as long as sufficient samples are taken, a spatially quantized image is as good as the original image (for the purposes required, i.e. at a given level of detail).

The precise answer to our question is provided by the *sampling theorem* (also referred to as the *Nyquist sampling theorem* or *Shannon's sampling theorem*). The sampling theorem, in relation to imaging, states: an analogue image can be reconstructed exactly from its digital form as long as the sampling frequency (i.e. number of samples per linear measure) is at least twice the highest frequency (i.e. variation per measure) present in the image. Thus we require:

$$\text{sampling interval} \leq \frac{1}{\text{Nyquist frequency}} \qquad (2.22)$$

where

$$\text{Nyquist frequency} = 2 \times (\text{Maximum frequency in image})$$

The sampling theorem is concerned with the number of samples needed to recreate the original image, not with the adequacy of the digitization for any particular type of processing or presentation. As far as the implementation in digitization hardware is concerned, the sampling interval is effectively set by:

- Δt (time between samples) in a scanning-based image capture process where a sensor is being swept over a region or scene (e.g. a desktop flat-bed scanner);

- the spatial distribution of the charge-coupled device (CCD) or complementary metal–oxide–semiconductor (CMOS) elements on the capture device itself (see Section 2.3.2.2).

Spatial quantization of the image causes aliasing to occur at edges and features within the image (Figure 2.17, left) and differing spatial quantization can affect the level of detail apparent within the image (Figure 2.17, right).

In addition to spatial sampling, we also have the colour or intensity sampling to consider – effectively *intensity quantization*. For each sample point (pixel) captured at a given spatial resolution we obtain a voltage reading on the image sensor (i.e. the CCD/CMOS or analogue-to-digital (A/D) converter) that relates to the amount and wavelength of light being projected through the camera lens to that point on the image plane. Depending on the sensitivity of the sensor to particular levels and wavelengths of light, these analogue voltage signals can be divided (i.e. discretized) into a number of bins each representing a specific level of intensity – pixel values. For an 8-bit grey-scale sensor this continuous voltage is divided into 2^8 bins, giving 256 possible pixel values and thus representing a grey scale ranging from 0 (black) to 255 (white).

Different levels of intensity quantization give different levels of image colour quality, as shown in Figure 2.18. Notably, a significant reduction in the level of quantization can be performed ($256 \rightarrow 16$) before any real effects become noticeable. For colour sensors each of the red, green and blue channels are similarly each quantized into an N-bit representation (typically 8-bits per channel, giving a 24-bit colour image). In total, 24-bit colour gives 16.7 million possible colour combinations. Although these representations may seem limited, given current computing abilities, it is worth noting that the human visual system can at most

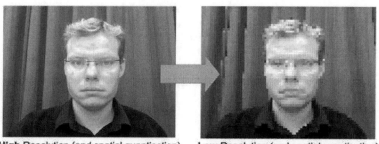

High Resolution (and spatial quantisation)   **Low** Resolution (and spatial quantisation)

Figure 2.17 The effects of spatial quantization in digital images

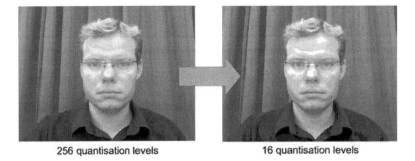

256 quantisation levels   16 quantisation levels

Figure 2.18 The effects of intensity quantization in digital images

determine between ~40 different levels of grey and it is estimated to see around 7–10 million distinct colours. This is illustrated in Figure 2.18, where a perceived difference in colour quality is only visible after a significant 256 → 16 level change. This limitation of human vision has implications for what 'we can get away with' in image compression algorithms (Section 1.3.2). Note that.. although the human visual system cannot determine the difference between these levels, the same is not always true for image processing. A shallow colour depth (i.e. low-intensity quantization) can lead to different objects/features appearing at the same quantization level despite being visually distinct in terms of colour. Again, aliasing in intensity quantization can occur (e.g. the 16 grey-levels example in Figure 2.18).

2.3.2.2 Digitization hardware

Both forms of quantization, spatial and intensity, are performed near simultaneously in the digitization hardware.

Traditionally, image capture was performed based on an analogue video-type camera connected to a computer system via an analogue to digital (A/D) converter. The converter required enough memory to store an image and a high-speed interface to the computer (commonly direct PCI or SCSI interface). When signalled by the computer, the frame grabber digitized a frame and stored the result in its internal memory. The data stored in the image memory (i.e. the digital image) can then be read into the computer over the communications interface. The schematic of such a system would typically be as per Figure 2.19, where we see an analogue link from the camera to the computer system itself and digitization is essentially internal to the hardware unit where the processing is performed.

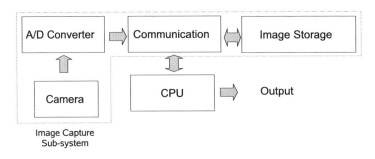

Image Capture
Sub-system

Figure 2.19 Schematic of capture-card-based system

Figure 2.20 Example of CCD sensor in a digital video camera

Such systems are now rarer and predominantly for high-speed industrial or specialist medical/engineering applications. By contrast, most modern frame-grabbing systems are directly integrated into the camera, with the A/D conversion being done at the sensor level in the electronics of the CCD or CMOS sensor inside the camera itself (i.e. digital capture, Figure 2.20). Here, analogue voltage readings occurring upon the sensor itself, in response to light being projected through the camera lens onto the sensor surface, are converted into a digitally sampled form. The digitization is carried out *in situ*, inside the sensor itself prior, to a digital transport link to the computer. Modern computer peripheral interfaces such as USB/FireWire (IEEE 1394)/GigE offer suitable high-speed digital interfaces to make digitization at sensor level and real-time transmission to the computer system a possibility. The image capture subsystem in Figure 2.19 is essentially reduced onto the capture device itself, either retaining only the most recent 'live' image to service requests from the computer (e.g. webcams) or, in the case of storage-based capture devices, retaining up to N digitized images in internal storage (e.g. digital camera/video). This is conceptually similar to what is believed to happen in the human eye, where basic 'processing' of the image is thought to occur in the retina before transmission onwards to the brain via the optic nerve. Increasingly, the advent of embedded processing is facilitating the onboard processing of images in so-called smart cameras. This trend in image capture and processing looks likely to continue in the near future.

2.3.2.3 *Resolution versus performance*
A final note on digitization relates to the spatial resolution of the image. A common misconception is that a greater resolution is somehow always a good thing. Image processing is by its very nature a very computationally demanding task. Just as in traditional computational analysis, where we discuss computational runtime in terms of $O(n)$ for a given algorithm operating on n items, in image processing we consider an $(n \times n)$ image (which generalizes to the differing $R \times C$ case).

With serial processing, any operation upon every image pixel (most common) is inherently quadratic in runtime, $O(n^2)$. Thus, the total number of operations increases rapidly as image size increases. Here, we see the limitation of increasing image resolution to achieve better image-processing results- *speed*. As a result, efficiency is a major concern when designing image-processing algorithms and is especially important for systems requiring the processing of video data with high temporal resolution in real time. In order to counter this high computational demand and to address the increasing need for real-time image-processing systems, specialist parallel computing facilities have traditionally be frequently used in image-processing tasks. Specialist hardware for conventional PCs, such as image-processing boards, were also commonplace and essentially performed user-defined image-processing tasks using fast hardware implementations independently of the PC itself.

With the ongoing increase in consumer-level image-processing power (Moore's law), the adaptation of processors to specialist multimedia tasks such as image processing (e.g. Intel MMX technologies) and now the advent of desktop-level multi-processor systems (e.g. Intel Core-Duo), the use of such specialist hardware has reduced in the image-processing industry. Although still used in industrial applications, the advent of embedded PC technology and robust embedded image-processing-capable operating systems look set to change this over the coming years. Reconfigurable, programmable hardware such as field-programmable gate arrays (FPGA) and the use of graphics processing unit (GPU) processing (image processing on the PC graphics card processor) also offer some of the benefits of a hardware solution with the additional flexibility of software based development/testing.

Modern PCs can cope well with real-time image-processing tasks and have memory capacities that realistically exceed any single image. The same is not always true of video stream processing, and performance evaluation and optimization is still an area of research.

2.3.3   Noise

The main barrier to effective image processing and signal processing in general is noise. By noise we mean a variation of the signal from its true value by a small (random) amount due to external or internal factors in the image-processing pipeline. These factors cannot be (easily) controlled and thus introduce random elements into the processing pipeline. Noise is the key problem in 99 % of cases where image processing techniques either fail or further image processing is required to achieve the required result. As a result, a large part of the image processing domain, and any image processing system pipeline, is dedicated to noise reduction and removal. A robust image processing system must be able to cope with noise. At each stage of image processing, capture and sampling noise is introduced (Figure 2.21).

Noise in digital images can originate from a variety of sources :

- *Capture noise* can be the result of variations in lighting, sensor temperature, electrical sensor noise, sensor nonuniformity, dust in the environment, vibration, lens distortion, focus limitations, sensor saturation (too much light), underexposure (too little light).

- *Sampling noise* As discussed previously (Section 2.3.2), limitations in sampling and intensity quantization are a source of noise in the form of representational aliasing. The sampled digital image is not a true representation of the analogue image, but an alias of the original.

Salt and Pepper (impulse) noise

Original Image

Figure 2.21 Examples of noise effects in digital imaging

- *Processing noise* Limitations in numerical precision (floating-point numbers), potential integer overflow and mathematical approximations (e.g. $\pi = 3.142\ldots$) are all potential sources of noise in the processing itself.

- *Image-encoding noise* Many modern image compression techniques (e.g. JPEG used intrinsically by modern digital cameras) are lossy compression techniques. By lossy we mean that they compress the image by removing visual information that represents detail not general perceivable to the human viewer. The problem is that this loss of information due to compression undermines image-processing techniques that rely on this information. This loss of detail is often referred to by the appearance of *compression artefacts* in the image. In general, loss = compression artefacts = noise. Modern image formats, such as PNG, offer lossless compression to counter this issue (see Section 1.3.2).

- *Scene occlusion* In the task of object recognition, objects are frequently obscured by other objects. This is known as occlusion. Occlusion poses a big problem for computer vision systems because *you don't know what you cannot see*. You may be able to infer recognition from a partial view of an object, but this is never as robust as full-view recognition. This limits available image information.

Ignoring systematic noise attributable to a specific cause (e.g. nonuniform lighting), noise in images can be characterized in a number of ways. The main two noise characterizations used in imaging are:

- *Salt and pepper noise* This is caused by the random introduction of pure white or black (high/low) pixels into the image (Figure 2.21). This is less common in modern image sensors, although can most commonly be seen in the form of camera sensor faults (hot pixels that are always at maximum intensity or dead pixels which are always black). This type of noise is also known as *impulse noise*.

- *Gaussian noise* In this case, the random variation of the image signal around its expected value follows the Gaussian or normal distribution (Figure 2.21). This is the most commonly used noise model in image processing and effectively describes most random noise encountered in the image-processing pipeline. This type of noise is also known as *additive noise*.

Ultimately, no digital image is a perfect representation of the original scene: it is limited in resolution by sampling and contains noise. One of the major goals of image processing is to limit the effects of these aspects in image visualization and analysis.

Exercises

The following exercises are designed to reinforce and develop the concepts and Matlab examples introduced in this chapter. Additional information on all of the Matlab functions represented in this chapter and throughout these exercises is available in Matlab from the function help browser (use *doc <function name>* at the Matlab command prompt, where <function name> is the function required)

Matlab functions: *rand()*, *fspecial()*, *filter2()*, *imresize()*, *tic()*, *toc()*, *imapprox()*, *rgb2ind()*, *gray2ind()*, *imnoise()*, *imtool()*, *hdrread()*, *tonemap()*, *getrangefromclass()*.

Exercise 2.1 Using Example 2.3 we can investigate the construction of a coherent image from a limited number of point samples via the concept of the PSF introduced in Section 2.1:

Example 2.3

Matlab code	What is happening?
A=imread('cameraman.tif');	%Read in an image
[rows dims]=size(A);	%Get image dimensions
Abuild=zeros(size(A));	%Construct zero image of equal size

%Randomly sample 1% of points only and convolve with Gaussian PSF

```
sub=rand(rows.*dims,1)<0.01;
Abuild(sub)=A(sub); h=fspecial('gaussian',[10 10],2);
B10=filter2(h,Abuild);
subplot(1,2,1), imagesc(Abuild); axis image; axis off;colormap(gray); title('Object points')
subplot(1,2,2), imagesc(B10); axis image; axis off;colormap(gray); title('Response of LSI system')
```

Here, we randomly select 1% (0.01) of the image pixels and convolve them using a Gaussian-derived PSF (see Section 4.4.4) applied as a 2-D image filter via the *filter2()* function. Experiment with this example by increasing the percentage of pixels randomly selected for PSF convolution to form the output image. At what point is a coherent image formed? What noise characteristics does this image show? Additionally, investigate the use of the *fspecial()* function and experiment with changing the parameters of the Gaussian filter used to approximate the PSF. What effect does this have on the convergence of the image towards a coherent image (i.e. shape outlines visible) as we increase the number of image pixels selected for convolution?

Exercise 2.2 The engineering of image formation poses limits on the spatial resolution (Section 1.2) of an image through the digitization process. At the limits of spatial resolution certain image details are lost. Using the Matlab example images 'football.jpg' and 'text.png', experiment with the use of the *imresize()* function both in its simplest form (using a single scale parameter) and using a specified image width and height (rows and columns). As you reduce image size, at what point do certain image details deteriorate beyond recognition/comprehension within the image? You may also wish to investigate the use of *imresize()* for enlarging images. Experiment with the effects of enlarging an image using all three of the available interpolation functions for the Matlab *imresize()* function. What differences do you notice? (Also try timing different resizing interpolation options using the Matlab *tic()*/*toc()* functions.)

Exercise 2.3 In addition to spatial resolution, the engineering of image formation also poses limits on the number of quantization (colour or grey-scale) levels in a given image. Using the Matlab function *imapprox()* we can approximate a given image represented in a given colour space (e.g. RGB or grey scale, Section 1.4.1.1) with fewer colours. This function operates on an indexed image (essentially an image with a reference look-up table for each of its values known as a colour map) – see 'indexed images' in Matlab Help. A conventional image (e.g. Matlab example 'peppers.png') can be converted to an indexed image and associated look-up table colour map using the Matlab *rgb2ind()* function (or *gray2ind()* for grey-scale images). Investigate the use of *imapprox()* and *rgb2ind()*/*gray2ind()* to reduce the number of colours in a colour image example and a greyscale image example. Experiment with different levels of colour reduction. How do your findings tally with the discussion of human colour perception in Section 2.3.2.1)?

Exercise 2.4 The Matlab *imnoise()* function can be used to add noise to images. Refer to Example 4.3 for details of how to use this function and experiment with its use as suggested in the associated Exercise 4.1.

Exercise 2.5 The Matlab function *imtool()* allows us to display an image and perform a number of different interactive operations upon it. Use this function to load the example image shown in Figure 2.16 (available as 'railway.png') and measure the length in pixels of the railway sleepers in the foreground (front) and background (rear) of the image. Based on the discussion of Section 2.3.1, what can we determine about the relative distance of the two sleepers you have measured to the camera? If we assume the standard length of a railway sleeper is 2.5 m, what additional information would be need to determine the absolute distance of these items from the camera? How could this be applied to other areas, such as estimating the distance of people or cars from the camera?

Exercise 2.6   High dynamic range (HDR) imaging is a new methodology within image capture (formation) whereby the image is digitized with a much greater range of quantization levels (Section 2.3.2.1) between the minimum and maximum colour levels. Matlab supports HDR imaging through the use of the ***hdrread()*** function. Use this function to load the Matlab example HDR image 'office.hdr' and convert it to a conventional RGB colour representation using the Matlab ***tonemap()*** function for viewing (with ***imshow()***). Use the Matlab ***getrangefromclass()*** function and ***imtool()*** to explore the value ranges of these two versions of the HDR image.

For further examples and exercises see http://www.fundipbook.com

3

Pixels

In this chapter we discuss the basis of all digital image processing: *the pixel*. We cover the types of tangible information that pixels can contain and discuss operations both on individual pixels (point transforms) and on distributions of pixels (global transforms).

3.1 What is a pixel?

The word pixel is an abbreviation of 'picture element'. Indexed as an (x, y) or column-row (c, r) location from the origin of the image, it represents the smallest, constituent element in a digital image and contains a numerical value which is the basic unit of information within the image at a given spatial resolution and quantization level. Commonly, pixels contain the colour or intensity response of the image as a small point sample of coloured light from the scene. However, not all images necessarily contain strictly visual information. An image is simply a 2-D signal digitized as a grid of pixels, the values of which may relate to other properties other than colour or light intensity. The information content of pixels can vary considerably depending on the type of image we are processing:

- *Colour/grey-scale images*   Commonly encountered images contain information relating to the colour or grey-scale intensity at a given point in the scene or image.

- *Infrared (IR)*   The visual spectrum is only a small part of the electromagnetic spectrum. IR offers us the capability of imaging scenes based upon either the IR light reflectance or upon the IR radiation they are emitting. IR radiation is emitted in proportion to heat generated/reflected by an object and, thus, IR imaging is also commonly referred to as thermal imaging. As IR light is invisible to the naked human eye, IR illumination and imaging systems offer a useful method for covert surveillance (as such, IR imaging commonly forms the basis for night-vision systems).

- *Medical imaging*   Many medical images contain values that are proportional to the absorption characteristics of tissue with respect to a signal projected through the body. The most common types are computed tomography (CT) and magnetic resonance imaging (MRI). CT images, like conventional X-rays, represent values that are directly proportional to the density of the tissue through which the signal passed. By contrast, magnetic resonance images exhibit greater detail but do not have a direct relationship to a single quantifiable property of the tissue. In CT and MRI our 2-D image format is commonly extended to a 3-D volume; the 3-D volume is essentially just a stack of 2-D images.

- *Radar/sonar imaging* A radar or sonar image represents a cross-section of a target in proportion to its distance from the sensor and its associated signal 'reflectivity'. Radar is commonly used in aircraft navigation, although it has also been used on road vehicle projects. Satellite-based radar for weather monitoring is now commonplace, as is the use of sonar on most modern ocean-going vessels. Ground-penetrating radar is being increasingly used for archaeological and forensic science investigations.

- *3-D imaging* Using specific 3-D sensing techniques such as stereo photography or 3-D laser scanning we can capture data from objects in the world around us and represent them in computer systems as 3-D images. 3-D images often correspond to depth maps in which every pixel location contains the distance of the imaged point from the sensor. In this case, we have explicit 3-D information rather than just a projection of 3-D as in conventional 2-D imaging. Depending on the capture technology, we may have only 3-D depth information or both 3-D depth and colour for every pixel location. The image depth map can be re-projected to give a partial view of the captured 3-D object (such data is sometimes called $2^1/_2$-D).

- *Scientific imaging* Many branches of science use a 2-D (or 3-D)-based discrete format for the capture of data and analysis of results. The pixel values may, in fact, correspond to chemical or biological sample densities, acoustic impedance, sonic intensity, etc. Despite the difference in the information content, the data is represented in the same form, i.e. a 2-D image. Digital image processing techniques can thus be applied in many different branches of scientific analysis.

Figure 3.1 shows some examples of images with different types of pixel information. These are just a few examples of both the variety of digital images in use and of the broad scale of application domains for digital image processing. In the colour/grey-scale images we will consider in this book the pixels will usually have integer values within a given quantization range (e.g. 0 to 255, 8-bit images), although for other forms of image information (e.g. medical, 3-D, scientific) floating-point real pixel values are commonplace.

3.2 Operations upon pixels

The most basic type of image-processing operation is a point transform which maps the values at individual points (i.e. pixels) in the input image to corresponding points (pixels) in an output image. In the mathematical sense, this is a one-to-one functional mapping from input to output. The simplest examples of such image transformations are arithmetic or logical operations on images. Each is performed as an operation between two images I_A and I_B or between an image and a constant value C:

$$I_{\text{output}} = I_A + I_B$$
$$I_{\text{output}} = I_A + C$$

(3.1)

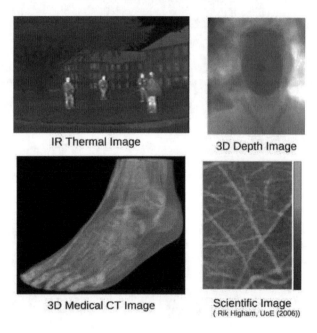

IR Thermal Image   3D Depth Image

3D Medical CT Image   Scientific Image
                                          (Rik Higham, UoE (2006))

Figure 3.1   Differing types of pixel information

In both cases, the values at an individual pixel location (i,j) in the output image are mapped as follows:

$$I_{\text{output}}(i,j) = I_A(i,j) + I_B(i,j)$$
$$I_{\text{output}}(i,j) = I_A(i,j) + C$$

(3.2)

To perform the operation over an entire image of dimension $C \times R$, we simply iterate over all image indices for $(i,j) = \{0 \ldots C{-}1, 0 \ldots R{-}1\}$ in the general case (N.B. in Matlab® pixel indices are $\{1 .. C, 1 .. R\}$).

3.2.1   Arithmetic operations on images

Basic arithmetic operations can be performed quickly and easily on image pixels for a variety of effects and applications.

3.2.1.1   Image addition and subtraction
Adding a value to each image pixel value can be used to achieve the following effects (Figure 3.2):

- *Contrast adjustment*   Adding a positive constant value C to each pixel location increases its value and, hence, its brightness.

- *Blending*   Adding images together produces a composite image of both input images. This can be used to produce blending effects using weighted addition.

Image Addition

Image Blending

Figure 3.2   Image contrast adjustment and blending using arithmetic addition

Image addition can be carried out in Matlab, as shown in Example 3.1.

Example 3.1

Matlab code	**What is happening?**
A=imread('cameraman.tif');	%Read in image
subplot(1,2,1), imshow(A);	%Display image
B = imadd(A, 100);	%Add 100 to each pixel value in image A
subplot(1,2,2), imshow(B);	%Display result image B

Comments

- Generally, images must be of the same dimension and of the same data type (e.g. 8-bit integer) for addition and subtraction to be possible between them.

- When performing addition operations we must also be aware of integer overflow. An 8-bit integer image can only hold integer values from 0 to 255. The Matlab functions *imadd,* *imsubtract* and *imabsdiff* avoid this problem by truncating or rounding overflow values.

Subtracting a constant value from each pixel (like addition) can also be used as a basic form of contrast adjustment. Subtracting one image from another shows us the difference

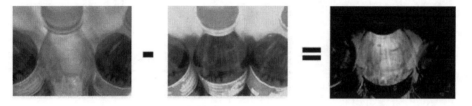

Figure 3.3 Image differencing using arithmetic subtraction

between images. If we subtract two images in a video sequence then we get a difference image (assuming a static camera) which shows the movement or changes that have occurred between the frames in the scene (e.g. Figure 3.3). This can be used as a basic form of change/movement detection in video sequences. Image subtraction can be carried out in Matlab as shown in Example 3.2.

Example 3.2

Matlab code	What is happening?
A=imread('cola1.png');	%Read in 1st image
B=imread('cola2.png');	%Read in 2nd image
subplot(1,3,1), imshow(A);	%Display 1st image
subplot(1,3,2), imshow(B);	%Display 2nd image
Output = imsubtract(A, B);	%Subtract images
subplot(1,3,3), imshow(Output);	%Display result

A useful variation on subtraction is the absolute difference $I_{\text{output}} = |I_A - I_B|$ between images (Example 3.3). This avoids the potential problem of integer overflow when the difference becomes negative.

Example 3.3

Matlab code	What is happening?
Output = imabsdiff(A, B);	%Subtract images
subplot(1,3,3), imshow(Output);	%Display result

3.2.1.2 Image multiplication and division

Multiplication and division can be used as a simple means of contrast adjustment and extension to addition/subtraction (e.g. reduce contrast to 25% = division by 4; increase contrast by 50% = multiplication by 1.5). This procedure is sometimes referred to as image colour scaling. Similarly, division can be used for image differencing, as dividing an image by another gives a result of 1.0 where the image pixel values are identical and a value not equal to 1.0 where differences occur. However, image differencing using subtraction is computationally more efficient. Following from the earlier examples, image multiplication and division can be performed in Matlab as shown in Example 3.4.

Example 3.4

Matlab code	What is happening?
Output = immultiply(A,1.5);	%Multiply image by 1.5
subplot(1,3,3), imshow(Output);	%Display result
Output = imdivide(A,4);	%Divide image by 4
subplot(1,3,3), imshow(Output);	%Display result

Comments

Multiplying different images together or dividing them by one another is not a common operation in image processing.

For all arithmetic operations between images we must ensure that the resulting pixel values remain within the available integer range of the data type/size available. For instance, an 8-bit image (or three-channel 24-bit colour image) can represent 256 values in each pixel location. An initial pixel value of 25 multiplied by a constant (or secondary image pixel value) of 12 will exceed the 0–255 value range. Integer overflow will occur and the value will ordinarily 'wrap around' to a low value. This is commonly known as saturation in the image space: the value exceeds the representational capacity of the image. A solution is to detect this overflow and avoid it by setting all such values to the maximum value for the image representation (e.g. truncation to 255). This method of handling overflow is implemented in the **imadd**, **imsubtract**, **immultiply** and **imdivide** Matlab functions discussed here. Similarly, we must also be aware of negative pixel values resulting from subtraction and deal with these accordingly; commonly they are set to zero. For three-channel RGB images (or other images with vectors as pixel elements) the arithmetic operation is generally performed separately for each colour channel.

3.2.2 Logical operations on images

We can perform standard logical operations between images such as NOT, OR, XOR and AND. In general, logical operation is performed between each corresponding bit of the image pixel representation (i.e. a bit-wise operator).

- *NOT (inversion)*   This inverts the image representation. In the simplest case of a binary image, the (black) background pixels become (white) foreground and vice versa. For grey-scale and colour images, the procedure is to replace each pixel value $I_{input}(i, j)$ as follows:

$$I_{output}(i, j) = \text{MAX} - I_{ouput}(i, j) \qquad (3.3)$$

where MAX is the maximum possible value in the given image representation. Thus, for an 8-bit grey-scale image (or for 8-bit channels within a colour image), MAX = 255.

In Matlab this can be performed as in Example 3.5.

Example 3.5

Matlab code	What is happening?
A=imread('cameraman.tif');	%Read in image
subplot(1,2,1), imshow(A);	%Display image
B = imcomplement(A);	%Invert the image
subplot(1,2,2), imshow(B);	%Display result image B

- *OR/XOR*   Logical OR (and XOR) is useful for processing binary-valued images (0 or 1) to detect objects which have moved between frames. Binary objects are typically produced through application of thresholding to a grey-scale image. Thresholding is discussed in Section 3.2.3.

- *AND*   Logical AND is commonly used for detecting differences in images, highlighting target regions with a binary mask or producing bit-planes through an image, as discussed in Section 1.2.1 (Figure 1.3). These operations can be performed in Matlab as in Example 3.6.

Example 3.6

Matlab code	What is happening?
A=imread('toycars1.png');	%Read in 1st image
B=imread('toycars2.png');	%Read in 2nd image
Abw=im2bw(A);	%convert to binary
Bbw=im2bw(B);	%convert to binary
subplot(1,3,1), imshow(Abw);	%Display 1st image
subplot(1,3,2), imshow(Bbw);	%Display 2nd image
Output = xor(Abw, Bbw);	%xor images images
subplot(1,3,3), imshow(Output);	%Display result

Comments

- note that the images are first converted to binary using the Matlab im2bw function (with an automatic threshold - Section 3.2.3).

- note that the resulting images from the im2bw function and the xor logical operation is of Matlab type 'logical'.

- the operators for AND is '&' whilst the operator for OR is '|' and are applied in infix notation form as A & B, A | B.

Combined operators, such as NAND, NOR and NXOR, can also be applied in a similar manner as image-processing operators.

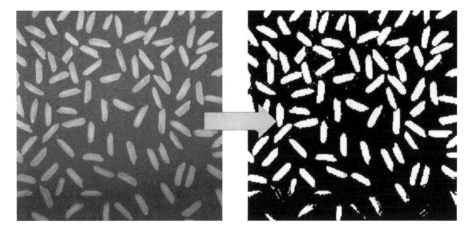

Figure 3.4   Thresholding for object identification

3.2.3   Thresholding

Another basic point transform is thresholding. Thresholding produces a binary image from a grey-scale or colour image by setting pixel values to 1 or 0 depending on whether they are above or below the threshold value. This is commonly used to separate or segment a region or object within the image based upon its pixel values, as shown in Figure 3.4.

In its basic operation, thresholding operates on an image I as follows:

```
for each pixel I(i,j) within the image I
      if I(i,j) > threshold
            I(i,j) = 1
      else
            I(i,j) = 0
      end
end
```

In Matlab, this can be carried out using the function ***im2bw*** and a threshold in the range 0 to 1, as in Example 3.7.

Example 3.7

Matlab code	What is happening?
I=imread('trees.tif');	%Read in 1st image
T=im2bw(I, 0.1);	%Perform thresholding
subplot(1,3,1), imshow(I);	%Display original image
subplot(1,3,2), imshow(T);	%Display thresholded image

The ***im2bw*** function automatically converts colour images (such as the input in the example) to grayscale and scales the threshold value supplied (from 0 to 1) according to the given quantization range of the image being processed. An example is shown in Figure 3.5.

For grey-scale images, whose pixels contain a single intensity value, a single threshold must be chosen. For colour images, a separate threshold can be defined for each channel (to

Figure 3.5   Thresholding of a complex image

correspond to a particular colour or to isolate different parts of each channel). In many applications, colour images are converted to grey scale prior to thresholding for simplicity. Common variations on simple thresholding are:

- the use of two thresholds to separate pixel values within a given range;

- the use of multiple thresholds resulting in a labelled image with portions labelled 0 to N;

- retaining the original pixel information for selected values (i.e. above/between thresholds) whilst others are set to black.

Thresholding is the 'work-horse' operator for the separation of image foreground from background. One question that remains is how to select a good threshold. This topic is addressed in Chapter 10 on image segmentation.

3.3   Point-based operations on images

The dynamic range of an image is defined as the difference between the smallest and largest pixel values within the image. We can define certain functional transforms or mappings that alter the effective use of the dynamic range. These transforms are primarily applied to improve the contrast of the image. This improvement is achieved by altering the relationship between the dynamic range of the image and the grey-scale (or colour) values that are used to represent the values.

In general, we will assume an 8-bit (0 to 255) grey-scale range for both input and resulting output images, but these techniques can be generalized to other input ranges and individual channels from colour images.

3.3.1   Logarithmic transform

The dynamic range of an image can be compressed by replacing each pixel value in a given image with its logarithm: $I_{\text{output}}(i,j) = \ln I_{\text{input}}(i,j)$, where $I(i,j)$ is the value of a pixel at a location (i,j) in image I and the function $\ln()$ represents the natural logarithm. In practice, as the logarithm is undefined for zero, the following general form of the logarithmic

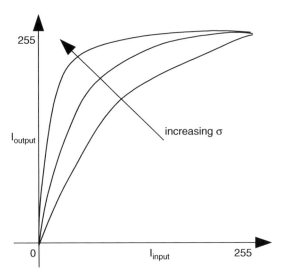

Figure 3.6 The Logarithmic Transform: Varying the parameter σ changes the gradient of the logarithmic function used for input to output

transform is used:

$$I_{\text{output}}(i,j) = c \ln[1 + (e^{\sigma}-1)I_{\text{input}}(i,j)] \tag{3.4}$$

Note that the scaling factor σ controls the input range to the logarithmic function, whilst c scales the output over the image quantization range 0 to 255. The addition of 1 is included to prevent problems where the logarithm is undefined for $I_{\text{input}}(i,j) = 0$. The level of dynamic range compression is effectively controlled by the parameter σ. As shown in Figure 3.6, as the logarithmic function is close to linear near the origin, the compression achieved is smaller for an image containing a low range of input values than one containing a broad range of pixel values.

The scaling constant c can be calculated based on the maximum allowed output value (255 for an 8-bit image) and the maximum value $\max(I_{\text{input}}(i,j))$ present in the input:

$$c = \frac{255}{\log[1 + \max(I_{\text{input}}(i,j))]} \tag{3.5}$$

The effect of the logarithmic transform is to increase the dynamic range of dark regions in an image and decrease the dynamic range in the light regions (e.g. Figure 3.7). Thus, the logarithmic transform maps the lower intensity values or dark regions into a larger number of greyscale values and compresses the higher intensity values or light regions in to a smaller range of greyscale values.

In Figure 3.7 we see the typical effect of being photographed in front of a bright background (left) where the dynamic range of the film or camera aperture is too small to capture the full range of the scene. By applying the logarithmic transform we brighten the foreground of this image by spreading the pixel values over a wider range and revealing more of its detail whilst compressing the background pixel range.

In Matlab, the logarithmic transform can be performed on an image as in Example 3.8. From this example, we can see that increasing the multiplication constant c increases the overall brightness of the image (as we would expect from our earlier discussion of multiplication in Section 3.4). A common variant which achieves a broadly similar result is the simple

Figure 3.7 Applying the logarithmic transform to a sample image

square-root transform (i.e. mapping output pixel values to the square root of the input) which similarly compresses high-value pixel ranges and spreads out low-value pixel ranges.

Example 3.8

Matlab code	What is happening?
I=imread('cameraman.tif');	%Read in image
subplot(2,2,1), imshow(I);	%Display image
Id=im2double(I);	
Output1=2*log(1 + Id);	
Output2=3*log(1 + Id);	
Output3=5*log(1 + Id);	
subplot(2,2,2), imshow(Output1);	%Display result images
subplot(2,2,3), imshow(Output2);	
subplot(2,2,4), imshow(Output3);	

3.3.2 Exponential transform

The exponential transform is the inverse of the logarithmic transform. Here, the mapping function is defined by the given base e raised to the power of the input pixel value:

$$I_{\text{output}}(i,j) = e^{I_{\text{input}}(i,j)} \tag{3.6}$$

where $I(i,j)$ is the value of a pixel at a location (i,j) in image I.

This transform enhances detail in high-value regions of the image (bright) whilst decreasing the dynamic range in low-value regions (dark) – the opposite effect to the logarithmic transform. The choice of base depends on the level of dynamic range compression required. In general, base numbers just above 1 are suitable for photographic image enhancement. Thus, we expand our exponential transform notation to include a variable base and scale to the appropriate output range as before:

$$I_{\text{output}}(i,j) = c[(1+\alpha)^{I_{\text{input}}(i,j)} - 1] \tag{3.7}$$

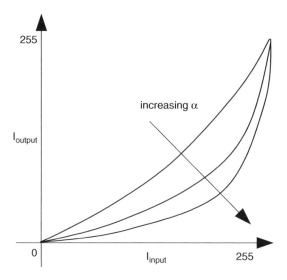

Figure 3.8 The Exponential Transform: Varying the parameter α changes the gradient of the exponential function used for input to output

Here, $(1 + \alpha)$ is the base and c is the scaling factor required to ensure the output lies in an appropriate range. As is apparent when $I_{input}(i, j) = 0$, this results in $I_{output}(i, j) = c$ unless we add in the -1 to counter this potential offset appearing in the output image. The level of dynamic range compression and expansion is controlled by the portion of the exponential function curve used for the input to output mapping; this is determined by parameter α. As shown in Figure 3.8, as the exponential function is close to linear near the origin, the compression is greater for an image containing a lower range of pixel values than one containing a broader range.

We can see the effect of the exponential transform (and varying the base) in Figure 3.9 and Example 3.9. Here, we see that the contrast of the background in the original image can be improved by applying the exponential transform, but at the expense of contrast in the darker areas of the image. The background is a high-valued area of the image (bright), whilst the darker regions have low pixel values. This effect increases as the base number is increased.

Figure 3.9 Applying the exponential transform to a sample image

---

Example 3.9

Matlab code	What is happening?
I=imread('cameraman.tif');	%Read in image
subplot(2,2,1), imshow(I);	%Display image
Id=im2double(I);	
Output1=4*(((1+0.3).^(Id)) - 1);	
Output2=4*(((1+0.4).^(Id)) - 1);	
Output3=4*(((1+0.6).^(Id)) - 1);	
subplot(2,2,2), imshow(Output1);	%Display result images
subplot(2,2,3), imshow(Output2);	
subplot(2,2,4), imshow(Output3);	

In Matlab, the exponential transform can be performed on an image as in Example 3.9. In this example we keep c constant whilst varying the exponential base parameter α, which in turn varies the effect of the transform on the image (Figure 3.9).

3.3.3 Power-law (gamma) transform

An alternative to both the logarithmic and exponential transforms is the 'raise to a power' or power-law transform in which each input pixel value is raised to a fixed power:

$$I_{output}(i,j) = c(I_{input}(i,j))^{\gamma} \qquad (3.8)$$

In general, a value of $\gamma > 1$ enhances the contrast of high-value portions of the image at the expense of low-value regions, whilst we see the reverse for $\gamma < 1$. This gives the power-law transform properties similar to both the logarithmic ($\gamma < 1$) and exponential ($\gamma > 1$) transforms. The constant c performs range scaling as before.

In Matlab, this transform can be performed on an image as in Example 3.10. An example of this transform, as generated by this Matlab example, is shown in Figure 3.10.

Example 3.10

Matlab code	What is happening?
I=imread('cameraman.tif');	%Read in image
subplot(2,2,1), imshow(I);	%Display image
Id=im2double(I);	
Output1=2*(Id.^0.5);	
Output2=2*(Id.^1.5);	
Output3=2*(Id.^3.0);	
subplot(2,2,2), imshow(Output1);	%Display result images
subplot(2,2,3), imshow(Output2);	
subplot(2,2,4), imshow(Output3);	

Figure 3.10   Applying the power-law transform to a sample image

3.3.3.1   *Application: gamma correction*

A common application of the power-law transform is gamma correction. Gamma correction is the term used to describe the correction required for the nonlinear output curve of modern computer displays. When we display a given intensity on a monitor we vary the analogue voltage in proportion to the intensity required. The problem that all monitors share is the nonlinear relationship between input voltage and output intensity. In fact, the monitor output intensity generally varies with the power of input voltage curve, as approximately $\gamma = 2.5$ (this varies with type of monitor, manufacturer, age, etc.). This means that, when you request an output intensity equal to say i, you in fact get an intensity of $i^{2.5}$. In order to counter this problem, we can preprocess image intensities using an inverse power-law transform prior to output to ensure they are displayed correctly. Thus, if the effect of the display gamma can be characterized as $I_{\text{output}} = (I_{\text{input}})^{\gamma}$ where γ is the r value for the output curve of the monitor (lower right), then pre-correcting the input with the inverse power-law transform, i.e. $(I_{\text{input}})^{1/\gamma}$, compensates to produce the correct output as

$$I_{\text{output}} = ((I_{\text{input}})^{1/\gamma})^{\gamma} = I_{\text{input}} \tag{3.9}$$

Modern operating systems have built in gamma correction utilities that allow the user to specify a γ value that will be applied in a power-law transform performed automatically on all graphical output. Generally, images are displayed assuming a γ of 2.2. For a specific monitor, gamma calibration can be employed to determine an accurate γ value.

We can perform gamma correction on an image in Matlab using the ***imadjust*** function to perform the power-law transform, as in Example 3.11. An example of the result is shown in Figure 3.11.

Figure 3.11 Gamma correction on a sample image

Example 3.11

Matlab code	What is happening?
A=imread('cameraman.tif');	%Read in image
subplot(1,2,1), imshow(A);	%Display image
B=imadjust(A,[0 1],[0 1],1./3);	%Map input grey values of image A in range 0–1 to %an output range of 0–1 with gamma factor of 1/3 %(i.e. r = 3).
	%Type ≫ *doc imadjust* for details of possible syntaxes
subplot(1,2,2), imshow(B);	%Display result.

3.4 Pixel distributions: histograms

An image histogram is a plot of the relative frequency of occurrence of each of the permitted pixel values in the image against the values themselves. If we normalize such a frequency plot, so that the total sum of all frequency entries over the permissible range is one, we may treat the image histogram as a discrete probability density function which defines the likelihood of a given pixel value occurring within the image. Visual inspection of an image histogram can reveal the basic contrast that is present in the image and any potential differences in the colour distribution of the image foreground and background scene components.

For a simple grey-scale image, the histogram can be constructed by simply counting the number of times each grey-scale value (0–255) occurs within the image. Each 'bin' within the histogram is incremented each time its value is encountered thus an image histogram can easily be constructed as follows -

```
initialize all histogram array entries to 0
for each pixel I(i,j) within the image I
     histogram(I(i,j)) = histogram(I(i,j)) + 1
end
```

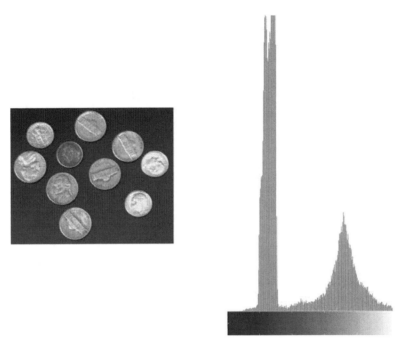

Figure 3.12 Sample image and corresponding image histogram

In Matlab we can calculate and display an image histogram as in Example 3.12. An example histogram is shown in Figure 3.12, where we see a histogram plot with two distinctive peaks: a high peak in the lower range of pixel values corresponds to the background intensity distribution of the image and a lower peak in the higher range of pixel values (bright pixels) corresponds to the foreground objects (coins).

Following on from Example 3.12, we can also individually query and address the histogram 'bin' values to display the pixel count associated with a selected 'bin' within the first peak; see Example 3.13.

Example 3.12

Matlab code	What is happening?
I=imread('coins.png');	%Read in image
subplot(1,2,1), imshow(I);	%Display image
subplot(1,2,2), imhist(I);	%Display histogram

Example 3.13

Matlab code	What is happening?
I=imread('coins.png');	%Read in image
[counts,bins]=imhist(I);	%Get histogram bin values
counts(60)	%Query 60th bin value

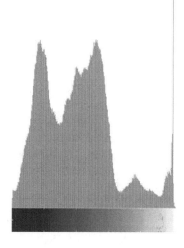

Figure 3.13 Sample image and corresponding image histogram

3.4.1 Histograms for threshold selection

In the example shown in Figure 3.12, we essentially see the image histogram plotted as a bar graph. The *x*-axis represents the range of values within the image (0–255 for 8-bit grayscale) and the *y*-axis shows the number of times each value actually occurs within the particular image. By selecting a threshold value between the two histogram peaks we can successfully separate the background/foreground of the image using the thresholding approach of Section 3.2.3 (a threshold of ∼120 is suitable for Figure 3.12). Generally, scenes with clear bimodal distributions work best for thresholding. For more complex scenes, such as Figure 3.13, a more complex histogram occurs and simple foreground/background thresholding cannot be applied. In such cases, we must resort to more sophisticated image segmentation techniques (see Chapter 10).

In Matlab, we can use the image histogram as the basis for calculation of an automatic threshold value. The function ***graythresh*** in Example 3.14 exploits the Otsu method, which chooses that threshold value which minimizes the interclass statistical variance of the thresholded black and white pixels.

Example 3.14

Matlab code	What is happening?
I=imread('coins.png');	%Read in image
level=graythresh(I);	%Get OTSU threshold
It=im2bw(I, level);	%Threshold image
imshow(It);	%Display it

3.4.2   Adaptive thresholding

Adaptive thresholding is designed to overcome the limitations of conventional, global thresholding by using a different threshold at each pixel location in the image. This local threshold is generally determined by the values of the pixels in the neighbourhood of the pixel. Thus, adaptive thresholding works from the assumption that illumination may differ over the image but can be assumed to be roughly uniform in a sufficiently small, local neighbourhood.

The local threshold value t in adaptive thresholding can be based on several statistics. Typically, the threshold is chosen to be either $t = \text{mean} + C$, $t = \text{median} + C$ or $\text{floor}((\text{max}-\text{min})/2) + C$ of the local $N \times N$ pixel neighbourhood surrounding each pixel. The choice of N is important and must be large enough to cover sufficient background and foreground pixels at each point but not so large as to let global illumination deficiencies affect the threshold. When insufficient foreground pixels are present, then the chosen threshold is often poor (i.e. the mean/median or average maximum/minimum difference of the neighbourhood varies largely with image noise). For this reason, we introduce the constant offset C into the overall threshold to set the threshold above the general image noise variance in uniform pixel areas.

In Example 3.15 we carry out adaptive thresholding in Matlab using the threshold $t = \text{mean} + C$ for a neighbourhood of $N = 15$ and $C = 20$. The result of this example is shown in Figure 3.14, where we see the almost perfect foreground separation of the rice grains

---

Example 3.15

Matlab code	What is happening?
I=imread('rice.png');	%Read in image
Im=imfilter(I,fspecial('average',[15 15]),'replicate');	%Create mean image
It=I-(Im + 20);	%Subtract mean image
	(+ constant C=20)
Ibw=im2bw(It,0);	%Threshold result at 0
	(keep +ve results only)
subplot(1,2,1), imshow(I);	%Display image
subplot(1,2,2), imshow(Ibw);	%Display result

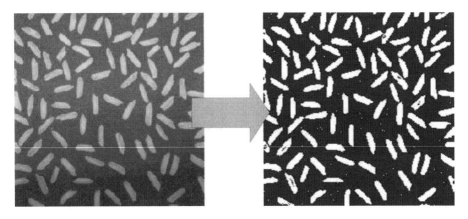

Figure 3.14   Adaptive thresholding applied to sample image

despite the presence of an illumination gradient over the image (contrast this to the bottom of Figure 3.4 using global thresholding). Further refinement of this result could be performed using the morphological operations of opening and closing discussed in Chapter 8.

Example 3.15 can also be adapted to use the threshold $t = \text{median} + C$ by replacing the line of Matlab code that constructs the mean image with one that constructs the median image; see Example 3.16. Adjustments to the parameters N and C are also required to account for the differences in the threshold approach used.

Example 3.16

Matlab code	What is happening?
Im=medfilt2(I,[N N]);	%Create median image

3.4.3   Contrast stretching

Image histograms are also used for contrast stretching (also known as normalization) which operates by stretching the range of pixel intensities of the input image to occupy a larger dynamic range in the output image.

In order to perform stretching we must first know the upper and lower pixel value limits over which the image is to be normalized, denoted a and b respectively. These are generally the upper and lower limits of the pixel quantization range in use (i.e. for an 8-bit image, $a = 255$ and $b = 0$ – see Section 1.2). In its simplest form, the first part of the contrast stretch operation then scans the input image to determine the maximum and minimum pixel values currently present, denoted c and d respectively. Based on these four values (a, b, c and d), the image pixel range is stretched according to the following formula:

$$I_{\text{output}}(i,j) = (I_{\text{input}}(i,j) - c)\left(\frac{a-b}{c-d}\right) + a \qquad (3.10)$$

for each pixel location denoted (i,j) in the input and output images. Contrast stretching is thus a further example of a point transform, as discussed earlier in this chapter.

In reality, this method of choosing c and d is very naive, as even a single outlier pixel (i.e. one that is unrepresentative of the general pixel range present within the image – e.g. salt and pepper noise) will drastically affect the overall enhancement result. We can improve upon this method by ensuring that c and d are truly representative of the image content and robust to such statistical outliers. Two such methods are presented here.

- *Method 1*   Compute the histogram of the input image and select c and d as the 5th and 95th percentile points of the cumulative distribution (i.e. 5% of the image pixels will be less than c and 5% greater than d).

- *Method 2*   Compute the histogram of the input image and find the most frequently occurring intensity value within the image. Let us assume that this peak value has a bin count of N. Select as a cut-off some fraction of N (e.g. 5%). Move away from the peak in either direction (left towards 0 one way, right towards 255) until the last values greater than the cut-off are reached. These values are c and d (see the histogram in Figure 3.12).

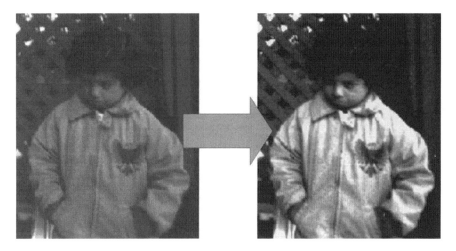

Figure 3.15   Contrast stretch applied to sample image

Method 2 is somewhat less robust to complex, multi-peak histograms or histograms that do not follow typical intensity gradient distributions.

In Example 3.17, we carry out contrast stretching in Matlab using method 1. The result of Example 3.17 is shown in Figure 3.15, where we see that the contrast is significantly improved (albeit slightly saturated). We can display the histograms before and after contrast stretching for this example using Example 3.18. These histogram distributions are shown in Figure 3.16. We can clearly see the restricted dynamic range of the original image and how the contrast-stretched histogram corresponds to a horizontally scaled version of the original.

Example 3.17

Matlab code	What is happening?
I=imread('pout.tif');	%Read in image
Ics=imadjust(I,stretchlim(I, [0.05 0.95]),[]);	%Stretch contrast using method 1
subplot(1,2,1), imshow(I);	%Display image
subplot(1,2,2), imshow(Ics);	%Display result

Comments

Here we use the *stretchlim()* function to identify the c and d pixel values at the 5th and 95th percentile points of the (normalized) histogram distribution of the image. The *imadjust()* function is then used to map the this range to the (default) maximum quantization range of the image.

Example 3.18

Matlab code	What is happening?
subplot(1,2,3), imhist(I);	%Display input histogram
subplot(1,2,4), imhist(Ics);	%Display output histogram

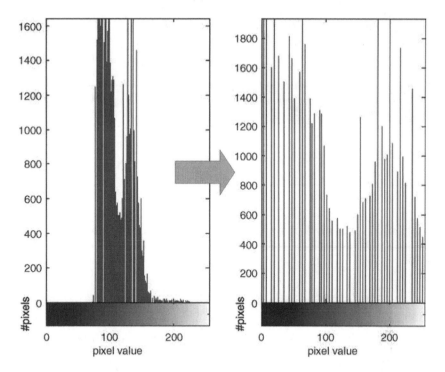

Figure 3.16 Before and after histogram distributions for Figure 3.15

As is evident in this example (Figure 3.15), image enhancement operations such as this are highly subjective in their evaluation, and parameter adjustments (i.e. values c and d) can subsequently make a significant difference to the resulting output.

3.4.4 Histogram equalization

The second contrast enhancement operation based on the manipulation of the image histogram is histogram equalization. This is one of the most commonly used image enhancement techniques.

3.4.4.1 Histogram equalization theory

Initially, we will assume a grey-scale input image, denoted $I_{input}(x)$. If the variable x is continuous and normalized to lie within the range $[0, 1]$, then this allows us to consider the normalized image histogram as a probability density function (PDF) $p_x(x)$, which defines the likelihood of given grey-scale values occurring within the vicinity of x. Similarly, we can denote the resulting grey-scale output image after histogram equalisation as $I_{output}(y)$ with corresponding PDF $p_y(y)$.

The essence of the histogram equalization problem is that we seek some transformation function $y = f(x)$ that maps between the input and the output grey-scale image values and which will transform the input PDF $p_x(x)$ to produce the desired output PDF $p_y(y)$. A standard result from elementary probability theory states that:

$$p_y(y) = p_x(x)\left|\frac{dx}{dy}\right| \tag{3.11}$$

which implies that the desired output PDF depends only on the known input PDF and the transformation function $y = f(x)$. Consider, then, the following transformation function, which calculates the area under the input probability density curve (i.e. integral) between 0 and an upper limit x:

$$y(x) = \int_0^x p_x(x')\,dx' \tag{3.12}$$

This is recognizable as the cumulative distribution function (CDF) of the random variable x. Differentiating this formula, applying Leibniz's rule[1] and substituting into our previous statement we obtain the following:

$$p_y(y) = p_x(x)\left|\frac{1}{p_x(x)}\right| \tag{3.13}$$

Finally, because $p_x(x)$ is a probability density and guaranteed to be positive $(0 \leq p_x(x) \leq 1)$, we can thus obtain:

$$p_y(y) = \frac{p_x(x)}{p_x(x)} = 1 \qquad 0 \leq y \leq 1 \tag{3.14}$$

The output probability density $p_y(y)$ is thus constant, indicating that all output intensity values y are equally probable and, thus, the histogram of output intensity values is equalized.

The principle of histogram equalization for the continuous case, therefore, is encapsulated by this mapping function $y = f(x)$ between the input intensity x and the output intensity y, which, as shown, is given by the cumulative integral of the input PDF (the cumulative distribution in statistics). However, we must note that the validity of the density transformation result given above depends on the satisfaction of two conditions. The function $y(x)$ must be (a) single valued and (b) monotonically increasing, i.e. $0 \leq y(x) \leq 1$ for $0 \leq x \leq 1$. Condition (a) first ensures that the inverse transformation $x^{-1}(y)$ exists and that the ascending pixel value order from black to white (dark to light, 0 to 255) is preserved. Note that, if the function were not monotonically increasing, the inverse transformation would not be single valued, resulting in a situation in which some output intensities map to more than one input intensity. The second requirement simply ensures that the output range maps the input.

3.4.4.2   *Histogram equalization theory: discrete case*
Intensity values in real digital images can occupy only a finite and discrete number of levels. We assume that there are S possible discrete intensity values denoted x_k, where $k = \{0, 1, 2, \dots, S{-}1\}$, and the input probability density for level x_k is $p_x(x_k)$. In this

---

[1] Leibniz's rule: the derivative of a definite integral with respect to its upper limit is equal to the integrand evaluated at the upper limit.

discrete case, the required general mapping $y = f(x)$ can be defined specifically for x_k as the following summation:

$$y(x_k) = \sum_{j=0}^{k} p_x(x_k) \tag{3.15}$$

This is essentially the cumulative histogram for the input image x, where the kth entry $y(x_k)$ is the sum of all the histogram bin entries up to and including k. Assuming, as is normally the case, that the permissible intensity values in the input and output image can be denoted simply by discrete integer values k, where $k = \{0, 1, 2, \ldots, S-1\}$, this can be written simply as:

$$y_k = \sum_{j=0}^{k} p_x(j) = \frac{1}{N} \sum_{j=0}^{k} n_j \tag{3.16}$$

where the population of the kth level is given by n_k and the total number of pixels in the image is N.

This definition makes the corresponding, computational procedure of histogram equalization a simple procedure to implement in practice. Unlike its continuous counterpart, however, the discrete transformation for $y = f(x)$ cannot in general produce a perfectly equalized (uniform) output histogram (i.e. in which all intensity values are strictly equally probable). In practice, it forms a good approximation to the ideal, driven and adapted to the cumulative histogram of the input image, that spreads the intensity values more evenly over the defined quantization range of the image.

3.4.4.3 *Histogram equalization in practice*

The two main attractions of histogram equalization are that it is a fully automatic procedure and is computationally simple to perform. The intensity transformation $y = f(x)$ we have defined in this section depends only on the readily available histogram of the input image.

Histogram modelling provides a means of modifying the dynamic range of an image such that its histogram distribution conforms to a given shape. In histogram equalization, we employ a monotonic, nonlinear mapping such that the pixels of the input image are mapped to an output image with a uniform histogram distribution. As shown in Section 3.4.4.2, this required mapping can be defined as the cumulative histogram $C(i)$ such that each entry in is the sum of the frequency of occurrence for each grey level up to and including the current histogram bin entry i. By its very nature $C()$ is a single-valued, monotonically increasing function. Its gradient will be in proportion to the current equalization of the image (e.g. a constant gradient of 1 will be a perfectly equalized image, as the increase at each step is constant).

In the idealized case, the resulting equalized image will contain an equal number of pixels each having the same grey level. For L possible grey levels within an image that has N pixels, this equates to the jth entry of the cumulative histogram $C(j)$ having the value jN/L in this idealized case (i.e. j times the equalized value). We can thus find a mapping between input pixel intensity values i and output pixel intensity values j as follows:

$$C(i) = j\frac{N}{L} \tag{3.17}$$

from which rearrangement gives:

$$j = \frac{L}{N}C(i) \tag{3.18}$$

which represents a mapping from a pixel of intensity value i in the input image to an intensity value of j in the output image via the cumulative histogram $C()$ of the input image.

Notably, the maximum value of j from the above is L, but the range of grey-scale values is strictly $j = \{0 \ldots (L-1)\}$ for a given image. In practice this is rectified by adding a -1 to the equation, thus also requiring a check to ensure a value of $j = -1$ is not returned:

$$j = \max\left(0, \frac{L}{N}C(i) - 1\right) \tag{3.19}$$

which, in terms of our familiar image-based notation, for 2-D pixel locations $i = (c, r)$ and $j = (c, r)$ transforms to

$$I_{\text{output}}(c, r) = \max\left(0, \frac{L}{N}C(I_{\text{input}}(c, r)) - 1\right) \tag{3.20}$$

where $C()$ is the cumulative histogram for the input image I_{input}, N is the number of pixels in the input image $(C \times R)$ and L is the number of possible grey levels for the images (i.e. quantization limit, Section 1.2). This effectively provides a compact look-up table for mapping pixel values in I_{input} to I_{output}. As an automatic procedure, this mapping, which constitutes histogram equalization, can be readily performed as in Example 3.19. The output result of Example 3.19 is shown in Figure 3.17, which can be compared with the input dynamic range of the image and corresponding input image histogram of Figures 3.15 (left) and 3.16 (left) respectively. Here (Figure 3.17), we can see the equalization effect on the dynamic range of the image and the corresponding equalization of the histogram

Example 3.19

Matlab code	What is happening?
I=imread('pout.tif');	%Read in image
Ieq=histeq(I);	
subplot(2,2,1), imshow(I);	%Display image
subplot(2,2,2), imshow(Ieq);	%Display result
subplot(2,2,3), imhist(I);	%Display histogram of image
subplot(2,2,4), imhist(Ieq);	%Display histogram of result

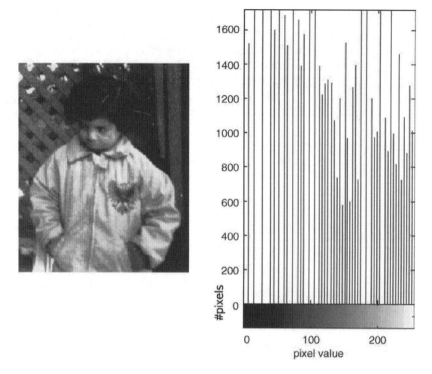

Figure 3.17   Histogram equalization applied to sample image

distribution over the full quantization range of the image (in comparison with the inputs of Figures 3.15 (left) and 3.16 (left)). Notably, the resulting histogram of Figure 3.17 does not contain all equal histogram entries, as discussed in Sections 3.4.4.1 and 3.4.4.2 – *this is a common misconception of histogram equalization.*

3.4.5   Histogram matching

Despite its attractive features, histogram equalization is certainly no panacea. There are many instances in which equalization produces quite undesirable effects. A closely related technique known as histogram matching (also known as *histogram specification*) is sometimes more appropriate and can be employed both as a means for improving visual contrast and for regularizing an image prior to subsequent processing or analysis.

The idea underpinning histogram matching is very simple. Given the original (input) image I_{input} and its corresponding histogram $p_x(x)$, we seek to effect a transformation $f(x)$ on the input intensity values such that the transformed (output) image I_{output} has a desired (target) histogram $p_z(z)$. Often, the target histogram will be taken from a model or 'typical' image of a similar kind.

3.4.5.1   Histogram matching theory
For the moment, we will gloss over the question of how to obtain the target histogram and simply assume it to be known. Our matching task, then, can be formulated as follows.

Let the input intensity levels be denoted by the variable $x(0 \leq x \leq 1)$ and the PDF (i.e. the continuous, normalized version of the histogram) of the input image be denoted by $p_x(x)$. Similarly, let the output (matched) intensity levels be denoted by the variable $z(0 \leq z \leq 1)$ and the corresponding PDF of the input image be denoted by $p_z(z)$. Our histogram-matching task can then be formulated as the derivation of a mapping $f(x)$ between the input intensities x and the output intensities z such that the mapped output intensity values have the desired PDF $p_z(z)$. We can approach this problem as follows.

The cumulative distribution (CDF) $C_x(x)$ of the input image is, by definition, given by the integral of the input PDF:

$$C_x(x) = \int_0^x p_{x'}(x') \, dx' \tag{3.21}$$

Similarly, the CDF of the output image $C_z(z)$ is given by the integral of the output PDF:

$$C_z(z) = \int_0^z p_{z'}(z') \, dz' \tag{3.22}$$

A key point in our reasoning is the recognition that both of these transforms are invertible. Specifically, an arbitrary PDF $f(x)$ is related to its CDF $p(X)$ by the relation

$$p(x) = \left. \frac{dP}{dX} \right|_{X=x} \tag{3.23}$$

Thus, knowledge of a CDF uniquely determines the PDF; and if the CDF is known, then the PDF is also known and calculable through this explicit differential. It follows that if we can define a mapping $f(x)$ between input intensities (x) and output intensities (z) such that the input and output CDFs are identical, we thereby guarantee that the corresponding PDFs will be the same. Accordingly, we demand that the CDFs defined previously, $C_x(x)$ and $C_z(z)$, be equal for the mapping $z = f(x)$:

$$C_z[f(x)] = C_x(x) \tag{3.24}$$

from which it follows that the required mapping $f()$ is

$$f(x) = C_z^{-1}[C_x(x)] \tag{3.25}$$

The definition of this mapping is fundamental to the ability to map input $C_x(x)$ to $C_z(z)$ and, hence, input PDF $p_x(x)$ to output PDF $p_z(z)$.

3.4.5.2 *Histogram matching theory: discrete case*
In general, the inverse mapping C^{-1} defined previously in Section 3.4.5.1 is not an analytic function and we must resort to numerical techniques to approximate the required mapping.

To effect a discrete approximation of the matching of two arbitrary image histograms, we proceed by first calculating the discrete CDF $C_x(k)$ of the input image:

$$C_x(k) = \sum_{j=0}^{k} p_x(j) = \frac{1}{N} \sum_{j=0}^{k} x_j \qquad (3.26)$$

where x_j is the population of the jth intensity level in the input image, N is the total number of pixels in that image and $k = \{0, 1, 2, \ldots, L-1\}$, where L is the number of possible grey levels for the image.

Second, in a similar manner, we calculate the discrete CDF $C_z(l)$ of the output image:

$$C_z(l) = \sum_{j=0}^{l} p_z(j) = \frac{1}{N} \sum_{j=0}^{l} z_j \qquad (3.27)$$

where z_j is the population of the jth intensity level in the output image, N is the total number of pixels and $l = \{0, 1, 2, \ldots, L-1\}$, where L is the number of possible grey levels for the image as before. Finally, we effect the discrete version of the earlier transform $f(x) = C_z^{-1}[C_x(x)]$ and must find the mapping defined by C_z^{-1} when the input to it is the CDF of the input image, $C_x()$. Like the earlier example on histogram equalization (Section 3.4.4), this mapping effectively defines a look-up table between the input and output image pixel values which can be readily and efficiently computed.

3.4.5.3 *Histogram matching in practice*

Histogram matching extends the principle of histogram equalization by generalizing the form of the target histogram. It is an automatic enhancement technique in which the required transformation is derived from a (user-) specified target histogram distribution.

In practice, the target histogram distribution t will be extracted from an existing reference image or will correspond to a specified mathematical function with the correct properties. In Matlab, histogram matching can be achieved as in Example 3.20. In this example, we specify a linearly ramped PDF as the target distribution t and we can see the resulting output together with the modified image histograms in Figure 3.18.

Example 3.20

Matlab code	What is happening?
I=imread('pout.tif');	
pz=0:255;	%Define ramp-like pdf as desired output histogram
Im=histeq(I, pz);	%Supply desired histogram to perform matching
subplot(2,3,1), imshow(I);	%Display image
subplot(2,3,2), imshow(Im);	%Display result
subplot(2,3,3), plot(pz);	%Display distribution t
subplot(2,3,4), imhist(I);	%Display histogram of image
subplot(2,3,5), imhist(Im);	%Display histogram of result

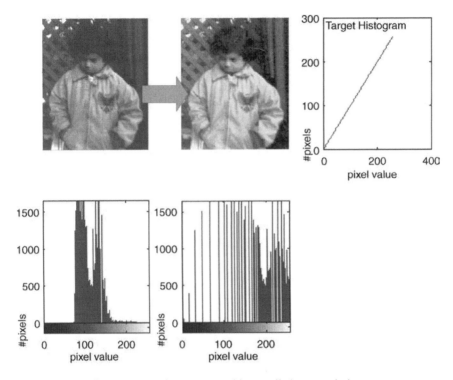

Figure 3.18 Histogram matching applied to sample image

It is evident from Figure 3.18 that the desired histogram of the output image (in this case chosen arbitrarily) does not produce an output with particularly desirable characteristics above that of the histogram equalization shown in Figure 3.17. In general, choice of an appropriate target histogram presupposes knowledge of what structures in the input image are present and need enhancing. One situation in which the choice of output histogram may be essentially fixed is when the image under consideration is an example of a given class (and in which the intensity distribution should thus be constrained to a specific form) but has been affected by extraneous factors. One example of this is a sequence of faces in frontal pose. The histogram of a given face may be affected by illumination intensity and by shadowing effects. A face captured under model/ideal conditions could supply the desired histogram to correct these illumination and shadowing effects prior to subsequent use of the transformed input image for facial recognition.

3.4.6   Adaptive histogram equalization

Sometimes the overall histogram of an image may have a wide distribution whilst the histogram of local regions is highly skewed towards one end of the grey spectrum. In such cases, it is often desirable to enhance the contrast of these local regions, but global histogram equalization is ineffective. This can be achieved by adaptive histogram equalization. The term adaptive implies that different regions of the image are processed differently (i.e. different look-up tables are applied) depending on local properties.

There are several variations on adaptive histogram equalization, but perhaps the simplest and most commonplace is the so-called sliding window (or tile-based) approach. In this method, the underlying image is broken down into relatively small contiguous 'tiles' or local $N \times M$ neighbourhood regions (e.g. 16×16 pixels). Each tile or inner window is surrounded by a larger, outer window which is used to calculate the appropriate histogram equalization look-up table for the inner window.

This is generally effective in increasing local contrast, but 'block artefacts' can occur as a result of processing each such region in isolation and artefacts at the boundaries between the inner windows can tend to produce the impression of an image consisting of a number of slightly incongruous blocks. The artefacts can generally be reduced by increasing the size of the outer window relative to the inner window.

An alternative method for adaptive histogram equalization (attributed to Pizer) has been applied successfully, the steps of which are as follows:

- A regular grid of points is superimposed over the image. The spacing of the grid points is a variable in this approach, but is generally a few tens of pixels.

- For each and every grid point, a rectangular window with twice the grid spacing is determined. A given window thus has a 50% overlap with its immediate neighbours to north, south, east and west.

- A histogram-equalized look-up table is calculated for each such window. Owing to the 50% overlap between the windows, every pixel within the image lies within four adjacent, rectangular windows or four neighbourhoods.

- The transformed value of a given image pixel is calculated as a weighted combination of the output values from the four neighbourhood look-up tables using the following bilinear formula:

$$I = (1-a)(1-b)I_1 + a(1-b)I_2 + (1-a)bI_3 + abI_4 \qquad (3.28)$$

for distance weights $0 \leq a, b \leq 1$ and histogram-equalized look-up table values I_1, I_2, I_3 and I_4 from each of the four adjacent neighbourhoods. This methodology is illustrated further in Figure 3.19. Note how this formula gives appropriate importance to the neighbourhoods to which the point of calculation I most fully belongs; e.g. a pixel located precisely on a grid point derives its equalized value from its surrounding neighbourhood only, whereas a point which is equidistant from its four nearest grid points is an equally weighted combination of all four surrounding neighbourhood values. It is important to remember the overlapping nature of the neighbourhoods when considering Figure 3.19.

A final extension to this method is the contrast-limited adaptive histogram equalization approach. In general terms, calculating local region histograms and equalizing the intensity values within those regions has a tendency to increase the contrast too much and amplify noise. Some regions within an image are inherently smooth with little real contrast and blindly equalizing the intensity histogram in the given region can have undesirable results.

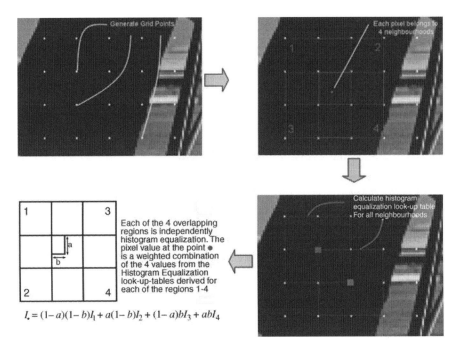

$$I_\bullet = (1-a)(1-b)I_1 + a(1-b)I_2 + (1-a)bI_3 + abI_4$$

Figure 3.19 Adaptive histogram equalization multi-region calculation approach

The main idea behind the use of contrast limiting is to place a limit l, $0 \leq l \leq 1$, on the overall (normalized) increase in contrast applied to any given region.

This extended method of contrast-limited adaptive histogram equalization is available in Matlab as in Example 3.21. The result of Example 3.21 is shown in Figure 3.20, where we see the differing effects of the specified contrast limit l and target distribution t on the resulting adaptively equalized image. Adaptive histogram equalization can sometimes effect significant improvements in local image contrast. However, owing to the fundamentally different nature of one image from the next, the ability to vary the number of tiles into which the image is decomposed and the specific form of the target probability distribution, there is little general theory to guide us. Obtaining an image with the desired contrast characteristics is thus, to a considerable degree, an art in which these parameters are varied on an experimental basis.

Example 3.21

Matlab code	What is happening?
I=imread('pout.tif');	%Read in image
I1=adapthisteq(I,'clipLimit',0.02,'Distribution','rayleigh');	
I2=adapthisteq(I,'clipLimit',0.02,'Distribution','exponential');	
I3=adapthisteq(I,'clipLimit',0.08,'Distribution','uniform');	
subplot(2,2,1), imshow(I); subplot(2,2,2), imshow(I2);	%Display orig. + output
subplot(2,2,3), imshow(I2); subplot(2,2,4), imshow(I3);	%Display outputs

Comments

- Here we use the ***adapthisteq()*** function to perform this operation with a contrast limit l (parameter 'clipLimit'=0.02/0.08) set accordingly. In addition, the Matlab implementation of adaptive histogram equalization also allows the user to specify a target distribution t for use with every region in the image (in the style of histogram matching, Section 3.4.5). Standard equalization is performed with the specification of (the default) *uniform* distribution.

- By default, the Matlab implementation uses an 8×8 division of the image into windows/neighbourhoods for processing. This, like the other parameters in this example, can be specified as named ('parameter name', value) pairs as inputs to the function. Please refer to the Matlab documentation on ***adapthisteq()*** for further details (*doc adapthisteq* at the Matlab command prompt).

Figure 3.20   Adaptive histogram equalization applied to a sample image

3.4.7 *Histogram operations on colour images*

Until this point, we have only considered the application of our contrast manipulation operations on single-channel, grey-scale images. Attempting to improve the contrast of colour images is a slightly more complex issue than for grey-scale intensity images. At first glance, it is tempting to consider application of histogram equalization or matching independently to each of the three channels (R,G,B) of the true colour image. However, the RGB values of a pixel determine both its intensity and its chromaticity (i.e. the subjective impression of colour). Transformation of the RGB values of the pixels to improve contrast will, therefore, in general, alter the chromatic (i.e. the colour hue) content of the image.

Figure 3.21 Adaptive histogram equalization applied to a sample colour image *(See colour plate section for colour version)*

The solution is first to transform the colour image to an alternative, perceptual colour model, such as HSV, in which the luminance component (intensity V) is decoupled from the chromatic (H and S) components which are responsible for the subjective impression of colour. In order to perform such histogram operations on colour images, we thus (a) transform the RGB component image to the HSV representation (hue, saturation, variance), (b) apply the histogram operation to the intensity component and finally (c) convert the result back to the RGB colour space as required. The HSV colour model space is not unique in this sense and there are actually several colour models (e.g. $L*a*b$) that we could use to achieve this. HSV is used as our example following on from its introduction in Chapter 1.

Using such a colour space conversion histogram operations can be applied in Matlab as in Example 3.22. The result of Example 3.22 is shown in Figure 3.21, where we see the application of histogram equalization (Section 3.4.4.3, Example 3.19) to an RGB colour image using an intermediate HSV colour space representation. Note that the subjective impression of the colour (chromaticity) has been maintained, yet the effect of the histogram equalization of the intensity component is evident in the result.

Example 3.22

Matlab code	What is happening?
I=imread('autumn.tif ');	%Read in image
Ihsv=rgb2hsv(I);	%Convert original to HSV image, I2
V=histeq(Ihsv(:,:,3));	%Histogram equalise V (3rd) channel of I2
Ihsv(:,:,3)=V;	%Copy equalized V plane into (3rd) channel I2
Iout=hsv2rgb(Ihsv);	%Convert I2 back to RGB form
subplot(1,2,1), imshow(I);	
subplot(1,2,2), imshow(Iout);	

Comments

- The functions **rgb2hsv** and **hsv2rgb** are used to convert between the RGB and HSV colour spaces.
- Individual colour channels of an image *I* can be referenced as 'I(:,:,n)' in Matlab syntax, where n is the numerical channel number in range $n = \{1,2,3\} = \{R,G,B\} = \{H,S,V\}$ and similarly for other colour space and image representations.

Exercises

The following exercises are designed to reinforce and develop the concepts and Matlab examples introduced in this chapter. Additional information on all of the Matlab functions represented in this chapter and throughout these exercises is available in Matlab from the function help browser (use *doc <function name>* at the Matlab command prompt, where *<function name>* is the function required)

   Matlab functions: *imresize, size, whos, imadd, imsubtract, immultiply, imagesc, imcomplement, imabsdiff, implay, uint8, horzcat, tic, toc, rgb2ycbcr*.

Exercise 3.1   Using the examples presented on arithmetic operations (Examples 3.1–3.4) we can investigate the use of these operators in a variety of ways. First, load the Matlab example images 'rice.png' and 'cameraman.tif'. Investigate the combined use of the Matlab *imresize()* and *size()* functions to resize one of the images to be the same size as the other. The Matlab command *whos* used in the syntax '*whos* v' will display information about the size and type of a given image or other variable *v*.

   Try adding both of these images together using the standard Matlab addition operator ' + ' and displaying the result. What happens? Now add them together by using the *imadd()* function but adding a third parameter to the function of 'uint16' or 'double' with quotes that forces an output in a 16-bit or floating-point double-precision data type. You will need to display the output using the *imagesc()* function. How do these two addition operator results differ and why do they differ?

   Repeat this exercise using the standard Matlab subtraction operator '−' and also by using the *imsubtract()* function.

Exercise 3.2   Building upon the logical inversion example presented in Example 3.6, investigate the application of the *imcomplement()* function to different image types and different applications. First load the example image 'peppers.png' and apply this operator. What effect do you see and to what aspect of traditional photography does the resulting image colouring relate?

   Image inversion is also of particular use in medical imaging for highlighting different aspects of a given image. Apply the operator to the example images 'mir.tif', 'spine.tif' and cell image 'AT3_1m4_09.tif'. What resulting effects do you see in the transformed images which may be beneficial to a human viewer?

Exercise 3.3   Use the sequence of cell images ('AT3_1m4_01.tif', 'AT3_1m4_02.tif', ... 'AT3_1m4_09.tif', 'AT3_1m4_10.tif') provided in combination with the Matlab *imabsdiff()* function and a Matlab *for* loop construct to display an animation of the differences between images in the sequence.

   You may wish to use an additional enhancement approach to improve the dynamic range of difference images that result from the *imabsdiff()* function. What is result of this differencing operation? How could this be useful?

   Hint. You may wish to set up an array containing each of the image file names 01 to 10. The animation effect can be achieved by updating the same figure for each set of differences (e.g. between the *k*th and (*k* − 1)th images in the sequence) or by investigating the Matlab *implay()* function to play back an array of images.

Exercise 3.4   Using a combination of the Matlab *immultiply()* and *imadd()* functions implement a Matlab function (or sequence of commands) for blending two images A and B into a single image with corresponding blending weights w_A and w_B such that output image C is

$$C = w_A A + w_B B \qquad\qquad (3.29)$$

Experiment with different example images and also with blending information from a sequence of images (e.g. cell images from Exercise 3.3). How could such a technique be used in a real application?

Exercise 3.5   Using the thresholding technique demonstrated in Example 3.7 and with reference to the histogram display functions of Example 3.12, manually select and apply a threshold to the example image 'pillsetc.png'. Compare your result with the adaptive threshold approaches shown in Examples 3.15 and 3.16. Which is better? (Also, which is easier?)

Repeat this procedure to isolate the foreground items in example images 'tape.png', 'coins.png' and 'eight.tif'. Note that images require to be transformed to grey scale (see Section 1.4.1.1) prior to thresholding.

Exercise 3.6   Looking back to Examples 3.15 and 3.16, investigate the effects of varying the constant offset parameter C when applying it to example images 'cameraman.tif' and 'coins.png'. How do the results for these two images differ?

Implement the third method of adaptive thresholding from Section 3.4.2 using the threshold $t = \text{floor}((\text{max}-\text{min})/2) + C$ method. Compare this approach against the examples already provided for thresholding the previous two images and other available example images.

Exercise 3.7   Read in the example images 'cameraman.tif' and 'circles.png' and convert the variable resulting from the latter into the unsigned 8-bit type of the former using the Matlab casting function *uint8()*. Concatenate these two images into a single image (using the Matlab function *horzcat()*) and display it. Why can you not see the circles in the resulting image? (Try also using the Matlab *imagesc* function).

Using the logarithmic transform (Example 3.8), adjust the dynamic range of this concatenated image so that both the outline of the cameraman's jacket and the outline of the circles are just visible (parameter $C > 10$ will be required). By contrast, investigate the use of both histogram equalization and adaptive histogram equalization on this concatenated image. Which approach gives the best results for overall image clarity and why is this approach better for this task?

Exercise 3.8   Consider the Matlab example image 'mandi.tif', where we can see varying lighting across both the background and foreground. Where is information not clearly visible within this image? What are the properties of these regions in terms of pixel values?

Consider the corresponding properties of the logarithmic and exponential transforms (Sections 3.3.1 and 3.3.2) and associated Examples 3.8 and 3.9. Experiment with both of these transforms and determine a suitable choice (with parameters) for enhancing this image. Note that this is large image example and processing may take a few seconds (use the Matlab functions *tic* and *toc* to time the operation).

Contrast the results obtained using these two transforms to applying histogram equalization contrast stretch (Section 3.4.4) or adaptive histogram equalization (Section 3.4.6) to this image.

Exercise 3.9 Based on the grey-scale gamma correction presented in Example 3.11, apply gamma correction to a colour image using the example 'autumn.tif'. Why can we (and why would we) apply this technique directly to the RGB image representation and not an alternative HSV representation as per the colour image histogram processing of Example 3.22?

Exercise 3.10 Based on Example 3.22, where we apply histogram equalization to a colour image, apply contrast stretching (Section 3.4.3, Example 3.17) to the colour example image 'westconcordaerial.png' using the same approach. Experiment with different parameter values to find an optimum for the visualization of this image. Compare the application of this approach with histogram equalization and adaptive histogram equalization on the same image. Which produces the best result? Which approach is more 'tunable' and which is more automatic in nature?

Repeat this for the Matlab image example 'peppers.png'. Do the results differ significantly for this example?

Exercise 3.11 Using the various contrast enhancement and dynamic range manipulation approaches presented throughout this chapter, investigate and make a selection of transforms to improve the contrast in the example image 'AT3_1m4_01.tif'. Once you have achieved a suitable contrast for this example image, extract the corresponding histogram distribution of the image (Example 3.13) and apply it to the other images in the sequence ('AT3_1m4_02.tif', ... 'AT3_1m4_09.tif', 'AT3_1m4_10.tif') using histogram matching (Example 3.20). Are the contrast settings determined for the initial example image suitable for all of the others? What if the images where not all captured under the same conditions?

Here, we see the use of histogram matching as a method for automatically setting the dynamic range adjustment on a series of images based on the initial determination of suitable settings for one example. Where else could this be applied?

Exercise 3.12 In contrast to the approach shown for colour histogram equalization shown in Example 3.22, perform histogram equalization on each of the (R,G,B) channels of an example colour image. Compare the results with that obtained using the approach of Example 3.22. What is the difference in the resulting images?

Repeat this for the operations of contrast stretching and adaptive histogram equalization. Investigate also using the YCbCr colour space (Matlab function *rgb2ybcr()*) for performing colour histogram operations on images. Which channel do you need to use? What else do you need to consider?

For further examples and exercises see http://www.fundipbook.com

4

Enhancement

The techniques we introduced at the end of Chapter 3 considered the manipulation of the dynamic range of a given digital image to improve visualization of its contents. In this chapter we consider more general image enhancement. We introduce the concept of image filtering based on localized image subregions (pixel neighbourhoods), outline a range of noise removal filters and explain how filtering can achieve edge detection and edge sharpening effects for image enhancement.

4.1 Why perform enhancement?

The basic goal of image enhancement is to process the image so that we can view and assess the visual information it contains with greater clarity. Image enhancement, therefore, is rather subjective because it depends strongly on the specific information the user is hoping to extract from the image.

The primary condition for image enhancement is that the information that you want to extract, emphasize or restore must exist in the image. Fundamentally, 'you cannot make something out of nothing' and the desired information must not be totally swamped by noise within the image. Perhaps the most accurate and general statement we can make about the goal of image enhancement is simply that the processed image should be more suitable than the original one for the required task or purpose. This makes the evaluation of image enhancement, by its nature, rather subjective and, hence, it is difficult to quantify its performance apart from its specific domain of application.

4.1.1 Enhancement via image filtering

The main goal of image enhancement is to process an image in some way so as to render it more visually acceptable or pleasing. The removal of noise, the sharpening of image edges and the 'soft focus' (blurring) effect so often favoured in romantic photographs are all examples of popular enhancement techniques. These and other enhancement operations can be achieved through the process of spatial domain filtering. The term spatial domain is arguably somewhat spurious, but is used to distinguish this procedure from frequency domain procedures (discussed in Chapter 5). Thus, spatial domain filtering simply indicates that the filtering process takes place directly on the actual pixels of the image itself.

Fundamentals of Digital Image Processing – A Practical Approach with Examples in Matlab
Chris Solomon and Toby Breckon
© 2011 John Wiley & Sons, Ltd

Therefore, we shall refer simply to filtering in this chapter without danger of confusion. Filters act on an image to change the values of the pixels in some specified way and are generally classified into two types: linear and nonlinear. Linear filters are more common, but we will discuss and give examples of both kinds.

Irrespective of the particular filter that is used, all approaches to spatial domain filtering operate in the same simple way. Each of the pixels in an image – the pixel under consideration at a given moment is termed the target pixel – is successively addressed. The value of the target pixel is then replaced by a new value which depends only on the value of the pixels in a specified neighbourhood around the target pixel.

4.2 Pixel neighbourhoods

An important measure in images is the concept of *connectivity*. Many operations in image processing use the concept of a local image neighbourhood to define a local area of influence, relevance or interest. Central to this theme of defining the local neighbourhood is the notion of pixel connectivity, i.e. deciding which pixels are connected to each other. When we speak of 4-connectivity, only pixels which are N, W, E, S of the given pixel are connected. However, if, in addition, the pixels on the diagonals must also be considered, then we have 8-connectivity (i.e. N, NW, W, NE, SE, E, SW, S are all connected see Figure 4.1).

In Figure 4.1 (right) we use this concept to determine whether region A and region B are connected and use a local connectivity model (here $N \times N = 3 \times 3$) to determine if these are separate or the same image feature. Operations performed locally in images, such as filtering and edge detection, all consider a given pixel location (i, j) in terms of its local pixel neighbourhood indexed as an offset $(i \pm k, j \pm k)$. The size and, hence, the scale of the neighbourhood can be controlled by varying the neighbourhood size parameter N from which an offset k (generally $\lfloor N/2 \rfloor$) is computed. In general, neighbourhoods may be $N \times M$ where $N \neq M$ for unequal influence of pixels in the horizontal and vertical directions. More frequently $N = M$ is commonplace and, hence, $N \times N$ neighbourhoods arise. The majority

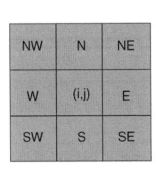

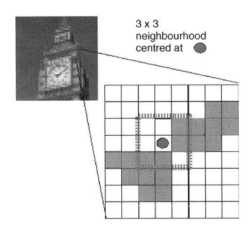

Figure 4.1 Image neighbourhood connectivity (left) and an example 3 × 3 neighbourhood centred at a specific image pixel location

of image processing techniques now use 8-connectivity by default, which for a reasonable size of neighbourhood is often achievable in real time on modern processors for the majority of operations. Filtering operations over a whole image are generally performed as a series of local neighbourhood operations using a sliding-window-based principle, i.e. each and every pixel in the image is processed based on an operation performed on its local $N \times N$ pixel neighbourhood (region of influence).

In Matlab® such an operation can be performed as in Example 4.1.

Example 4.1

Matlab code	What is happening?
A=imread('cameraman.tif');	%Read in image
subplot(1,2,1), imshow(A);	%Display image
func=@(x) max(x(:));	%Set filter to apply
B=nlfilter(A,[3 3],func);	%Apply over 3 × 3 neighbourhood
subplot(1,2,2), imshow(B);	%Display result image B

Comments

Here we specify *func()* as the *max()* filter function to apply over each and every 3 × 3 neighbourhood of the image. This replaces every input pixel in the output image with the maximum pixel value of the input pixel neighbourhood. You may wish to experiment with the effects of varying the neighbourhood dimensions and investigating the Matlab *min()* and *mean()* (for the latter, a type conversion will be required to display the double output type of the Matlab *mean()* function as an 8-bit image – specify the filter function as *uint8(mean())*).

4.3   Filter kernels and the mechanics of linear filtering

In linear spatial filters the new or filtered value of the target pixel is determined as some linear combination of the pixel values in its neighbourhood. Any other type of filter is, by definition, a nonlinear filter. The specific linear combination of the neighbouring pixels that is taken is determined by the filter kernel (often called a mask). This is just an array/sub-image of exactly the same size as the neighbourhood[1] containing the weights that are to be assigned to each of the corresponding pixels in the neighbourhood of the target pixel. Filtering proceeds by successively positioning the kernel so that the location of its centre pixel coincides with the location of each target pixel, each time the filtered value being calculated by the chosen weighted combination of the neighbourhood pixels. This filtering procedure can thus be visualized as sliding the kernel over all locations of interest in the original image (i, j), multiplying the pixels underneath the kernel by the corresponding weights w, calculating the new values by summing the total and copying them to the same locations in a new (filtered) image f (e.g. Figure 4.2).

The mechanics of linear spatial filtering actually express in discrete form a process called convolution, an important physical and mathematical phenomenon which we will develop

---

[1] A point of detail for the purist. With linear filters it is really the kernel that comes first and thus *defines* the neighbourhood. For some nonlinear filters (e.g. order filters) the order is reversed because we must define the region over which we wish to do the ranking and cannot write down a kernel.

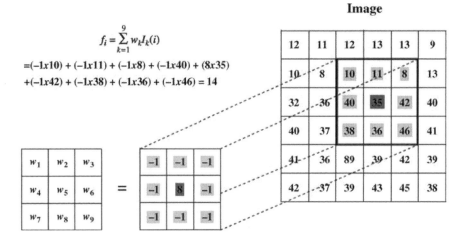

Figure 4.2 The mechanics of image filtering with an $N \times N = 3 \times 3$ kernel filter

further in Chapter 5. For this reason, many filter kernels are sometimes described as convolution kernels, it then being implicitly understood that they are applied to the image in the linear fashion described above. Formally, we can express the action of convolution between a kernel and an image in two equivalent ways. The first addresses the row and column indices of the image and the kernel:

$$f(x,y) = \sum_{i=I_{\min}}^{I_{\max}} \sum_{j=J_{\min}}^{J_{\max}} w(i,j)I(x+i, y+j) \tag{4.1}$$

Here, the indices $i = 0, j = 0$ correspond to the centre pixel of the kernel which is of size $(I_{\max} - I_{\min} + 1, J_{\max} - J_{\min} + 1)$. A second, equivalent approach is to use linear indices:

$$f_i = \sum_{k=1}^{N} w_k I_k(i) \tag{4.2}$$

In this case, $I_k(i)$ represents the neighbourhood pixels of the ith image pixel, where k is a linear index running over the neighbourhood region according to a row-wise (as in Figure 4.2) or column-wise convention. Here, w_k are the corresponding kernel values and f_i represents the filtered value resulting from original input value $I_k(i)$. The former notation is more explicit, whereas the latter is neater and more compact; but neither is generally recommended over the other. Figure 4.2 illustrates this basic procedure, where the centre pixel of the kernel and the target pixel in the image are indicated by the dark grey shading. The kernel is 'placed' on the image so that the centre and target pixels match. The filtered value of the target pixel f_i is then given by a linear combination of the neighbourhood pixel values, the specific weights being determined by the kernel values w_k. In this specific case the target pixel value of original value 35 is filtered to an output value of 14.

The steps in linear (convolution) filtering can be summarized as follows:

(1) Define the filter kernel.

(2) Slide the kernel over the image so that its centre pixel coincides with each (target) pixel in the image.

(3)   Multiply the pixels lying beneath the kernel by the corresponding values (weights) in the kernel above them and sum the total.

(4)   Copy the resulting value to the same locations in a new (filtered) image.

In certain applications, we may apply such a linear filter to a selected region rather than the entire image; we then speak of region-based filtering. We may also take a slightly more sophisticated approach in which the filter itself can change depending on the distribution of pixel values in the neighbourhood, a process termed adaptive filtering (e.g. see adaptive thresholding, discussed in Section 3.4.2).

Filtering at the boundaries of images also poses challenges. It is reasonable to ask what we should do when a target pixel lies close to the image boundary such that the convolution kernel overlaps the edge of the image. In general, there are three main approaches for dealing with this situation:

(1)   Simply leave unchanged those target pixels which are located within this boundary region.

(2)   Perform filtering on only those pixels which lie within the boundary (and adjust the filter operation accordingly).

(3)   'Fill in' in the missing pixels within the filter operation by mirroring values over the boundary.

Resulting undesirable edge artefacts are generally difficult to overcome and, in general, (2) or (3) is the preferred method. In certain instances, it is acceptable to 'crop' the image – meaning that we extract only a reduced-size image in which any edge pixels which have not been adequately filtered are removed entirely.

In Matlab, linear convolution filtering can be performed as in Example 4.2.

Example 4.2

Matlab code	What is happening?
A=imread('peppers.png');	%Read in image
subplot(1,2,1), imshow(A);	%Display image
k=fspecial('motion', 50, 54);	%Create a motion blur convolution kernel
B=imfilter(A, k, 'symmetric');	%Apply using symmetric mirroring at edges
subplot(1,2,2), imshow(B);	%Display result image B

Comments

Here we specify the *fspecial()* function to construct a kernel that will mimic the effect of motion blur (of specified length and angle) onto the image. Option 3) from our earlier discussion is used to deal with image edges during filtering. You may wish to investigate the use of other kernel filters that can be generated with the *fspecial()* function and the edge region filtering options available with the *imfilter()* function. Type *doc imfilter* at the Matlab prompt for details). How do they effect the image filtering result?

4.3.1 Nonlinear spatial filtering

Nonlinear spatial filters can easily be devised that operate through exactly the same basic mechanism as we described above for the linear case. The kernel mask slides over the image in the same way as the linear case, the only difference being that the filtered value will be the result of some nonlinear operation on the neighbourhood pixels. For example, employing the same notation as before, we could define a quadratic filter:

$$f_i = \sum_{k=1}^{N} w_{k1} I_k^2(i) + w_{k2} I_k(i) + w_{k3} \tag{4.3}$$

In this case, the action of the filter will be defined by the three weights which specify the contribution of the second, first- and zeroth-order terms. Nonlinear filters of this kind are not common in image processing. Much more important are order (or statistical) filters (discussed shortly), which operate by ranking the pixels in the specified neighbourhood and replacing the target pixel by the value corresponding to a chosen rank. In this case, we cannot write down a kernel and an equation of the form of our linear convolution is not applicable. In the following sections, we will present and discuss some of the more important examples of both linear and nonlinear spatial filters.

4.4 Filtering for noise removal

One of the primary uses of both linear and nonlinear filtering in image enhancement is for noise removal. We will now investigate the application of a number of different filters for removing typical noise, such as additive 'salt and pepper' and Gaussian noise (first introduced in Section 2.3.3). However, we first need to consider generation of some example images with noise added so that we can compare the effectiveness of different approaches to noise removal.

In Matlab, this can be achieved as in Example 4.3. The results of the noise addition from Example 4.3 are shown in Figure 4.3. These images will form the basis for our comparison of noise removal filters in the following sections of this chapter.

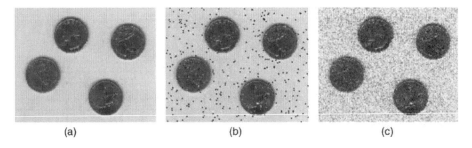

(a) (b) (c)

Figure 4.3 (a) Original image with (b) 'salt and pepper' noise and (c) Gaussian noise added

Example 4.3

Matlab code	What is happening?
I=imread('eight.tif');	%Read in image
subplot(1,3,1), imshow(I);	%Display image
Isp=imnoise(I,'salt & pepper',0.03);	%Add 3% (0.03) salt and pepper noise
subplot(1,3,2), imshow(Isp);	%Display result image Isp
Ig=imnoise(I,'gaussian',0.02);	%Add Gaussian noise (with 0.02 variance)
subplot(1,3,3), imshow(Ig);	%Display result image Ig

Comments
Here we use the ***imnoise()*** function to add both 'salt and pepper' noise and Gaussian noise to the input image. The strength is specified using the percentage density and variance (with zero mean) respectively. The reader may wish to experiment with the effects of changing these noise parameters and also explore the other noise effects available using this function (type *doc imnoise* at the Matlab prompt for details).

4.4.1 Mean filtering

The mean filter is perhaps the simplest linear filter and operates by giving equal weight w_K to all pixels in the neighbourhood. A weight of $W_K = 1/(NM)$ is used for an $N \times M$ neighbourhood and has the effect of smoothing the image, replacing every pixel in the output image with the mean value from its $N \times M$ neighbourhood. This weighting scheme guarantees that the weights in the kernel sum to one over any given neighbourhood size. Mean filters can be used as a method to suppress noise in an image (although the median filter which we will discuss shortly usually does a better job). Another common use is as a preliminary processing step to smooth the image in order that some subsequent processing operation will be more effective.

In Matlab, the mean filter can be applied as in Example 4.4 with the results as shown in Figure 4.4.

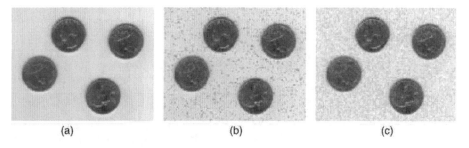

(a)	(b)	(c)

Figure 4.4 Mean filter (3 × 3) applied to the (a) original, (b) 'salt and pepper' noise and (c) Gaussian noise images of Figure 4.3

Example 4.4

Matlab code	What is happening?
k=ones(3,3)/9;	%Define mean filter
I_m=imfilter(I,k);	%Apply to original image
Isp_m=imfilter(Isp,k);	%Apply to salt and pepper image
Ig_m=imfilter(Ig,k);	%Apply to Gaussian image
subplot(1,3,1), imshow(I_m);	%Display result image
subplot(1,3,2), imshow(Isp_m);	%Display result image
subplot(1,3,3), imshow(Ig_m);	%Display result image

Comments

Here we define a 3 × 3 mean filter and apply it to the three images generated in Example 4.3 and shown in Figure 4.3. Experiment with using larger neighbourhood sizes for this filter. What effect does it have on the resulting image?

We can see that the mean filtering is reasonably effective at removing the Gaussian noise (Figure 4.4c), but at the expense of a loss of high-frequency image detail (i.e. edges). Although a significant portion of the Gaussian noise has been removed (compared with Figure 4.3c), it is still visible within the image. Larger kernel sizes will further suppress the Gaussian noise but will result in further degradation of image quality. It is also apparent that mean filtering is not effective for the removal of 'salt and pepper' noise (Figure 4.4b). In this case, the large deviation of the noise values from typical values in the neighbourhood means that they perturb the average value significantly and noise is still very apparent in the filtered result. In the case of 'salt and pepper' noise, the noisy high/low pixel values thus act as outliers in the distribution. For this reason, 'salt and pepper' noise is best dealt with using a measure that is robust to statistical outliers (e.g. a median filter, Section 4.4.2).

In summary, the main drawbacks of mean filtering are (a) it is not robust to large noise deviations in the image (outliers) and (b) when the mean filter straddles an edge in the image it will cause blurring. For this latter reason, the mean filter can also be used as a general low-pass filter. A common variation on the filter, which can be partially effective in preserving edge details, is to introduce a threshold and only replace the current pixel value with the mean of its neighbourhood if the magnitude of the change in pixel value lies below this threshold.

4.4.2 Median filtering

Another commonly used filter is the median filter. Median filtering overcomes the main limitations of the mean filter, albeit at the expense of greater computational cost. As each pixel is addressed, it is replaced by the statistical median of its $N \times M$ neighbourhood rather than the mean. The median filter is superior to the mean filter in that it is better at preserving sharp high-frequency detail (i.e. edges) whilst also eliminating noise, especially isolated noise spikes (such as 'salt and pepper' noise).

The median m of a set of numbers is that number for which half of the numbers are less than m and half are greater; it is the midpoint of the sorted distribution of values. As the median is a pixel value drawn from the pixel neighbourhood itself, it is more robust to outliers and does not create a new unrealistic pixel value. This helps in preventing edge blurring and loss of image detail.

By definition, the median operator requires an ordering of the values in the pixel neighbourhood at every pixel location. This increases the computational requirement of the median operator.

Median filtering can be carried out in Matlab as in Example 4.5. The results of Example 4.5 are shown in Figure 4.5, where we can now see the effectiveness of median filtering on the two types of noise ('salt and pepper' and Gaussian) introduced to the images in Example 4.3/Figure 4.3.

Example 4.5

Matlab code	What is happening?
I_m=medfilt2(I,[3 3]);	%Apply to original image
Isp_m=medfilt2(Isp,[3 3]);	%Apply to salt and pepper image
Ig_m=medfilt2(Ig,[3 3]);	%Apply to Gaussian image
subplot(1,3,1), imshow(I_m);	%Display result image
subplot(1,3,2), imshow(Isp_m);	%Display result image
subplot(1,3,3), imshow(Ig_m);	%Display result image

Comments

Here we define a 3×3 median filter *medfilt2()* and apply it to the three images generated in Example 4.3 and shown in Figure 4.3. Experiment with using larger neighbourhood sizes for this filter and the effect it has on the resulting image.

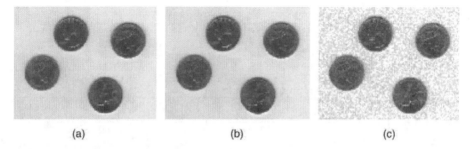

(a)   (b)   (c)

Figure 4.5 Median filter (3×3) applied to the (a) original, (b) 'salt and pepper' noise and (c) Gaussian noise images of Figure 4.3

In this result (Figure 4.5c), we again see the removal of some Gaussian noise at the expense of a slight degradation in image quality. By contrast, the median filter is very good at removing 'salt and pepper'-type noise (Figure 4.5b), where we see the removal of this high/low impulse-type noise with minimal degradation or loss of detail in the image. This is a key advantage of median filtering.

4.4.3 Rank filtering

The median filter is really just a special case of a generalized order (or rank) filter. The general order filter is a nonlinear filter comprising the following common steps:

(1) Define the neighbourhood of the target pixel ($N \times N$).

(2) Rank them in ascending order (first is lowest value, $(N \times N)^{th}$ is highest value).

(3) Choose the order of the filter (from 1 to N).

(4) Set the filtered value to be equal to the value of the chosen rank pixel.

Order filters which select the maximum and minimum values in the defined neighbourhood are (unsurprisingly) called maximum and minimum filters. We can use the Matlab order-filter function as shown in Example 4.6. The results of Example 4.6 are shown in Figure 4.6, where we can now see the result of maximum filtering on the two types of noise ('salt and pepper' and Gaussian) introduced to the images in Figure 4.3. Notably, the Gaussian noise has been largely removed (Figure 4.6c), but at the expense of image detail quality (notably the lightening of the image background). The nature of the 'salt and pepper'-type noise causes its high values to be amplified by the use of a maximum filter.

Example 4.6

Matlab code	What is happening?
I_m=ordfilt2(I,25,ones(5,5));	%Apply to original image
Isp_m=ordfilt2(Isp,25,ones(5,5));	%Apply to salt and pepper image
Ig_m=ordfilt2(Ig,25,ones(5,5));	%Apply to Gaussian image
subplot(1,3,1), imshow(I_m); %	%Display result image
subplot(1,3,2), imshow(Isp_m);	%Display result image
subplot(1,3,3), imshow(Ig_m);	%Display result image

Comments
Here we define a 5 × 5 max filter and apply it to the three images generated in Example 4.3 and shown in Figure 4.3. Experiment with using larger neighbourhood sizes for this filter, varying the rank of the filter (second parameter of *ordfilt2()* function) and the effect it has on the resulting image.

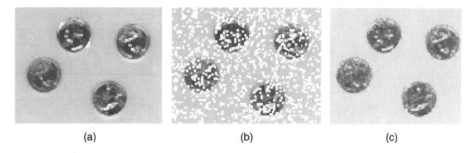

(a) (b) (c)

Figure 4.6 Order filtering (max, order = 25, 5 × 5) applied to the (a) original, (b) 'salt and pepper' noise and (c) Gaussian noise images of Figure 4.3

A variation on simple order filtering is *conservative smoothing*, in which a given pixel is compared with the maximum and minimum values (excluding itself) in the surrounding $N \times N$ neighbourhood and is replaced only if it lies outside of that range. If the current pixel value is greater than the maximum of its neighbours, then it is replaced by the maximum. Similarly, if it is less than the minimum, then it is replaced by the minimum.

4.4.4 Gaussian filtering

The Gaussian filter is a very important one both for theoretical and practical reasons. Here, we filter the image using a discrete kernel derived from a radially symmetric form of the continuous 2-D Gaussian function defined as follows:

$$f(x, y) = \frac{1}{2\pi\sigma^2} \exp\left(-\frac{x^2 + y^2}{2\sigma^2}\right) \tag{4.4}$$

Discrete approximations to this continuous function are specified using two free parameters:

(1) the desired size of the kernel (as an $N \times N$ filter mask);

(2) the value of σ, the standard deviation of the Gaussian function.

As is always the case with linear convolution filters (Section 4.3), there is a trade-off between the accurate sampling of the function and the computational time required to implement it. Some examples of discrete Gaussian filters, with varying kernel and standard deviation sizes, are shown in Figure 4.7.

Applying the Gaussian filter has the effect of smoothing the image, but it is used in a way that is somewhat different to the mean filter (Section 4.4.1). First, the degree of smoothing is controlled by the choice of the standard deviation parameter σ, not by the absolute value of the kernel size (which is the case with the mean filter). Second, the Gaussian function has a rather special property, namely that its Fourier transform is also a Gaussian function, which makes it very convenient for the frequency-domain analysis of filters (Chapter 5).

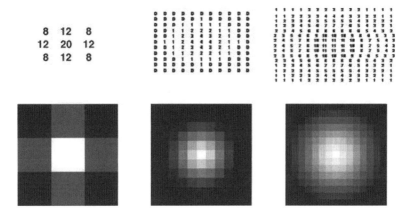

Figure 4.7   Gaussian filter kernels $3 \times 3 \sigma = 1$, 11×11 $\sigma = 2$ and 21×21 $\sigma = 4$ (The numerical values shown are unnormalised)

A Gaussian function with a large value of σ is an example of a so-called low-pass filter in which the high spatial frequency content (i.e. sharp edge features) of an image is suppressed. To understand this properly requires a background in the Fourier transform and frequency-domain analysis, subjects that are developed in Chapter 5.

We can apply the Gaussian filter in Matlab as in Example 4.7. The results of Example 4.7 are shown in Figure 4.8. In all cases, the smoothing effect of the filter degrades high-frequency (edge) detail as expected (e.g. Figure 4.8a), but it also removes to some degree the noise present in both Figure 4.8b and c.

Example 4.7

Matlab code	What is happening?
k=fspecial('gaussian', [5 5], 2);	%Define Gaussian filter
I_g=imfilter(I,k);	%Apply to original image
Isp_g=imfilter(Isp,k);	%Apply to salt and pepper image
Ig_g=imfilter(Ig,k);	%Apply to Gaussian image
subplot(1,3,1), imshow(I_g);	%Display result image
subplot(1,3,2), imshow(Isp_g);	%Display result image
subplot(1,3,3), imshow(Ig_g);	%Display result image

Comments
Here we define a 5×5 Gaussian filter kernel with $\sigma = 2$ using the Matlab *fspecial()* function and apply it to the three images generated in Example 4.3 and shown in Figure 4.3. The reader can experiment by trying different kernel sizes and different σ values to understand the effect it has on the resulting image.

Gaussian smoothing (or filtering) commonly forms the first stage of an edge-detection algorithm (e.g. the Canny edge detector, discussed in Chapter 10), where it is used as a means of noise suppression.

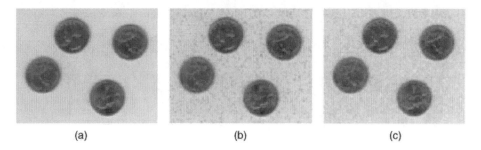

(a) **(b)** **(c)**

Figure 4.8 Gaussian filtering (5 × 5 with σ = 2) applied to the (a) original, (b) 'salt and pepper' noise and (c) Gaussian noise images of Figure 4.3

4.5 Filtering for edge detection

In addition to noise removal, the other two main uses of image filtering are for (a) feature extraction and (b) feature enhancement. We will next look at the use of image filtering in both of these areas through the detection of edges within images. An edge can be considered as a discontinuity or gradient within the image. As a result, the consideration of derivative filters is central to edge detection in image processing.

4.5.1 Derivative filters for discontinuities

The mean or averaging filter sums the pixels over the specified neighbourhood and, as we have seen, this has the effect of smoothing or blurring the image. In effect, this is just integration in discrete form. By contrast, derivative filters can be used for the detection of discontinuities in an image and they play a central role in sharpening an image (i.e. enhancing fine detail). As their name implies, derivative filters are designed to respond (i.e. return significant values) at points of discontinuity in the image and to give no response in perfectly smooth regions of the image, i.e. they detect *edges*.

One of the most important aspects of the human visual system is the way in which it appears to make use of the outlines or edges of objects for recognition and the perception of distance and orientation. This feature has led to one theory for the human visual system based on the idea that the visual cortex contains a complex of feature detectors that are tuned to the edges and segments of various widths and orientations. Edge features, therefore, can play an important role in the analysis of the image.

Edge detection is basically a method of segmenting an image into regions based on discontinuity, i.e. it allows the user to observe those features of an image where there is a more or less abrupt change in grey level or texture, indicating the end of one region in the image and the beginning of another. Enhancing (or amplifying) the presence of these discontinuities in the image allows us to improve the perceived image quality under certain conditions. However, like other methods of image analysis, edge detection is sensitive to noise.

Edge detection makes use of differential operators to detect changes in the gradients of the grey or colour levels in the image. Edge detection is divided into two main categories:

Table 4.1 Derivative operators: their formal (continuous) definitions and corresponding discrete approximations

2-D derivative measure	Continuous case	Discrete case
$\dfrac{\partial f}{\partial x}$	$\lim\limits_{\Delta x \to 0} \dfrac{f(x+\Delta x, y)-f(x,y)}{\Delta x}$	$f(x+1,y)-f(x,y)$
$\dfrac{\partial f}{\partial y}$	$\lim\limits_{\Delta y \to 0} \dfrac{f(x,y+\Delta y)-f(x,y)}{\Delta y}$	$f(x,y+1)-f(x,y)$
$\nabla f(x,y)$	$\left[\dfrac{\partial f}{\partial x}, \dfrac{\partial f}{\partial y}\right]$	$[f(x+1,y)-f(x,y), f(x,y+1)$ $-f(x,y)]$
$\dfrac{\partial^2 f}{\partial x^2}$	$\lim\limits_{\Delta x \to 0} \dfrac{(\partial f/\partial x)(x+\Delta x, y)-(\partial f/\partial x)f(x,y)}{\Delta x}$	$f(x+1,y)-2f(x,y)+f(x\text{-}1,y)$
$\dfrac{\partial^2 f}{\partial y^2}$	$\lim\limits_{\Delta y \to 0} \dfrac{(\partial f/\partial x)(x, y+\Delta y)-(\partial f/\partial x)(x,y)}{\Delta y}$	$f(x,y+1)-2f(x,y)+f(x,y-1)$
$\nabla^2 f(x,y)$	$\dfrac{\partial^2 f}{\partial x^2} + \dfrac{\partial^2 f}{\partial y^2}$	$f(x+1,y)+f(x-1,y)-4f(x,y)$ $+f(x,y+1)+f(x,y-1)$

first-order edge detection and second-order edge detection. As their names suggest, first-order edge detection is based on the use of first-order image derivatives, whereas second-order edge detection is based on the use of second-order image derivatives (in particular the Laplacian). Table 4.1 gives the formal definitions of these derivative quantities in both continuous and corresponding discrete forms for a 2-D image $f(x,y)$.

Before we discuss the implementation of these operators, we note two things:

(1) Differentiation is a linear operation and a discrete approximation of a derivative filter can thus be implemented by the kernel method described in Section 4.3. From the discrete approximations given in Table 4.1, we must, therefore, devise appropriate filter kernels to represent each of the derivative operators (see Section 4.5.2).

(2) A very important condition we must impose on such a filter kernel is that its response be zero in completely smooth regions. This condition can be enforced by ensuring that the weights in the kernel mask sum to zero.

Although they are relatively trivial to implement, the discrete representations given in Table 4.1 in kernel form are not generally the filter kernels of choice in practice. This is because the detection of edges (which is the main application of derivative filters) is generally assisted by an initial stage of (most often Gaussian) smoothing to suppress noise. Such noise might otherwise elicit a large response from the edge-detector kernel and dominate the true edges in the image. The smoothing operation and the edge-response operator can actually be combined into a single kernel, an example of which we will see in Section 4.6). More generally, noise suppression is a key part of more advanced edge-detection approaches, such as the Canny method (see Chapter 10).

Although a conceptually simple task, effective and robust edge detection is crucial in so many applications that it continues to be a subject of considerable research activity. Here, we will examine some basic edge-detection filters derived directly from the discrete derivatives.

4.5.2 First-order edge detection

A number of filter kernels have been proposed which approximate the first derivative of the image gradient. Three of the most common (their names being taken from their original authors/designers in the early image-processing literature) are shown in Figure 4.9, where we see the Roberts, Prewitt and Sobel edge-detector filter kernels. All three are implemented as a combination of two kernels: one for the x-derivative and one for the y-derivative (Figure 4.9).

The simple 2×2 Roberts operators (commonly known as the Roberts cross) were one of the earliest methods employed to detect edges. The Roberts cross calculates a simple, efficient, 2-D spatial gradient measurement on an image highlighting regions corresponding to edges. The Roberts operator is implemented using two convolution masks/kernels, each designed to respond maximally to edges running at $\pm 45°$ to the pixel grid, (Figure 4.9 (left)) which return the image x-derivative and y-derivative, G_x and G_y respectively. The magnitude $|G|$ and orientation θ of the image gradient are thus given by:

$$|G| = \sqrt{G_x^2 + G_y^2}$$

$$\theta = \tan^{-1}\left(\frac{G_y}{G_x}\right) + \frac{1}{4}\pi \qquad (4.5)$$

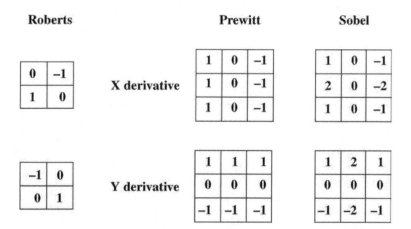

Figure 4.9 First-order edge-detection filters

This gives an orientation $\theta = 0$ for a vertical edge which is darker on the left side in the image. For speed of computation, however, $|G|$ is often approximated as just the sum of the magnitudes of the x-derivative and y-derivative, G_x and G_y.

The Roberts cross operator is fast to compute (due to the minimal size of the kernels), but it is very sensitive to noise. The Prewitt and Sobel edge detectors overcome many of its limitations but use slightly more complex convolution masks (Figure 4.9 (centre and right)).

The Prewitt/Sobel kernels are generally preferred to the Roberts approach because the gradient is not shifted by half a pixel in both directions and extension to larger sizes (for filter neighbourhoods greater than 3×3) is not readily possible with the Roberts operators. The key difference between the Sobel and Prewitt operators is that the Sobel kernel implements differentiation in one direction and (approximate) Gaussian averaging in the other (see Gaussian kernels, Figure 4.7). The advantage of this is that it smoothes the edge region, reducing the likelihood that noisy or isolated pixels will dominate the filter response.

Filtering using these kernels can be achieved in Matlab as shown in Example 4.8. The edge filter output of Example 4.8 is shown in Figure 4.10, where we can see the different responses of the three Roberts, Prewitt and Sobel filters to a given sample image. As expected, stronger and highly similar edge responses are obtained from the more sophisticated Prewitt and Sobel approaches (Figure 4.10 (bottom)). The Roberts operator is notably susceptible to image noise, resulting in a noisy and less distinct edge magnitude response (Figure 4.10 (top right)).

Example 4.8

Matlab code	What is happening?
I=imread('circuit.tif');	%Read in image
IEr = edge(I,'roberts');	%Roberts edges
IEp = edge(I,'prewitt');	%Prewitt edges
IEs = edge(I,'sobel');	%Sobel edges
subplot(2,2,1), imshow(I);	%Display image
subplot(2,2,2), imshow(IEr);	%Display image
subplot(2,2,3), imshow(IEp);	%Display image
subplot(2,2,4), imshow(IEs);	%Display image

Comments

Here we use the Matlab ***edge()*** function to apply the Roberts, Prewitt and Sobel edge detectors. As an extension, this function also facilitates the use of an additional third, threshold parameter in the form ***edge(Image, 'filter name', threshold)***. This exploits the concept of thresholding to select a subset of edge filter responses based on the magnitude of the filter response. The ***edge()*** function can also be used to return the individual G_x and G_y components of a given filter mask and automatically select a magnitude threshold (in a similar manner to Example 3.15; see *doc edge* in Matlab command prompt). Experiment with these parameters and the effects that can be achieved on the example images. You may also wish to investigate the use of the Matlab ***tic()***/***toc()*** functions for timing different edge-filtering operations.

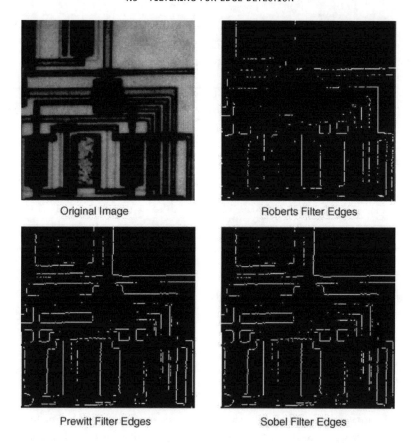

Original Image   Roberts Filter Edges

Prewitt Filter Edges   Sobel Filter Edges

Figure 4.10   Roberts, Prewitt and Sobel edge-magnitude responses

4.5.2.1   Linearly separable filtering

The Sobel and Prewitt filters are examples of linearly separable filters. This means that the filter kernel can be expressed as the matrix product of a column vector with a row vector. Thus, the filter kernels shown in Figure 4.9 can be expressed as follows:

$$\begin{bmatrix} 1 & 0 & -1 \\ 1 & 0 & -1 \\ 1 & 0 & -1 \end{bmatrix} = \begin{bmatrix} 1 \\ 1 \\ 1 \end{bmatrix} \begin{bmatrix} 1 & 0 & -1 \end{bmatrix} \quad \text{and} \quad \begin{bmatrix} 1 & 0 & -1 \\ 2 & 0 & -2 \\ 1 & 0 & -1 \end{bmatrix} = \begin{bmatrix} 1 \\ 2 \\ 1 \end{bmatrix} \begin{bmatrix} 1 & 0 & -1 \end{bmatrix}$$

$$(4.6)$$

An important consequence of this is that the 2-D filtering process can actually be carried out by two sequential 1-D filtering operations. Thus, the rows of the image are first filtered with the 1-D row filter and the resulting filtered image is then filtered column-wise by the 1-D column filter. This effects a computational saving in terms of reducing the amount of arithmetic operations required for a given convolution with a filter kernel. The saving is modest in the 3×3 case (a reduction to six multiplications/additions compared with nine for the 2-D version), but is considerably greater if we are considering larger kernel sizes. In general, linear separable filters result in a saving of order $2N$ operations as opposed to order N^2 for nonseparable, 2-D convolution.

4.5.3   Second-order edge detection

In general, first-order edge filters are not commonly used as a means of image enhancement. Rather, their main use is in the process of edge detection as a step in image segmentation procedures. Moreover, as we shall see in our discussion of the Canny method in Chapter 10, although the application of a derivative operator is a vital step, there is actually considerably more to robust edge detection than the simple application of a derivative kernel. A much more common means of image enhancement is through the use of a second-order derivative operator:- the Laplacian.

4.5.3.1   Laplacian edge detection
A very popular second-order derivative operator is the Laplacian:

$$\nabla^2 f(x,y) = \frac{\partial^2 f}{\partial x^2} + \frac{\partial^2 f}{\partial y^2} \tag{4.7}$$

The discrete form is given from Table 4.1 as:

$$\nabla^2 f = f(x+1,y) + f(x-1,y) - 4f(x,y) + f(x,y+1) + f(x,y-1) \tag{4.8}$$

This can easily be implemented in a 3×3 kernel filter, as shown in Figure 4.11A. However, if we explore an image applying this operator locally, then we expect the response to be greatest (as the Laplacian is a second-order derivative operator) at those points in the image where the local gradient changes most rapidly. One of the potential shortcomings of applying the mask in the form given by Figure 4.11A is the relative insensitivity to features lying in approximately diagonal directions with respect to the image axes. If we imagine rotating the axes by $45°$ and superimposing the rotated Laplacian on the original, then we can construct a filter that is invariant under multiple rotations of $45°$ (Figure 4.11B).

Figure 4.12 compares the response of the first-order Sobel and second-order Laplacian derivative filters. Note how the first-order gradient operator tends to produce 'thick edges', whereas the Laplacian filter tends to produce finer edges in response to the change in gradient rather than the image gradient itself. The Laplacian operator can be applied in Matlab as in Example 4.9.

The second-order derivative property that allows the Laplacian to produce a fine edge response corresponding to a change in gradient, rather than the less isolated response of the first-order edge filters, makes it suitable as the first stage of digital edge enhancement.

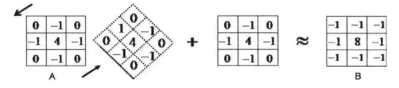

Figure 4.11   Construction of the Laplacian discrete kernel

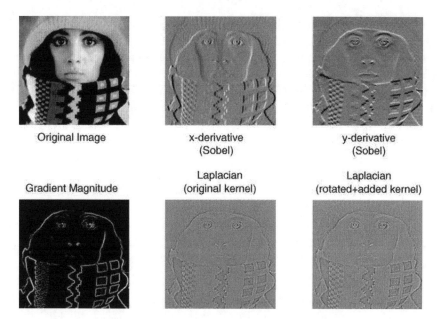

| Original Image | x-derivative (Sobel) | y-derivative (Sobel) |

| Gradient Magnitude | Laplacian (original kernel) | Laplacian (rotated+added kernel) |

Figure 4.12 Comparison of first-order derivative (Sobel) and second-order (Laplacian) filters

However, as the Laplacian kernels approximate a second derivative over the image they are in fact very sensitive to noise.

Example 4.9

Matlab code	What is happening?
I=rgb2gray(imread('peppers.png'));	%Read in image (in grey scale)
k=fspecial('laplacian');	%Create Laplacian filter
IEl=imfilter(double(I),k,'symmetric');	%Laplacian edges
subplot(1,2,1), imagesc(I);	%Display image
subplot(1,2,2), imagesc(IEl);	%Display image
colormap('gray');	

Comments

Here we first construct the Laplacian filter (in a similar manner to Example 4.2) and then apply it to the image using the Matlab *imfilter()* function. Note the inline use of the *rgb2gray()* function to load the (colour) example image as grey scale. In addition, we perform the Laplacian operation on a floating-point version of the input image (function *double()*) and, as the Laplacian operator returns both positive and negative values, use function *imagesc()* and *colormap()* to correctly scale and display the image as shown in Figure 4.12.

4.5.3.2 Laplacian of Gaussian

To counter this high noise sensitivity of the Laplacian filter, the standard Laplacian kernel (Figure 4.11) is commonly combined with the Gaussian kernel (Figure 4.7) to produce a robust filtering method. These two kernels could be applied sequentially to the image as two separate convolution operations – first smoothing with the Gaussian kernel and then with the Laplacian. However, as convolution is associative (Section 4.3), we can combine the kernels by convolving the Gaussian smoothing operator with the Laplacian operator to produce a single kernel: the Laplacian of Gaussian (LoG) filter. This single kernel is then applied to the image in a single pass. This offers a significant computational saving by approximately halving the calculations required.

The response of the filter will be zero in areas of uniform image intensity, whilst it will be nonzero in an area of transition. At a given edge, the operator will return a positive response on the darker side and negative on the lighter side.

We can see this effect in Figure 4.12 and the results of Example 4.9. If we wish to apply the LoG operator in Matlab, then we can do so by replacing the 'laplacian' parameter of the *fspecial()* function with 'log'. The result should be a smoothed version of the result in Example 4.9. Additional input parameters to the *fspecial()* function allow the level of smoothing to be controlled by varying the width of the Gaussian.

4.5.3.3 *Zero-crossing detector*

The zero-crossing property of the Laplacian (and LoG) also permits another method of edge detection: the zero-crossing method. We use a zero-crossing detector to locate pixels at which the value of the Laplacian passes through zero (i.e. points where the Laplacian changes sign). This occurs at 'edges' in the image where the intensity of the image changes rapidly (or in areas of intensity change due to noise). It is best to think of the zero-crossing detector as a type of feature detector rather than as a specific edge detector. The output from the zero-crossing detector is usually a binary image with single-pixel thickness lines showing the positions of the zero-crossing points.

The starting point for the zero-crossing detector is an image which has been filtered using the LoG filter (to overcome the effects of noise). The zero crossings that result are strongly influenced by the size of the Gaussian used for the smoothing stage of this operator. As the smoothing is increased, then fewer and fewer zero-crossing contours will be found, and those that do remain will correspond to features of larger and larger scale in the image.

We can use a zero-crossing detector with the LoG filter in Matlab as in Example 4.10. Figure 4.13 shows edge transitions detected using the zero-crossings concept applied to the LoG filter (from Example 4.10). The effect of noise on the second derivative despite the use of Gaussian smoothing is evident.

Example 4.10

Matlab code	What is happening?
I=rgb2gray(imread('peppers.png'));	%Read in image (in grey scale)
k=fspecial('log', [10 10], 3.0);	%Create LoG filter
IEzc = edge(I, 'zerocross', [], k);	%Zero-crossing edges (auto-thresholded)
subplot(1,2,1), imshow(I);	%Display image
subplot(1,2,2), imshow(IEzc);	%Display image

> **Comments**
> As is evident from Figure 4.13, the results are still quite noisy even with a broad 10×10 Gaussian ($\sigma = 3.0$) as specified in the above example. The reader may wish to experiment with different Gaussian parameters (note that the filter k can itself be visualized as an image using the *imagesc()* function).

Figure 4.13   Edges detected as the zero crossings of the LOG operator

A couple of points to note with this operator are:

- In the general case, all edges detected by the zero-crossing detector are in the form of closed curves in the same way that contour lines on a map are always closed. In the Matlab implementation, if a threshold of zero is specified then this is always the case (NB: Example 4.10 uses an autoselected threshold (parameter specified as '[ ]') on which edges to keep). The only exception to this is where the curve goes off the edge of the image. This can have advantages for later processing.

- As we have seen, the LoG filter is quite susceptible to noise if the standard deviation of the smoothing Gaussian is small. One solution to this is to increase Gaussian smoothing to preserve only strong edges. An alternative is to look at the gradient of the zero crossing and only keep zero crossings where this is above a certain threshold (i.e. use the third derivative of the original image). This will tend to retain only stronger edges but as the third derivative is also highly sensitive to noise this greatly amplifies any high-frequency noise in the image.

4.6 Edge enhancement

In the final part of this chapter we look at the use of second-order edge detection (Section 4.5.3) as a method for edge enhancement (commonly known as image sharpening).

4.6.1 Laplacian edge sharpening

We have seen that the Laplacian responds only to the fine detail in the image (i.e. those image regions where the change in gradient is significant) but has a zero response to constant regions and regions of smooth gradient in the image. If, therefore, we take the original image and add or subtract the Laplacian, then we may expect to enhance the fine detail in the image artificially. It is common practice just to subtract it from the original, truncating any values which exhibit integer overflow in the common 8-bit representation. Using the Laplacian definition of Section 4.5.3.1, we can define this as follows:

$$I_{\text{output}}(x, y) = I_{\text{in}}(x, y) - \nabla^2 I_{\text{in}}(x, y) \qquad (4.9)$$

Using Matlab, Laplacian image sharpening can be achieved as in Example 4.11. The output from Example 4.11 is shown in Figure 4.14, where we can see the original image, the Laplacian 'edges' and the sharpened final output. Note that we can see the enhancement of edge contrast in this filtering result, but also an increase in image noise in the sharpened image.

Example 4.11

Matlab code
```
A=imread('cameraman.tif');
h=fspecial(laplacian, [10 0], 3.0);
B=imfilter(A,h);
C=imsubtract(A,B);
subplot(1,3,1), imshow(A);
subplot(1,3,2), imagesc(B); axis image; axis off
subplot(1,3,3), imshow(C);
```

What is happening?
%Read in image
%Generate 3 × 3 Laplacian filter
%Filter image with Laplacian kernel
%Subtract Laplacian from original.

%Display original, Laplacian and
%enhanced image

Comments
In this example, because the images are not first converted to floating-point format (data type double), the Laplacian filtered image is automatically truncated into the 8-bit form. The reader may experiment by first converting both images A and B to floating point (***double()***), performing the calculation in floating-point image arithmetic and displaying such images as per Example 4.9.

Original Image　　　Laplacian "edges"　　　Sharpened Image

Figure 4.14 Edge sharpening using the Laplacian operator

| Original Image | LoG "edges" | Sharpened Image |

Figure 4.15 Edge sharpening using the LoG operator

In order to overcome this problem we replace the Laplacian operator in Example 4.11 with the LoG operator (Section 4.5.3.2). The reader may wish to experiment using this variation on Example 4.11, applying different-size Gaussian kernels to investigate the difference in image sharpening effects that can be achieved. The example of Figure 4.14 is again shown in Figure 4.15 using the alternative LoG operator where we can note the reduced levels of noise in the intermediate LoG edge image and the final sharpened result.

4.6.2 The unsharp mask filter

An alternative edge enhancement filter to the Laplacian-based approaches (Section 4.6.1) is the unsharp mask filter (also known as boost filtering). Unsharp filtering operates by subtracting a smoothed (or unsharp) version of an image from the original in order to emphasize or enhance the high-frequency information in the image (i.e. the edges). First of all, this operator produces an edge image from the original image using the following methodology:

$$I_{edges}(c, r) = I_{original}(c, r) - I_{smoothed}(c, r) \qquad (4.10)$$

The smoothed version of the image is typically obtained by filtering the original with a mean (Section 4.4.1) or a Gaussian (Section 4.4.4) filter kernel. The resulting difference image is then added onto the original to effect some degree of sharpening:

$$I_{enhanced}(c, r) = I_{original}(c, r) + k(I_{edges}(c, r)) \qquad (4.11)$$

using a given constant scaling factor k that ensures the resulting image is within the proper range and the edges are not 'oversharp' in the resulting image. Generally, $k = 0.2$–0.7 is acceptable, depending on the level of sharpening required. It is this secondary stage that gives rise to the alternative name of boost filtering.

To understand this approach to sharpening we need to consider two facts. First, the smooth, relatively unchanging regions of the original image will not be changed significantly by the smoothing filter (for example, a constant region which is already perfectly smooth will be completely unaffected by the smoothing filter, e.g. Figures 4.4 and 4.8). By contrast, edges and other regions in the image in which the intensity changes rapidly will be affected significantly. If we subtract this smoothed image $I_{smoothed}$ from the original image $I_{original}$, then we get a resulting image I_{edges} with higher values (differences) in the areas affected significantly (by the smoothing) and low values in the areas where little change occurred. This broadly corresponds to a smoothed edge map of the image. The result of subtracting this smoothed image (or some multiple thereof) from the original is an image of higher value pixels in the areas of high contrast change (e.g. edges) and lower pixel values in the areas of uniformity I_{edges}. This resulting image can then be added back onto the original, using a specified scaling factor k, to enhance areas of rapid intensity change within the image whilst leaving areas of uniformity largely unchanged. It follows that the degree of enhancement will be determined by the amount of smoothing that is imposed before the subtraction takes place and the fraction of the resulting difference image which is added back to the original. Unsharp filtering is essentially a reformulation of techniques referred to as high boost – we are essentially boosting the high-frequency edge information in the image.

This filtering effect can be implemented in Matlab as shown in Example 4.12. The result of Example 4.12 is shown in Figure 4.16, where we see the original image, the edge difference image (generated by subtraction of the smoothed image from the original) and a range of enhancement results achieved with different scaling factors k. In the examples shown, improvement in edge detail (sharpness) is visible but the increasing addition of the edge image (parameter k) increases both the sharpness of strong edges and noise apparent within the image. In all cases, image pixels exceeding the 8-bit range 0–255 were truncated.

Example 4.12

Matlab code	**What is happening?**
A=imread('cameraman.tif');	%Read in image
Iorig=imread('cameraman.tif');	%Read in image
g=fspecial('gaussian',[5 5],1.5);	%Generate Gaussian kernel
subplot(2,3,1), imshow(Iorig);	%Display original image
Is=imfilter(Iorig,g);	%Create smoothed image by filtering
Ie=(Iorig - Is);	%Get difference image
subplot(2,3,2), imshow(Ie);	%Display unsharp difference
Iout=Iorig + (0.3).*Ie;	%Add k * difference image to original
subplot(2,3,3), imshow(Iout);	
Iout=Iorig + (0.5).*Ie;	%Add k * difference image to original
subplot(2,3,4), imshow(Iout);	
Iout=Iorig + (0.7).*Ie;	%Add k * difference image to original
subplot(2,3,5), imshow(Iout);	
Iout=Iorig + (2.0).*Ie;	%Add k * difference image to original
subplot(2,3,6), imshow(Iout);	

Comments

In this example we perform all of the operations using unsigned 8-bit images (Matlab type **uint8**) and we compute the initial smoothed image using a 5×5 Gaussian kernel ($\sigma = 1.5$). Try changing the levels of smoothing, using an alternative filter such as the mean for the smoothing operator and varying the scaling factor k.

Original Image	Edge Image	k = 0.3
k = 0.5	k = 0.7	k = 2.0

Figure 4.16   Edge sharpening using unsharp mask filter

Exercises

The following exercises are designed to reinforce and develop the concepts and Matlab examples introduced in this chapter. Additional information on all of the Matlab functions presented in this chapter and throughout these exercises is available in Matlab from the function help browser (use *doc <function name>* at the Matlab command prompt where *<function name>* is the function required).

Matlab functions: ***imnoise, plot, tic, toc, min, max, mean, function, imresize, for, colfilt, roipoly, roifilt, imfilter, fspecial.***

Exercise 4.1   Based on Example 4.3, experiment with the Matlab ***imnoise()*** function for adding different levels of 'salt and pepper' and Gaussian noise to images. Use the 'peppers.png' and 'eight.tif' example images as both a colour and grey-scale example to construct a series of image variables in Matlab with varying levels and types of noise (you may also wish to investigate the other noise types available from this function, type *doc imnoise* at the Matlab prompt). Based on the filtering topics presented in Section 4.4, investigate the usefulness of each of the mean, median and Gaussian filtering for removing different types and levels of image noise (see Examples 4.4, 4.5 and 4.7 for help).

Exercise 4.2   Building on Example 4.2, experiment with the use of different neighbour-hood dimensions and the effect on the resulting image output. Using the plotting functions of Matlab (function ***plot()***) and the Matlab timing functions ***tic*** and ***toc***, create a plot of neighbourhood dimension N against operation run-time for applying the *min()*, *max()* and *mean()* functions over an example image. Does the timing increase linearly or not? Why is this?

Exercise 4.3   Building on the local pixel neighbourhood definition described Section 4.2, write a Matlab function (see Matlab help for details on functions or type *doc function* at the Matlab function) to extract the $N \times N$ pixel neighbourhood of a given specified pixel location (x, y) and copy the pixel values to a new smaller $N \times N$ image. You may want to take a look at the syntax for the Matlab for loop function (*doc for*) in the first instance.

   Combine your extraction program with the Matlab *imresize()* function to create a region extraction and magnification program.

Exercise 4.4   The Matlab *colfilt()* function utilizes the highly efficient matrix operations of Matlab to perform image filtering operations on pixel neighbourhoods. Using the Matlab timing functions *tic()/toc()*, time the operation performed in Example 4.2 with and without the use of this optimizing parameter. Vary the size of the neighbourhood over which the operation is performed (and the operation: *min()/max()/mean()*). What do you notice? Is the increase in performance consistent for very small neighbourhoods and very large neighbourhoods?

Exercise 4.5   Based on Example 4.4 for performing mean filtering on a given image, record the execution time for this operation (using Matlab timing functions *tic()/toc()*) over a range of different neighbourhood sizes in the range 0–25. Use Matlab's plotting facilities to present your results as a graph. What do you notice? How does the execution time scale with the increase in size of the neighbourhood? (*Hint*. To automate this task you may want to consider investigating the Matlab *for* loop constructor).

   As an extension you could repeat the exercise for differently sized images (or a range of image sizes obtained from *imresize()*) and plot the results. What trends do you notice?

Exercise 4.6   Repeat the first part of Exercise 4.5, but compare the differences between mean filtering (Example 4.4) and median filtering (Example 4.5). How do the trends compare? How can any differences be explained?

Exercise 4.7   A region of interest (ROI) within an image is an image sub-region (commonly rectangular in nature) over which localized image-processing operations can be performed. In Matlab an ROI can be selected interactively by first displaying an image (using *imshow()*) and then using the *roipoly()* function that returns an image sub-region defined as a binary image the same size as the original (zero outside the ROI and one outside the ROI).

   Investigate the use of the *roifilt()* function for selectively applying both Gaussian filtering (Example 4.7) and mean filtering (Example 4.4) to an isolated region of interest within one of the available example images.

   You may also wish to investigate combining the ROI selection function with your answer to Exercise 4.3 to extract a given ROI for histogram equalization (Section 3.4.4) or edge-detection processing (Section 4.5) in isolation from the rest of the image.

Exercise 4.8   Several of the filtering examples in this chapter make use of the Matlab *imfilter()* function (Example 4.2), which itself has a parameter of how to deal with image boundaries for a given filtering operation. Select one of the filtering operations from this chapter and experiment with the effect of selecting different boundary options with the *imfilter()* function. Does it make a difference to the result? Does it make a difference to the amount of time the filtering operation takes to execute? (Use *tic()*/*toc()* for timing.)

Exercise 4.9   Implement a Matlab function (*doc function*) to perform the conservative smoothing filter operation described towards the end of Section 4.4.3. (*Hint.* You may wish to investigate the Matlab *for* loop constructor.) Test this filter operation on the noise examples generated in Example 4.3. How does it compare to mean or median filtering for the different types of noise? Is it slower or faster to execute? (Use *tic()*/*toc()* for timing.)

Exercise 4.10   Considering the Roberts and Sobel edge detectors from Example 4.8, apply edge detection to three-channel RGB images and display the results (e.g. the 'peppers.png' and 'football.jpg' images). Display the results as a three-channel colour image and as individual colour channels (one per figure).

   Note how some of the edge responses relate to distinct colour channels or colours within the image. When an edge is visible in white in the edge-detected three-channel image, what does this mean? You may also wish to consider repeating this task for the HSV colour space we encountered in Chapter 1. How do the results differ in this case?

Exercise 4.11   Building upon the unsharp operator example (Example 4.12), extend the use of this operator to colour images (e.g. the 'peppers.png' and 'football.jpg' images). How do the areas of image sharpening compare with the areas of edge-detection intensity in Exercise 4.10?

   The unsharp operator can also be constructed by the use of an 'unsharp' parameter to the Matlab *fspecial()* function for us with *imfilter()*. Compare the use of this implementation with that of Example 4.12. Are there any differences?

For further examples and exercises see http://www.fundipbook.com

5

Fourier transforms and frequency-domain processing

5.1 Frequency space: a friendly introduction

Grasping the essentials of frequency-space methods is very important in image processing. However, if the author's experience is anything to go by, Fourier methods can prove something of a barrier to students. Many discussions of the Fourier transform in the literature simply begin by defining it and then offer a comprehensive but rather formal mathematical presentation of the related theorems. The significant risk in this approach is that all but the mathematically inclined or gifted are quickly turned off, either feeling inadequate for the task or not understanding why it is all worth the trouble of struggling with. The *real significance* of Fourier methods, the small number of really central concepts, simply gets lost in the mathematical details. We will try to avoid this overly formal approach. Of course, we cannot simply ignore the mathematics – it is central to the whole subject – but we will certainly aim to stress the key points and to underline the real significance of the concepts we discuss.

Frequency-space analysis is a widely used and powerful methodology centred around a particular mathematical tool, namely the *Fourier transform*.[1] We can begin simply by saying that the Fourier transform is a particular type of integral transform that enables us to view imaging and image processing from an alternative viewpoint by transforming the problem to another space. In image processing, we are usually concerned with 2-D spatial distributions (i.e. functions) of intensity or colour which exist in *real space* – i.e. a 2-D Cartesian space in which the axes define units of length. The Fourier transform operates on such a function to produce an entirely equivalent form which lies in an abstract space called *frequency space*. Why bother? In the simplest terms, frequency space is useful because it can make the solution of otherwise difficult problems *much easier* (Figure 5.1).

Fourier methods are sufficiently important that we are going to break the pattern and digress (for a time) from image processing to devote some time to understanding some of the key concepts and mathematics of Fourier transforms and frequency space. Once this

---

[1] Strictly, frequency-space analysis is not exclusively concerned with the Fourier transform. In its most general sense, it also covers the use of similar transforms such as the Laplace and Z transform. However, we will use frequency space and Fourier space synonymously.

Fundamentals of Digital Image Processing – A Practical Approach with Examples in Matlab
Chris Solomon and Toby Breckon
© 2011 John Wiley & Sons, Ltd

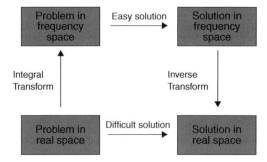

Figure 5.1   Frequency-space methods are used to make otherwise difficult problems easier to solve

foundation has been laid, we will move on to see how they can be used to excellent effect both to look at image processing from a new perspective and to carry out a variety of applications.

5.2   Frequency space: the fundamental idea

We will begin our discussion of Fourier methods by summarizing, without any attempt at rigour, some of the key concepts. To stay general, we will talk for the moment of the Fourier analysis of signals rather than images.

(1)   *The harmonic content of signals.*   The fundamental idea of Fourier analysis is that any signal, be it a function of time, space or any other variables, may be expressed as a weighted linear combination of *harmonic* (i.e. sine and cosine) functions having different periods or frequencies. These are called the (spatial) frequency components of the signal.

(2)   *The Fourier representation is a complete alternative.*   In the Fourier representation of a signal as a weighted combination of harmonic functions of different frequencies, the assigned weights constitute the *Fourier spectrum*. This spectrum extends (in principle) to infinity and any signal can be reproduced to arbitrary accuracy. Thus, the Fourier spectrum is a complete and valid, alternative representation of the signal.

(3)   *Fourier processing concerns the relation between the harmonic content of the output signal to the harmonic content of the input signal.*   In frequency space, signals are considered as combinations of harmonic signals. Signal processing in frequency space (analysis, synthesis and transformation of signals) is thus concerned with the constituent harmonic content and how these components are preserved, boosted or suppressed by the processing we undertake.

These first three concepts are summarized in Figure 5.2.

(4)   *The space domain and the Fourier domain are reciprocal.*   In the Fourier representation of a function, harmonic terms with high frequencies (short periods) are needed to construct small-scale (i.e. sharp or rapid) changes in the signal. Conversely, smooth features in the signal can be represented by harmonic terms with low frequencies (long periods). The two domains are thus *reciprocal* – small in the space domain maps to

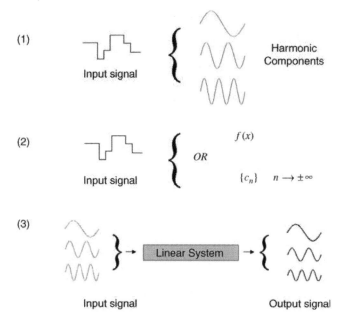

Figure 5.2   A summary of three central ideas in Fourier (frequency) domain analysis. (1) Input signals are decomposed into harmonic components. (2) The decomposition is a complete and valid representation. (3) From the frequency-domain perspective, the action of any linear system on the input signal is to modify the amplitude and phase of the input components

large in the Fourier domain and large in the space domain maps to small in the Fourier domain.

Students often cope with Fourier series quite well but struggle with the Fourier transform. Accordingly, we make one more key point.

(5)   *The Fourier series expansion and the Fourier transform have the same basic goal.* Conceptually, the Fourier series expansion and the Fourier transform do *the same thing*. The difference is that the Fourier series breaks down a periodic signal into harmonic functions of *discrete* frequencies, whereas the Fourier transform breaks down a nonperiodic signal into harmonic functions of *continuously varying* frequencies. The maths is different but the idea is the same.

We will expand on these basic ideas in what follows. We begin our discussion in the next section with Fourier series.

5.2.1   The Fourier series

Key point 1

Any *periodic* signal may be expressed as a weighted combination of sine and cosine functions having different periods or frequencies.

This is Fourier's basic hypothesis. The process of breaking down a *periodic* signal as a sum of sine and cosine functions is called a *Fourier decomposition* or *Fourier expansion*. If the signal is something that varies with time, such as a voltage waveform or stock-market share price, the harmonic functions that build the signal are *1-D* and have a *temporal* frequency. In such a 1-D case, Fourier's basic theorem says that a periodic signal[2] $V(t)$ having period T can be constructed exactly as an infinite sum of harmonic functions, a Fourier series, as follows:

$$V(t) = \sum_{n=0}^{\infty} a_n \cos\left(\frac{2\pi nt}{T}\right) + \sum_{n=1}^{\infty} b_n \sin\left(\frac{2\pi nt}{T}\right)$$

$$= \sum_{n=0}^{\infty} a_n \cos(\omega_n t) + \sum_{n=1}^{\infty} b_n \sin(\omega_n t)$$

(5.1)

An arbitrary periodic 1-D function of a *spatial* coordinate x $f(x)$ having spatial period λ can be represented in exactly the same way:

$$V(x) = \sum_{n=0}^{\infty} a_n \cos\left(\frac{2\pi nx}{\lambda}\right) + \sum_{n=1}^{\infty} b_n \sin\left(\frac{2\pi nx}{\lambda}\right)$$

$$= \sum_{n=0}^{\infty} a_n \cos(k_n x) + \sum_{n=1}^{\infty} b_n \sin(k_n x)$$

(5.2)

Let us first make some simple observations on the Fourier expansion of the *spatial* function $f(x)$ expressed by Equation (5.2).

(1) The infinite series of harmonic functions in the expansion, namely $\cos(k_n x)$ and $\sin(k_n x)$, are called the Fourier *basis functions*.

(2) We are dealing with a function that varies in space and the (inverse) periodicity, determined by $k_n = 2\pi n/\lambda$, is thus called the *spatial frequency*.

(3) The coefficients a_n and b_n indicate how much of each basis function (i.e. harmonic wave of the given spatial frequency) is required to build $f(x)$. The complete set of coefficients $\{a_n \text{ and } b_n\}$ are said to constitute the *Fourier* or *frequency spectrum* of the spatial function. The function $f(x)$ itself is called the spatial domain representation.

(4) To reproduce the original function $f(x)$ *exactly*, the expansion must extend to an infinite number of terms. In this case, as $n \rightarrow \infty$, the spatial frequencies $k_n \rightarrow \infty$ and the number of coefficients $\{a_n \text{ and } b_n\}$ describing the Fourier spectrum also approach infinity.

---

[2] Strictly, the signal must satisfy certain criteria to have a valid Fourier expansion. We will not digress into these details but assume that we are dealing with such functions here.

> ## Key point 2
>
> The Fourier spectrum is a valid and complete alternative representation of a function.

This point is essential, namely that in knowing the coefficients a_n and b_n we have a complete representation of the function just as valid as $f(x)$ itself. This is so because we can rebuild $f(x)$ with arbitrary precision by carrying out the summation in Equation (5.2). Figure 5.3 shows how a simple 1-D periodic function (only one cycle of the function is shown) – a step function – can be increasingly well approximated by a Fourier series representation as more terms in Equation (5.2) are added. If we continue this procedure for ever, the Fourier spectrum $\{a_n, b_n\}$ can reproduce the function exactly and can thus be rightly considered an *alternative* (frequency-domain) representation.

If we examine the synthesis of the periodic square wave in Figure 5.3, we can observe our key point number 4 in the Fourier decomposition/synthesis of a function. The *low spatial frequencies* (corresponding to the lower values of n) build the 'basic' smooth shape, whereas the high spatial frequencies are required to reproduce the sharp transitions in the function (the edges). The synthesis of a step function by a Fourier series is just one simple example,

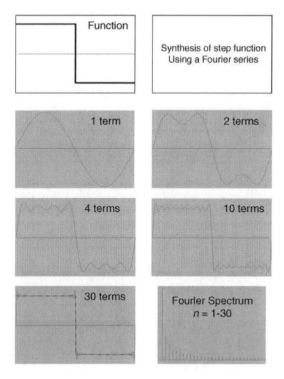

Figure 5.3   The synthesis of a step function of period λ using a Fourier series. The resulting spectrum is the frequency-domain representation of the spatial function determining the contribution of each sine wave of frequency $\sin(k_n x) = \sin(2\pi n x/\lambda)$

but this is in fact a basic 'ground rule' of the Fourier representation of *any* function (true in one or more dimensions).

It is also worth emphasizing that although the ability to synthesize a given function *exactly* requires harmonic (i.e. sine and cosine) frequencies extending to infinite spatial frequency, it is often the case that a good approximation to a function can be obtained using a *finite and relatively small number of spatial frequencies*. This is evident from the synthesis of the step function. When we approximate a spatial function by a Fourier series containing a finite number of harmonic terms N, we then effectively define a so-called spatial *frequency cut-off* $k_{CO} = 2\pi N/\lambda$. The loss of the high spatial frequencies in a signal generally results in a loss of fine detail.

5.3   Calculation of the Fourier spectrum

Fourier's basic theorem states that we can synthesize periodic functions using the sinusoidal basis functions, but we have so far glossed over the question of *how we actually calculate the Fourier spectrum* (i.e. the expansion coefficients in Equation (5.2)). Fortunately, this is easy. By exploiting the *orthogonality* properties of the Fourier basis,[3] we can obtain simple formulae for the coefficients:

$$a_n = \frac{2}{\lambda} \int_{-\lambda/2}^{\lambda/2} f(x)\cos(k_n x)\, dx$$

$$b_n = \frac{2}{\lambda} \int_{-\lambda/2}^{\lambda/2} f(x)\sin(k_n x)\, dx$$

(5.3)

where $k_n = 2\pi n/\lambda$. Note that we get the coefficients in each case by integrating over one full spatial period of the function.[4]

5.4   Complex Fourier series

Fourier series can actually be expressed in a more convenient and compact, complex form. Thus, a periodic, 1-D spatial function $f(x)$ is expressed as a weighted sum of *complex exponential* (harmonic) functions:

$$f(x) = \sum_{n=-\infty}^{\infty} c_n \exp\left(\frac{i2\pi n x}{\lambda}\right)$$

$$= \sum_{n=-\infty}^{\infty} c_n \exp(ik_n x)$$

(5.4)

---

[3] A proof of these formulae is offered on the book website http://www.fundipbook.com/materials/.
[4] We can of course add *any constant value* to our chosen limits of $-\lambda/2$ and $\lambda/2$. Zero and λ are commonly quoted, but all that matters is that the periodic function is integrated over a full spatial period λ.

where n may assume all integer values from $-\infty$ to $+\infty$. In an entirely analogous way to the real form of the expansion, we may exploit the orthogonality relations between complex exponential functions to obtain the Fourier expansion coefficients:[5]

$$c_n = \frac{1}{\lambda} \int_{-\lambda/2}^{\lambda/2} f(x)\exp(ik_n x)\, dx \qquad (5.5)$$

Note that the c_n are, in general, *complex* numbers. In using the complex representation we stress that nothing essential changes. We are still representing $f(x)$ as an expansion in terms of the real sine and cosine functions. The sines and cosines can actually be made to reappear by virtue of grouping the positive and negative exponentials with the same magnitude of the index n but of opposite sign. A strictly real function can then be constructed from the complex Fourier basis because the expansion coefficients (the c_n) are also, in general, complex. It is relatively straightforward to show that (see the exercises on the book's website[6]) that the complex coefficients c_n are related to the real coefficients a_n and b_n in the real Fourier series (Equation (5.3)) by

$$c_k = a_k + ib_k$$
$$c_{-k} = a_k - ib_k \qquad \text{for } k = 0, 1 \ldots \to \infty \qquad (5.6)$$

To expand the discussion of Fourier methods to deal with images, there are two main differences from the formalism we have presented so far that we must take into account:

(1) images are not, in general, periodic functions.

(2) images are typically *2-D* (and sometimes higher dimensional) spatial functions of finite support.[7]

The extension of the Fourier hypothesis from periodic functions to deal with non-periodic functions is really just the extension from a Fourier series to a Fourier transform. First, we will consider the issue of periodicity and only after extend to two dimensions. In what follows (just to keep it simple) we will consider a *1-D* spatial function $f(x)$, though an analogous argument can be made for two or more dimensions.

5.5 The 1-D Fourier transform

To move from a Fourier series to a Fourier transform, we first express our function $f(x)$ as a complex Fourier series:

[5,6] See http://www.fundipbook.com/materials/.
[7] The *support* of a function is the region over which it is nonzero. Obviously, any real digital image has finite support.

$$f(x) = \sum_{n=-\infty}^{\infty} c_n \exp(ik_n x) \qquad k_n = \frac{2\pi n}{\lambda} \qquad (5.7)$$

Multiplying top and bottom by $\lambda \to \infty$, this can also be written as

$$f(x) = \frac{1}{\lambda} \sum_{n=-\infty}^{\infty} (\lambda c_n) \exp(ik_n x) \qquad (5.8)$$

We now let the *period* of the function $\lambda \to \infty$. In other words, the trick here is to consider that *a nonperiodic function* may be considered as *a periodic function having infinite spatial period*. Note that the individual spatial frequencies in the summation expressed by Equation (5.8) are given by $k_n = 2\pi n/\lambda$ and the frequency interval between them is thus given by $\Delta k = 2\pi/\lambda$. As the spatial period $\lambda \to \infty$, the frequency interval Δk becomes infinitesimally small and λc_n tends towards a *continuous* function $F(k_x)$. It is in fact possible to show[8] that the limit of Equation (5.8) as $\lambda \to \infty$ is the *inverse Fourier transform*:

$$f(x) = \int_{-\infty}^{\infty} F(k_x) \exp(ik_x x) \, \mathrm{d}k_x \qquad (5.9)$$

Note that:

- the specific weights assigned to the harmonic (complex exponential) functions are given by the function $F(k_x)$;

- the frequencies k_x are now *continuous* and range over all possible values;

- The summation for $f(x)$ becomes an integral and the Fourier spectrum $F(k_x)$ is now a *function* as opposed to a set of discrete values.

The orthogonality properties of the complex Fourier basis (the $\exp(ik_x x)$ functions) enable us to calculate the weighting function $F(k_x)$ in Equation (5.9). This is, in fact, the *Fourier transform* of $f(x)$:

$$F(k_x) = \frac{1}{2\pi} \int_{-\infty}^{\infty} f(x) \exp(-ik_x x) \, \mathrm{d}x \qquad (5.10)$$

It is vital to emphasize that the essential meaning and purpose of the Fourier transform is really no different from that of the Fourier series. The Fourier transform of a function also fundamentally expresses its decomposition into *a weighted set of harmonic functions*. Moving from the Fourier series to the Fourier transform, we move from function synthesis using weighted combinations of harmonic functions having *discrete* frequencies (a *summation*) to weighted, infinitesimal, combinations of *continuous* frequencies (an *integral*).

---

[8] See book website http://www.fundipbook.com/materials/ for details.

Table 5.1   Comparison of the synthesis of spatial functions using the real Fourier series, the complex Fourier series and the Fourier transform

	Real Fourier series	Complex Fourier series	Fourier transform
Spatial frequencies	$k_n = \dfrac{2\pi n}{\lambda}$ $n = 1, 2, \ldots, \infty$	$k_n = \dfrac{2\pi n}{\lambda}$ $n = \pm 1, \pm 2, \ldots, \pm \infty$	k_x $-\infty \le k_x \le \infty$
Basis functions	$\sin k_n x, \ \cos k_n x$ $n = 1, 2, \ldots, \infty$	$\exp(ik_n x)$ $n = 0, \pm 1, \pm 2, \ldots, \pm \infty$	$\exp(ik_x x)$ $-\infty \le k_x \le \infty$
Spectrum	Coefficients $\{a_n, b_n\}$ $n = 0, 1, 2, \ldots, \infty$	Coefficients $\{c_n\}$ $n = 0, \pm 1, \pm 2, \ldots, \pm \infty$	Function $F(k_x)$ $-\infty \le k_x \le \infty$

The synthesis of functions using the real Fourier series, the complex Fourier series and the Fourier transform are summarized in Table 5.1.

The Fourier transform $F(k_x)$ is a *complex function* and we can, therefore, also write the Fourier spectrum in polar form as the product of the Fourier modulus and (the exponential of) the Fourier phase:

$$F(k_x) = |F(k_x)| \exp i\varphi(k_x)$$

where $\quad |F(k_x)|^2 = [\mathrm{Re}\{F(k_x)\}]^2 + [\mathrm{Im}\{F(k_x)\}]^2 \quad \varphi(k_x) = \tan^{-1}\left[\dfrac{\mathrm{Im}\{F(k_x)\}}{\mathrm{Re}\{F(k_x)\}}\right] \quad (5.11)$

This form is useful because it helps us to see that the Fourier transform $F(k_x)$ defines both the 'amount' of each harmonic function contributing to $f(x)$ (through $|F(k_x)|$) and the relative placement/position of the harmonic along the axis through the associated complex phase term $\exp(i\varphi(k_x))$. As we shall see in Figure 5.11, most of what we will loosely call the 'visual information' in a Fourier transform is actually contained in the phase part.

5.6   The inverse Fourier transform and reciprocity

Examination of Equations (5.9) and (5.10) shows that there is a close similarity between the Fourier transform and its inverse. It is in fact arbitrary whether we define the forward Fourier transform with the negative form of the exponential function or the positive form, but the convention we have chosen here is the normal one and we will use this throughout. There is also a certain freedom with regard to the 2π factor which can be placed on either the forward or reverse transforms or split between the two.[9] The factor of 2π also appears in

---

[9] Some authors define both forward and reverse transforms with a normalization factor of $(2\pi)^{-1/2}$ outside.

FT for width a FT for width $2a$ FT for width $a/2$

Figure 5.4   The Fourier transform of the rectangle function $F(k_x) = (a/2\pi)\{[\sin(k_xa/2)]/(k_xa/2)\}$

the 2-D Fourier transform, which is presented in the next section, but is actually *remarkably unimportant in image processing* – it enters simply as an overall scaling factor and thus has no effect on the spatial structure/content of an image. For this reason, it is quite common to neglect it entirely.

Rather like the Fourier series expansion of a square wave, the Fourier transform of the rectangle function is easy to calculate and informative. The 1-D rectangle function is defined as

$$\mathrm{rect}\left(\frac{x}{a}\right) = 1 \quad |x| \le \frac{a}{2}$$

$$= 0 \quad |x| \ge \frac{a}{2}$$

By substituting directly into the definition we can show that its Fourier transform is given by

$$F(k_x) = \frac{a}{2\pi}\frac{\sin(k_xa/2)}{k_xa/2} = \frac{a}{2\pi}\mathrm{sinc}\left(\frac{k_xa}{2}\right)$$

where $\mathrm{sinc}\,\theta = \sin\theta/\theta$. This function is plotted in Figure 5.4 for three values of the rectangle width a, over a spatial frequency range $-8\pi/a \le k_x \le 8\pi/a$. Note the reciprocity between the extent of the spatial function and the extent of its Fourier transform (the scaling factor of $2a/\pi$ has been omitted to keep the graphs the same height).

Note that a *decrease* in the width of the rectangle in the spatial domain results in the Fourier transform spreading out in the frequency domain. Similarly, an increase of the extent of the function in the spatial domain results in the frequency-domain representation shrinking. This reciprocal behaviour is a central feature of the frequency-domain representation of functions.

Important examples of 1-D Fourier transforms and Fourier transform relations are provided in Tables 5.2 and 5.3. We also offer some exercises (with worked solutions) on the book's website[10] which we strongly recommend to the reader as a means to consolidate the basic concepts discussed so far and to gaining further insight into the basic behaviour and properties of the Fourier transform.

---

[10] http://www.fundipbook.com/materials/.

Table 5.2 Fourier transforms of some important functions

Function name	Space domain $f(x)$ $f(x)$	Frequency domain $F(k)$ $F(k) = \int_{-\infty}^{\infty} f(x)e^{-ikx}\, dx$
Impulse (delta function)	$\delta(x)$	1
Constant	1	$\delta(k)$
Gaussian	$\exp\left(-\dfrac{x^2}{2\sigma^2}\right)$	$(\sigma\sqrt{2\pi})\exp\left(-\dfrac{\sigma^2 k^2}{2}\right)$
Rectangle	$\mathrm{rect}\left(\dfrac{x}{L}\right) = \prod\left(\dfrac{x}{L}\right) \equiv \begin{cases} 1 & \|x\| \le L/2 \\ 0 & \text{elsewhere} \end{cases}$	$L\,\mathrm{sinc}\left(\dfrac{kL}{2\pi}\right)$
Triangle	$\Lambda\left(\dfrac{x}{W}\right) \equiv \begin{cases} 1-(\|x\|/W) & \|x\| \le W \\ 0 & \text{elsewhere} \end{cases}$	$W\,\mathrm{sinc}^2\left(\dfrac{kW}{2\pi}\right)$
Sinc	$\mathrm{sinc}(Wx) \equiv \dfrac{\sin(Wx)}{Wx}$	$\dfrac{1}{W}\mathrm{rect}\left(\dfrac{k}{W}\right)$
Exponential	$e^{-a\|x\|} \qquad a>0$	$\dfrac{2a}{a^2+k^2}$
Complex exponential	$\exp(ik_0 x)$	$\delta(k-k_0)$
Decaying exponential	$\exp(-ax)u(x) \qquad \mathrm{Re}\{a\}>0$	$\dfrac{1}{a+ik}$
Impulse train	$\sum\limits_{n=-\infty}^{\infty} \delta(x-nx_s)$	$\dfrac{2\pi}{x_s}\sum\limits_{k=-\infty}^{\infty} \delta\left[k\left(1-\dfrac{2\pi}{x_s}\right)\right]$
Cosine	$\cos(k_0 x + \theta)$	$e^{i\theta}\delta(k-k_0) + e^{-i\theta}\delta(k+k_0)$
Sine	$\sin(k_0 x + \theta)$	$-i[e^{i\theta}\delta(k-k_0) - e^{-i\theta}\delta(k+k_0)]$
Unit step	$u(x) \equiv \begin{cases} 1 & x \ge 0 \\ 0 & x<0 \end{cases}$	$\pi\delta(k) + \dfrac{1}{jik}$
Signum	$\mathrm{sgn}(x) \equiv \begin{cases} 1 & x \ge 0 \\ -1 & x<0 \end{cases}$	$\dfrac{2}{ik}$
Sinc^2	$\mathrm{sinc}^2(Bx)$	$\dfrac{1}{B}\Lambda\left(\dfrac{\omega}{2\pi B}\right)$
Linear decay	$1/x$	$-i\pi\mathrm{sgn}(k)$

5.7 The 2-D Fourier transform

In an entirely analogous way to the 1-D case, a function of two spatial variables $f(x,y)$ can be expressed as a weighted superposition of 2-D harmonic functions.[11] The basic concept is

---

[11] Certain functions do not, in the strict sense, possess Fourier transforms, but this is a mathematical detail which has no real practical importance in image processing and we will fairly assume that we can obtain the Fourier transform of any function of interest.

Table 5.3   Some important properties of the 2-D Fourier transform: $f(x, y)$ has a Fourier transform $F(k_x, k_y)$; $g(x, y)$ has a Fourier transform $G(k_x, k_y)$

Property	Spatial domain	Frequency domain	Comments
Addition theorem	$f(x,y)+g(x,y)$	$F(k_x,k_y)+G(k_x,k_y)$	Fourier transform of the sum equals sum of the Fourier transforms.
Similarity theorem	$f(ax,by)$	$\dfrac{1}{\|ab\|}F\left(\dfrac{k_x}{a},\dfrac{k_y}{b}\right)$	The frequency-domain function scales in inverse proportion to the spatial-domain function.
Shift theorem	$f(x-a,y-b)$	$\exp[-i(k_x a + k_y b)]F(k_x,k_y)$	Shift a function in space and its transform is multiplied by a pure phase term.
Convolution theorem	$f(x,y)*g(x,y)$	$F(k_x,k_y)G(k_x,k_y)$	Transform of a convolution is equal to the product of the individual transforms.
Separable product	$f(x,y)=h(x)g(y)$	$F(k_x,k_y)=H(k_x)G(k_y)$	If a 2-D function separates into two 1-D functions, then so does its Fourier transform
Differentiation	$\dfrac{\partial}{\partial x^m}\dfrac{\partial}{\partial y^n}f(x,y)$	$(ik_x)^m(ik_y)^n F(k_x,k_y)$	Calculation of image derivatives is trivial using the image Fourier transform
Rotation	$f(x\cos\theta + y\sin\theta,$ $-x\sin\theta + y\cos\theta)$	$F(k_x\cos\theta + k_y\sin\theta,$ $-k_x\sin\theta + k_y\cos\theta)$	Rotate a function by θ in the plane and its Fourier transform rotates by θ
Parseval's theorem	$\displaystyle\int_{-\infty}^{\infty}\int_{-\infty}^{\infty}\|f(x,y)\|^2\,dx\,dy$	$\displaystyle\int_{-\infty}^{\infty}\int_{-\infty}^{\infty}\|F(k_x,k_y)\|^2\,dk_x\,dk_y$	Fourier transformation preserves image 'energy'
Laplacian	$-\nabla^2 f(x,y)$	$(k_x^2+k_y^2)F(k_x,k_y)$	Calculation of the image Laplacian is trivial using the image Fourier transform
Multiplication	$f(x,y)\cdot g(x,y)$	$F(k_x,k_y)**G(k_x,k_y)$	Product of functions is a convolution in the Fourier domain

graphically illustrated in Figure 5.5, in which examples of 2-D harmonic functions are combined to synthesize $f(x,y)$. Mathematically, the reverse or *inverse* 2-D Fourier transform is defined as

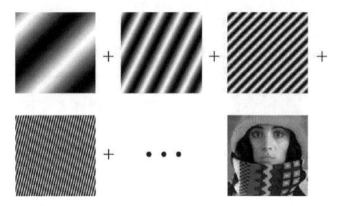

Figure 5.5 The central meaning of the 2-D Fourier transform is that some scaled and shifted combination of the 2-D harmonic basis functions (some examples are shown) can synthesize an arbitrary spatial function

$$f(x,y) = \frac{1}{\sqrt{2\pi}} \int\limits_{-\infty}^{\infty} \int\limits_{-\infty}^{\infty} F(f_x, f_y) \exp[2\pi i (f_x x + f_y y)] \, df_x \, df_y$$

$$= \frac{1}{\sqrt{2\pi}} \int\limits_{-\infty}^{\infty} \int\limits_{-\infty}^{\infty} F(k_x, k_y) \exp[i(k_x x + k_y y)] \, dk_x \, dk_y$$

(5.12)

where the weighting function $F(k_x, k_y)$ is called the *2-D Fourier transform* and is given by

$$F(f_x, f_y) = \frac{1}{\sqrt{2\pi}} \int\limits_{-\infty}^{\infty} \int\limits_{-\infty}^{\infty} f(x,y) \exp[-2\pi i (f_x x + f_y y)] \, dx \, dy$$

$$F(k_x, k_y) = \frac{1}{\sqrt{2\pi}} \int\limits_{-\infty}^{\infty} \int\limits_{-\infty}^{\infty} f(x,y) \exp[-i(k_x x + k_y y)] \, dx \, dy$$

(5.13)

Considering the 2-D Fourier transform purely as an integral to be 'worked out', it can at first appear rather formidable.[12] Indeed, the number of functions whose Fourier transform can be *analytically* calculated is relatively small and, moreover, they tend to be rather simple functions (some are listed in Table 5.2). We stress two things, however. First, Fourier transforms of even very complicated functions *can* be calculated accurately and quickly on a computer. In fact, the development and digital implementation of the Fourier transform for the computer (known as the fast Fourier transform (FFT)) revolutionized the world of

---

[12] One of the authors freely confesses that he never understood it at all as an undergraduate, despite encountering it in several different courses at various times.

image processing and indeed scientific computing generally. Second, the real value of the frequency-domain representation lies not just in the ability to calculate such complex integrals numerically, but much more in the *alternative and complementary viewpoint* that the frequency domain provides on the processes of image formation and image structure.

Just as for the 1-D case, there are a relatively small number of simple 2-D functions whose Fourier transform can be analytically calculated (and a number of them are, in fact, separable into a product of two 1-D functions). It is instructive, however, to work through some examples, and these are provided with worked solutions on the book's website.[13]

There are a number of theorems relating operations on a 2-D spatial function to their effect on the corresponding Fourier transform which make working in the frequency domain and swapping back to the spatial domain much easier. Some of these are listed in Table 5.3. Again their formal proofs are provided on the book's website for the interested reader.

5.8 Understanding the Fourier transform: frequency-space filtering

In spatial-domain image processing, we are basically concerned with how imaging systems and filters of various kinds affect the individual pixels in the image. In frequency-domain image processing, however, we consider imaging systems and filtering operations from an alternative perspective – namely, *how they affect the constituent harmonic components that make up the input.* Figure 5.6 depicts the basic ideas of spatial and frequency-space

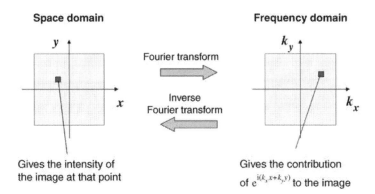

Figure 5.6 The relation between the space domain and the frequency domain. The value at a point (x, y) in the space domain specifies the intensity of the image at that point. The (complex) value at a point (k_x, k_y) in the frequency domain specifies the contribution of the harmonic function $\exp[i(k_x x + k_y y)]$ to the image

---

[13] See http://www.fundipbook.com/materials/.

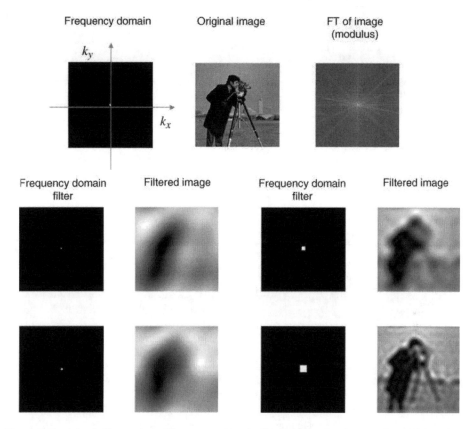

Figure 5.7 A basic illustration of frequency-domain filtering. The Fourier transform of the original image is multiplied by the filters indicated above (white indicates spatial frequency pairs which are preserved and black indicates total removal). An inverse Fourier transform then returns us to the spatial domain and the filtered image is displayed to the right. (The Matlab code for this figure can be found at http://www.fundipbook.com/materials/.)

representation. In the spatial domain, we refer to the pixel locations through a Cartesian (x, y) coordinate system. In the frequency-space representation the value at each coordinate point in the (k_x, k_y) system tells us the contribution that the harmonic frequency component $\exp[i(k_x x + k_y y)]$ makes to the image.

Figure 5.7 demonstrates the basic idea behind *frequency-domain filtering*, in which certain harmonic components are removed from the image. The basic filtering procedure involves three steps: (i) calculate the Fourier transform; (ii) suppress certain frequencies in the transform through multiplication with a *filter* function (in this case, the filter is set to zero at certain frequencies but is equal to one otherwise); and then (iii) calculate the inverse Fourier transform to return to the spatial domain. In Figure 5.7, the image and (the modulus) of its corresponding Fourier transform are displayed alongside the image which results when we remove selected groups of harmonic frequencies from the input. This removal is achieved by multiplication of the filter function image (where white is 1, black is 0) with the Fourier transform of the image. Components of increasing spatial frequency (low to high) are thus transmitted as we increase the size of the filter function centred on the Fourier image.

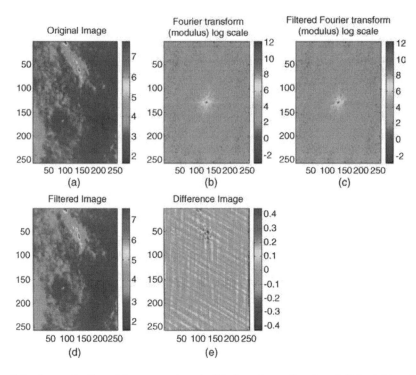

Figure 5.8   An application of frequency-domain filtering. Proceeding from left to right and top to bottom, we have: (a) the original image with striping effect apparent; (b) the Fourier modulus of the image (displayed on log scale); (c) the Fourier modulus of the image after filtering (displayed on log scale); (d) the filtered image resulting from recombination with the original phase; (e) the difference between the original and filtered images (*See colour plate section for colour version*)

Figure 5.8 shows a real-life example of frequency-domain filtering undertaken by one of the authors. The Infrared Astronomical Satellite (IRAS) was a joint project of the USA, UK and the Netherlands.[14] The IRAS mission performed a sensitive all-sky survey at wavelengths of 12, 25, 60 and 100 µm. As a result of this survey, we were presented with acquired images which suffered from striping artefacts – a result of the difficulty in properly aligning and registering the strips recorded as the detector effectively made 1-D sweeps across the sky. The stripes which are visible in the image can also be clearly seen in the Fourier transform of the image, where they show up as discrete groups of spatial frequencies which are enhanced over the background. Note that we show the absolute value of the Fourier transform on a *log* scale to render them visible. Clearly, if we can suppress these 'rogue' frequencies in frequency space, then we may expect to largely remove the stripes from the original image.

Simply setting all the identified rogue frequencies to zero is one option, but is not really satisfactory. We would expect a certain amount of the rogue frequencies to be naturally present in the scene and such a tactic will reduce the overall power spectrum of the image.

---

[14] See http://irsa.ipac.caltech.edu/IRASdocs/iras.html.

The basic approach taken to filtering was as follows:

(1) Calculate the Fourier transform and separate it into its modulus and phase.

(2) Identify the 'rogue frequencies'. Whilst, in principle, this could be achieved manually on a single image, we sought an automated approach that would be suitable for similarly striped images. We must gloss over the details here, but the approach relied on use of the Radon transform to identify the directionality of the rogue frequencies.

(3) Replace the rogue frequencies in the Fourier modulus with values which were statistically reasonable based on their neighbouring values.

(4) Recombine the filtered modulus with the original phase and perform an inverse Fourier transform.

5.9 Linear systems and Fourier transforms

From the frequency-domain perspective, the action of a linear imaging system on an input can be easily summarized:

> Any input image can be decomposed into a weighted sum of harmonic functions. The action of a linear system in general will be to preserve or alter the magnitude and phase of the input harmonics.

Broadly speaking, we may thus view a linear imaging system as something that operates on the constituent input harmonics of the image and can assess its quality by its ability to (more or less) faithfully transmit the input harmonics to the output. This basic action of a linear system in the frequency domain is illustrated in Figure 5.9. One measure for characterizing the performance of a linear, shift-invariant imaging system is through the *optical transfer function* (OTF). To understand this fully, we must first introduce the convolution theorem.

5.10 The convolution theorem

As we showed earlier in this chapter, the importance of the convolution integral originates from the fact that many situations involving the process of physical measurement with an imperfect instrument can be accurately described by convolution. The process of convolution (a rather messy integral in the spatial domain) has a particularly simple and convenient form in the frequency domain; this is provided by the famous *convolution theorem* – probably the single most important theorem in the field of image processing.

Consider two 2-D functions $f(x, y)$ and $h(x, y)$ having Fourier transforms respectively denoted by $F(k_x, k_y)$ and $H(k_x, k_y)$. Symbolically denoting the operation of taking a 2-D Fourier transform by **F**, the *first form* of the convolution theorem states that:

INPUT HARMONICS   OUTPUT HARMONICS

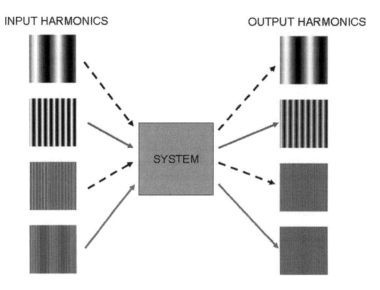

Figure 5.9   The basic action of a linear system can be understood by how well it transmits to the output each of the constituent harmonic components that make up the input. In this illustration, the lower frequencies are faithfully reproduced, but higher frequencies are suppressed.

$$\mathbf{F}\{f(x,y) * * h(x,y)\} = F(k_x, k_y)H(k_x, k_y) \qquad (5.14)$$

The Fourier transform of the convolution of the two functions is equal to the product of the individual transforms.

Thus, the processing of convolving two functions in the spatial domain can be equivalently carried out by *simple multiplication of their transforms in the frequency domain*. This first form of the convolution theorem forms the essential basis for the powerful methods of *frequency-domain filtering*. Thus, rather than attempt to operate directly on the image itself with a chosen spatial-domain filter (e.g. edge detection or averaging), we alternatively approach the problem by considering *what changes we would like to effect on the spatial frequency content of the image*. Filter design is often much easier in the frequency domain, and the alternative viewpoint of considering an image in terms of its spatial frequency content often allows a more subtle and better solution to be obtained.

For completeness, we note that there is a second form of the convolution theorem, which states that

$$\mathbf{F}\{f(x,y)h(x,y)\} = F(k_x, k_y) * * H(k_x, k_y) \qquad (5.15)$$

The Fourier transform of the product of the two functions is equal to the convolution of their individual transforms.

This form does not find quite such widespread use in digital image processing as it does not describe the basic image formation process. However, it finds considerable use in the fields of Fourier optics and optical image processing, where we are often interested in the effect of devices (diffraction gratings, apertures, etc.) which act multiplicatively on the incident light field.

The convolution theorem[15] lies at the heart of both frequency-domain enhancement techniques and the important subject of image restoration (a subject we will develop in more detail in Chapter 6).

5.11   The optical transfer function

Consider for a moment the specific imaging scenario in which $f(x, y)$ corresponds to the input distribution, $h(x, y)$ to the respective system PSF and $g(x, y)$ is the image given by their convolution:

$$g(x, y) = f(x, y) * * h(x, y) \tag{5.16}$$

Taking the Fourier transform of both sides, we can use the first form of the convolution theorem to write the right-hand side:

$$\mathbf{F}\{g(x, y)\} = \mathbf{F}\{f(x, y) * * h(x, y\} $$
$$G(k_x, k_y) = F(k_x, k_y)H(k_x, k_y) \tag{5.17}$$

Thus, the Fourier spectrum of the output image $G(k_x, k_y)$ is given by the product of the input Fourier spectrum $F(k_x, k_y)$ with a multiplicative filter function $H(k_x, k_y)$. $H(k_x, k_y)$ is called the OTF. The OTF is the frequency-domain equivalent of the PSF. Clearly, the OTF derives its name from the fact that it determines how the individual spatial frequency pairs (k_x, k_y) are transferred from input to output. This simple interpretation makes the OTF the most widely used measure of the quality or fidelity of a linear shift-invariant imaging system.

$$\mathbf{F}\{f(x, y) * h(x, y\} = \underbrace{G(k_x, k_y)}_{\substack{\text{output} \\ \text{Fourier} \\ \text{spectrum}}} = \underbrace{F(k_x, k_y)}_{\substack{\text{input} \\ \text{Fourier} \\ \text{spectrum}}} \underbrace{H(k_x, k_y)}_{\text{OTF}} \tag{5.18}$$

This multiplicative property of the OTF on the input spectrum is particularly convenient whenever we consider complex imaging systems comprising multiple imaging elements (e.g. combinations of lenses and apertures in a camera or telescope). If the kth element is characterized by its PSF $h_k(x, y)$, then the overall image is given by a sequence of convolutions of the input with the PSFs. Taking Fourier transforms and using the convolution theorem, this can be equivalently expressed by sequential *multiplication* of the OTFs in the frequency domain- a much easier calculation:

$$\mathbf{F}\{h_1(x, y) * h_2(x, y) * \cdots h_N(x, y)\} = H_1(k_x, k_y)H_2(k_x, k_y) \cdots H_N(k_x, k_y) \tag{5.19}$$

---

[15] For proof, go to http://www.fundipbook.com/materials/.

A detailed discussion of OTFs and their measurement is outside our immediate scope, but we stress two key points:

(1) The OTF is normalized to have a maximum transmission of unity. It follows that an *ideal imaging system* would be characterized by an OTF given by $H(k_x, k_y) = 1$ for all spatial frequencies.

(2) As the Fourier transform of the PSF, the OTF is, in general, a *complex* function:

$$H(k_x, k_y) = |H(k_x, k_y)| \exp[i\varphi(k_x, k_y)]$$

The square modulus of the OTF is a real function known as the modulation transfer function (MTF); this gives the magnitude of transmission of the spatial frequencies. However, it is the phase term $\varphi(k_x, k_y)$ which controls the position/placement of the harmonics in the image plane. Although the MTF is a common and useful measure of image quality, the phase transmission function $\varphi(k_x, k_y)$ is also crucial for a complete picture.

Figure 5.10 (generated using the Matlab® code in Example 5.1) shows on the left an image of the old BBC test card.[16] The test card has a number of features designed to allow

---

Example 5.1

Matlab code	What is happening?
A=imread('BBC_grey_testcard.png');	%Read in test card image
FA=fft2(A); FA=fftshift(FA);	%Take FFT and centre it
PSF=fspecial('gaussian',size(A),6);	%Define PSF
OTF=fft2(PSF); OTF=fftshift(OTF);	%Calculate corresponding OTF
Afilt=ifft2(OTF.*FA);	%Calculate filtered image
Afilt=fftshift(Afilt);	
subplot(1,4,1);imshow(A,[]);	%Display results
colormap(gray);	
subplot(1,4,2); imagesc(log(1+(PSF)));	
axis image; axis off;	
subplot(1,4,3); imagesc(log(1+abs	
(OTF))); axis image; axis off;	
subplot(1,4,4); imagesc(abs(Afilt)); axis	
image; axis off;	
PSF=fspecial('gaussian',size(A),6);	%Define PSF
OTF=fft2(PSF); OTF=fftshift(OTF);	%Calculate corresponding OTF
rlow=(size(A,1)./2)-3; rhigh=	%Define range to be altered
(size(A,1)./2)+3;	
clow=(size(A,2)./2)-3; chigh=	
(size(A,2)./2)+3;	

---

[16] For those readers too young to remember this, the test card was displayed when you switched on your TV set before transmission of programmes had started. This was back in the days when television often did not start until 11 am.

```
Fphase=angle(OTF);                              %Extract Fourier phase
Fphase(rlow:rhigh,clow:chigh)=                  %Add random component to selected phase
    Fphase(rlow:rhigh,clow:chigh) +
    0.*pi.*rand;

OTF=abs(OTF).*exp(i.*Fphase);                   %Recombine phase and modulus
Afilt=ifft2(OTF.*FA);                           %Calculate filtered image
    Afilt=fftshift(Afilt);
psfnew=abs(fftshift((otf2psf(OTF))));           %Calculate corresponding PSF
subplot(1,4,2); imagesc(log(1 + psfnew));
    axis image; axis off; colormap(gray);
subplot(1,4,3); imagesc(log(1 +
    abs(OTF))); axis image; axis off;
subplot(1,4,4); imagesc(abs(Afilt));
    axis image; axis off;

PSF=fspecial('motion',30,30);                   %Define motion PSF
OTF=psf2otf(PSF,size(A));                        %Calculate corresponding OTF
    OTF=fftshift(OTF);
Afilt=ifft2(OTF.*FA);                           %Calculate filtered image
subplot(1,4,1);imshow(A,[]);                     %Display results
subplot(1,4,2); imshow(log(1 + PSF),[]);
subplot(1,4,3); imshow(log(1 + abs
    (OTF)),[])
subplot(1,4,4); imshow(abs(Afilt),[]);
```

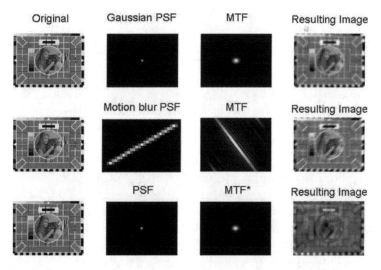

Figure 5.10   The effect of the OTF on the image quality. Column 1: the original image; column 2: the system PSF; column 3: the corresponding system MTF; column 4: the resulting images after transfer by the system OTF. Note the image at bottom right has the same MTF as that at the top right, but the phase has been shifted and this has significant consequences on the image quality

assessment of image quality. For example, hard edges, sinusoidal patterns of both low and high frequency and adjacent patches of slightly different intensities are all evident. These features allow systematic measurements to be made on the OTF and also the image contrast to be assessed. The top row shows the effect on the test card image of transmission through a system with an OTF well approximated by a 2-D Gaussian function. The PSF (also a Gaussian) corresponding to the OTF is also shown. Note that the overall effect on the image is an isotropic blurring and most of the fine structures in the original image are unresolved. The middle row shows the OTF corresponding to a *motion* of the source or detector during image acquisition. The corresponding PSF, a long thin line, is displayed (on a finer scale) indicating a motion of approximately 30 pixels at an angle of 30°. Note that the MTF extends to high spatial frequencies in the direction orthogonal to the motion but drops rapidly in the direction of the motion, and this substantiates what we have already discussed concerning the reciprocal behaviour in the space and spatial frequency domain in which short signals in the space domain become extended in the frequency domain and vice versa. The bottom row shows the effect of an OTF whose MTF is identical to that in the first row but to which *a random phase term has been added to some of the low-frequency components of the image*. The drastic effect on the image quality is apparent.

5.12   Digital Fourier transforms: the discrete fast Fourier transform

In our discussion, we have been a little disingenuous through our silence on a key computational issue. We commented earlier that Fourier transforms can be carried out quickly and effectively on a digital computer thanks to the development of a Fourier transform which works on discrete or digital data – the discrete Fast Fourier Transform (FFT) – but have offered no discussion of it. For example, we have implicitly assumed in the examples presented so far that the Fourier transform of an image (which has a finite number of pixels and thus spatial extent) *also has finite extent*. Moreover, we have implicitly made use of the fact that an $N \times N$ digital image in the spatial domain will transform to a corresponding $N \times N$ frequency-domain representation. This is true, but certainly not obvious. In fact, we have already seen that the continuous Fourier transform of a function having finite spatial extent (e.g. a delta or rectangle function) generally has *infinite* extent in the frequency domain. Clearly, then, there are issues to be resolved. In this chapter we will not attempt to offer a comprehensive treatment of the discrete FFT. This is outside our immediate concern and many excellent treatments exist already.[17] In the simplest terms, we can say that the discrete FFT is just an adaptation of the integral transform we have studied which preserves its essential properties when we deal with discrete (i.e. digital) data. The following section can be considered optional on a first reading, but it is included to ensure the reader is aware of some practical aspects of working with the discrete FFT.

---

[17] For example, see *The Fast Fourier Transform* by E. Oran Brigham or *Fourier Analysis and Imaging* by Ron Bracewell.

5.13 Sampled data: the discrete Fourier transform

Digital images are by definition *discrete*. A digital image which consists of a (typically) large but finite number of pixels must, by definition, have *finite support*. This immediately suggests one potential problem for using Fourier transforms in digital image processing, namely: how can we possibly represent a Fourier transform having *infinite support in a finite array of discrete pixels?* The answer is that we cannot, but there is, fortunately, an answer to this problem. The *discrete Fourier transform* (DFT) – and its inverse – calculates a frequency-domain representation of finite support from a discretized signal of finite support.

The theorems that we have presented and the concepts of frequency space that we have explored for the continuous Fourier transform are identical and carry directly over to the DFT. However, there is one important issue in using the DFT on discrete data that we need to be fully aware of.[18] This relates to the *centring of the DFT*, i.e. the centre of the Fourier array corresponding to the spatial frequency pair $(0,0)$. Accordingly, we offer a brief description of the DFT in the following sections.

Consider a function $f(x,y)$ which we wish to represent discretely by an $M \times N$ array of sampled values. In general, the samples are taken at arbitrary but regularly spaced intervals Δx and Δy along the x and y axes. We will employ the notation $f(x,y)$ for the discrete representation on the understanding that this actually corresponds to samples at coordinates $x_0 + x\Delta x$ and $y_0 + y\Delta y$:

$$f(x,y) \triangleq f(x_0 + x\Delta x, y_0 + y\Delta y) \tag{5.20}$$

where x_0 and y_0 are the chosen starting point for the sampling and the indices x and y assume integer values $x = 0, 1, 2, \ldots, M-1$ and $y = 0, 1, 2, \ldots, N-1$ respectively.

The 2-D (forward) DFT of the $M \times N$ array $f(x,y)$ is given by the expression

$$F(u,v) = \frac{1}{\sqrt{MN}} \sum_{x=0}^{M-1} \sum_{y=0}^{N-1} f(x,y) \exp\left[-2\pi i \left(\frac{ux}{M} + \frac{vy}{N}\right)\right] \tag{5.21}$$

Note that

- the DFT is also of dimension $M \times N$;

- the spatial frequency indices also assume integer values $u = 0, 1, 2, \ldots, M-1$ and $v = 0, 1, 2, \ldots, N-1$.

The discrete transform $F(u,v)$ actually corresponds to sampling spatial frequency pairs $(u\Delta u, v\Delta v)$, i.e. $\triangleq F(u,v)F(u\Delta u, v\Delta v)$, where the sampling intervals in the frequency domain $\{\Delta u, \Delta v\}$ are related to the spatial sampling interval by

$$\Delta u = \frac{1}{M\Delta x} \qquad \Delta v = \frac{1}{N\Delta v} \tag{5.22}$$

---

[18] It is easy to be caught out by this issue when using computer FFT routines, including the Matlab functions *fft*, *fft2* and *fftn*.

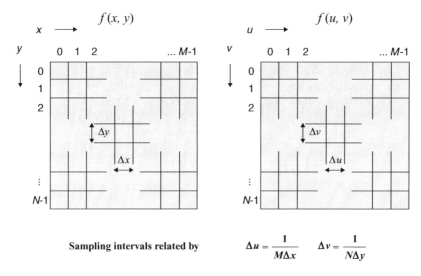

Sampling intervals related by $\Delta u = \dfrac{1}{M\Delta x}$ $\Delta v = \dfrac{1}{N\Delta y}$

Figure 5.11 A discretely sampled image $f(x,y)$ and its DFT $F(u,v)$ have the same number of pixels. The relationship between the sampling intervals in the two domains is given in the diagram

The inverse DFT (reverse DFT) is defined in a similar manner as

$$f(x,y) = \frac{1}{\sqrt{MN}} \sum_{u=0}^{M-1} \sum_{v=0}^{N-1} F(u,v) \exp\left[+2\pi i \left(\frac{ux}{M} + \frac{vy}{N} \right) \right] \tag{5.23}$$

Note that the only difference in the forward and reverse transforms is the sign in the exponential. It is possible to show that the forward and reverse DFTs are *exact inverses* of each other. Equations (5.21) and (5.23) thus represent an exact transform relationship which maintains finite support in both the spatial and frequency domains as required. Figure 5.11 illustrates graphically the basic sampling relationship between a 2-D digital image represented by $M \times N$ pixels and its DFT.

5.14 The centred discrete Fourier transform

The definition of the DFT in Equation (5.21) and the diagram in Figure 5.11 indicate that the spatial frequency coordinates run from the origin at the top left corner of the array, increasing as we move across right and down. It is usual practice *to centre the DFT* by shifting its origin to the centre of the array.

For clarity, we repeat the definition for the DFT given earlier of the 2-D *discrete* array $f(x,y)$:

$$F(u,v) = \frac{1}{\sqrt{MN}} \sum_{x=0}^{M-1} \sum_{y=0}^{N-1} f(x,y) \exp\left[-2\pi i \left(\frac{ux}{M} + \frac{vy}{N} \right) \right] \tag{5.24}$$

We now shift the frequency coordinates to new values $u' = u-(M/2)$, $v' = v-(N/2)$ so that $(u', v') = (0, 0)$ is at the centre of the array. The centred Fourier transform $F(u', v')$ is defined as

$$F(u', v') = F\left(u - \frac{M}{2}, v - \frac{N}{2}\right)$$

$$= \frac{1}{\sqrt{MN}} \sum_{x=0}^{M-1} \sum_{y=0}^{N-1} f(x, y) \exp\left(-2\pi i \left\{ \frac{[u-(M/2)]x}{M} + \frac{[v-(N/2)]y}{N} \right\}\right)$$

Factoring out the exponential terms which are independent of u and v and using $e^{i\pi} = -1$, we have

$$F(u', v') = F\left(u - \frac{M}{2}, v - \frac{N}{2}\right)$$

$$= \frac{1}{\sqrt{MN}} \sum_{x=0}^{M-1} \sum_{y=0}^{N-1} [(-1)^{x+y} f(x, y)] \exp\left[-2\pi i \left(\frac{ux}{M} + \frac{vy}{N} \right)\right] \tag{5.25}$$

which is, by definition, the DFT of the product $(-1)^{x+y} f(x, y)$. Thus, we have one simple way to achieve a centred Fourier transform:

Centred DFT: method 1

If $f(x, y)$ is an $M \times N$ array, then its *centred DFT* is given by the DFT of $(-1)^{x+y} f(x, y)$:

$$F(u', v') \equiv F\left(u - \frac{M}{2}, v - \frac{N}{2}\right) = \mathbf{F}_T\{(-1)^{x+y} f(x, y)\} \tag{5.26}$$

where $\mathbf{F}_T$ symbolically represents the DFT operation.

However, rather than premultiply our function by the factor $(-1)^{x+y}$, we can also achieve a centred DFT 'retrospectively'. Assume that we have calculated the DFT $F(u, v)$ of $f(x, y)$ using the standard definition. It is possible to show, that the *centred DFT* is given by a diagonal swapping of the quadrants in the DFT (Example 5.2, Figure 5.12).

Centred DFT: method 2

If $f(x, y)$ is an $M \times N$ array with DFT $F(u, v)$, its *centred DFT* is given by swapping the first quadrant of $F(u, v)$ with the third and swapping the second quadrant with the fourth.

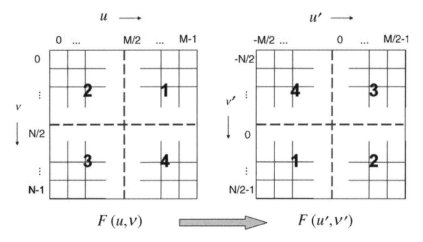

Figure 5.12 The centred DFT (right) can be calculated from the uncentred DFT (left) by dividing the array into four rectangles with two lines through the centre and diagonally swapping the quadrants. Using the labels indicated, we swap quadrant 1 with quadrant 3 and quadrant 2 with quadrant 4

Example 5.2

Matlab code	What is happening?
A=imread('cameraman.tif');	%Read in image
FT=fft2(A); FT_centred=fftshift(FT);	%Take FT, get centred version too
subplot(2,3,1), imshow(A);	%Display image
subplot(2,3,2), imshow(log(1 + abs(FT)),[]);	%Display FT modulus (log scale)
subplot(2,3,3), imshow(log(1 + abs(FT_centred)),[]);	%Display centred FT modulus (log scale)
Im1=abs(ifft2(FT)); subplot(2,3,5), imshow(Im1,[]);	%Inverse FFT and display
Im2=abs(ifft2(FT_centred)); subplot(2,3,6), imshow(Im1,[]);	%Inverse FFT and display
figure;	
[xd,yd]=size(A); x=-xd./2:xd./2-1; y=-yd./2:yd./2-1;	
[X,Y]=meshgrid(x,y); sigma=32;	
arg=(X.^2 + Y.^2)./sigma.^2; frqfilt=exp(-arg);	%Construct freq domain filter
imfilt1=abs(ifft2(frqfilt.*FT));	%Centred filter and noncentred spectrum
imfilt2=abs(ifft2(frqfilt.*FT_centred));	%image – centred filter on centred spectrum
subplot(1,3,1), imshow(frqfilt,[]);	%Display results
subplot(1,3,2), imshow(imfilt1,[]);	
subplot(1,3,3), imshow(imfilt2,[]);	

Comments

Matlab functions: *fft2*, *ifft2*, *fftshift*.

This example illustrates the role played by *fftshift*, which centres the Fourier transform such that the zeroth spatial frequency pair is at the centre of the array.

Why should we centre the DFT? The answer is that we do not *have* to do this, but there are two good reasons in its favour. First, shifting the origin to the centre makes the discrete frequency-space range $(-M/2 \rightarrow M/2-1; -N/2 \rightarrow N/2-1)$ more akin to the continuous space in which we have an equal distribution of positive and negative frequency components $(-\infty \leq k_x \leq \infty; -\infty \leq k_y \leq \infty)$. Second, the construction of frequency-domain filters for suppressing or enhancing groups of spatial frequencies is generally facilitated by using a centred coordinate system.

For further examples and exercises see http://www.fundipbook.com

6

Image restoration

Building upon our consideration of the Fourier (frequency) domain in the previous chapter, we now progress to explore a practical use of these methods in the field of image restoration.

6.1 Imaging models

Image restoration is based on the attempt to improve the quality of an image through knowledge of the physical processes which led to its formation. As we discussed in Chapter 5, we may consider image formation as a process which transforms an *input distribution* into an *output distribution*.[1] The input distribution represents the ideal, i.e. it is the 'perfect' image to which we do not have direct access but which we wish to recover or at least approximate by appropriate treatment of the imperfect or corrupted output distribution. Recall that, in 2-D linear imaging systems, the relationship between the input distribution $f(x', y')$ and the measured output distribution $g(x, y)$ is represented as a linear superposition integral. For linear, shift invariant (LSI) systems, this reduces to the form of a convolution:

$$g(x, y) = \iint f(x', y')h(x-x', y-y')\mathrm{d}x'\mathrm{d}y' + n(x, y)$$

$$g(x, y) = f(x, y) * *h(x, y) + n(x, y)$$

(6.1)

where $**$ is used to denote 2-D convolution. In Equation (6.1), the quantity $h(x-x', y-y')$ is the Point Spread Function (PSF) or impulse response and $n(x, y)$ is an additive noise term. These two factors are responsible for the imperfect output distribution which is obtained.

The image restoration task is (in principle at least) simple:

> Estimate the input distribution $f(x', y')$ using the measured output $g(x, y)$ and any knowledge we may possess about the PSF $h(x-x', y-y')$ and the noise $n(x, y)$.

Recovery of the input distribution $f(x', y')$ from Equation (6.1) is known as *deconvolution*. Image restoration has now evolved into a fascinating but quite complex field of research. The simple observation that *any* approach to image restoration will be *ultimately*

---

[1] These are also often referred to as the *object* and the *image*.

limited by our explicit or implicit knowledge of the PSF and the noise process is helpful in establishing firm ground at the beginning.

Equation (6.1) is not the only model which describes image formation – some situations require a more complex model – but it is certainly by far the most important in practice. Some imaging situations need to be described by more complex models, such as *spatially variant* and *nonlinear* models. However, these are relatively rare and, in any case, can be best understood by first developing a clear understanding of approaches to restoration for the LSI model. Accordingly, Equation (6.1) will form the basis for our discussion of image restoration.

6.2 Nature of the point-spread function and noise

We have said that the quality of our image restoration will be limited by our knowledge of the PSF and the noise. What, then, are the natures of these two entities? Although important exceptions do exist in practice,[2] the system PSF is typically *a fixed or deterministic quantity* that is determined by the physical hardware which constitutes the overall imaging system and which may, therefore, be considered to be unchanging with time. For example, in a simple optical imaging system, the overall PSF will be determined by the physical nature and shape of the lenses; in a medical diagnostic imaging system, such as the Anger camera (which detects gamma-rays emitted from the body of a patient), the PSF is determined by a combination of a mechanical collimator and a scintillator–photomultiplier system which allows the detection of the origin of an emitted gamma photon with a certain limited accuracy.

Thus, given certain physical limitations relating to the nature of the detected flux in the imaging system (i.e. optical, infrared, sonar, etc.), the PSF is typically something over which we initially have some control in principle but which becomes fixed once the system has been designed and engineered. Precisely because the PSF is the result of a design and engineering process, it is something of which we can usually expect to have knowledge and which will assist in approaching the restoration problem.

By contrast, the noise term in Equation (6.1) is typically *stochastic* in nature and produces a random and unwanted fluctuation on the detected signal. The key characteristic of noise processes is that we usually have no control over them and we cannot predict the noise which will be present in a given instant. Noise originates from the physical nature of detection processes and has many specific forms and causes. Whatever the specific physical process that gives rise to the formation of an image, the common characteristic is the unpredictable nature of the signal fluctuations. However, although we cannot know the values of a specific realization of the noise, we can often understand and model its statistical properties. The variety of noise models and behaviour are often a major factor in the subtly different approaches to image restoration. Figure 6.1 summarizes the basic image formation/restoration model described by Equation (6.1).

Image *formation* results from convolution of the input with the system PSF and the addition of random noise. Linear *restoration* employs a linear filter whose specific form

---

[2] The PSF of a telescope system viewing light waves which propagate through atmospheric turbulence is one important example in astronomical imaging where the PSF may be considered as *random* in nature and changes significantly over a short time-scale.

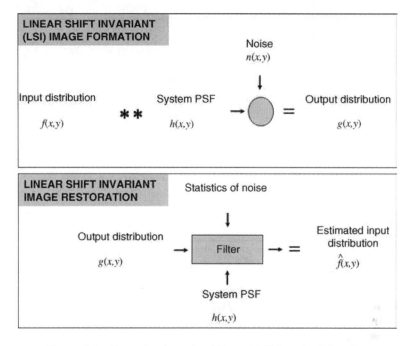

Figure 6.1   The main elements of linear (shift invariant) imaging

depends on the system PSF, our knowledge of the statistics of the noise and, in certain cases, known or assumed statistical properties of the input distribution.

6.3   Restoration by the inverse Fourier filter

Consider again the LSI imaging equation presented in Equation (6.1):

$$g(x,y) = \iint f(x',y')h(x-x',y-y')\mathrm{d}x'\mathrm{d}y' + n(x,y)$$

in which the quantity $h(x-x',y-y')$ is the PSF, $n(x,y)$ is an additive noise term and $f(x',y')$ is the quantity we wish to restore.

If we Fourier transform both sides of this equation, then we may apply the *convolution theorem* of Chapter 5 to obtain

$$\mathbf{F}_\mathrm{T}\{g(x,y)\} = \mathbf{F}_\mathrm{T}\{f(x,y) * *h(x,y) + n(x,y)\}$$
$$G(k_x,k_y) = F(k_x,k_y)H(k_x,k_y) + N(k_x,k_y) \tag{6.2}$$

Convolution in the spatial domain has now become a simple multiplication in the frequency domain. Let us consider the ideal situation in which the additive noise term in our imaging equation $n(\mathbf{x})$ is negligible, i.e. $n(\mathbf{x}) = 0$.

In this case a trivial solution results. We simply divide both sides of Equation (6.2) by $H(K_x, K_y)$ and then take the inverse Fourier transform of both sides. Thus:

$$F(k_x, k_y) = \frac{G(k_x, k_y)}{H(k_x, k_y)} = Y(k_x, k_y)G(k_x, k_y) \tag{6.3}$$

and so

$$f(x, y) = F^{-1}\{Y(k_x, k_y)G(k_x, k_y)\} \tag{6.4}$$

where F^{-1} denotes the operation of inverse Fourier transformation. The frequency-domain filter

$$Y(k_x, k_y) = \frac{1}{H(k_x, k_y)} \tag{6.5}$$

where $H(k_x, k_y)$ is the system optical transfer function (OTF) is called the *inverse filter*.

In practice, the straight inverse filter rarely works satisfactorily and should only ever be used with extreme caution. The reason for this can be understood by examining Equation (6.5) and considering the simple fact that *the OTF of any imaging system will generally not extend to the diffraction limit*. This means that any attempt to recover spatial frequency pairs (k_x, k_y) which may exist in the input but at which the OTF of the system has dropped to an effective value of zero will be disastrous. The magnitude of the multiplicative filter $Y(k_x, k_y)$ at these frequencies will then tend to $\sim 1/0 \rightarrow \infty$.

One way round this is to use a truncated inverse filter in which one effectively monitors the value of the OTF $H(k_x, k_y)$ and sets the value of the filter function $Y(k_x, k_y)$ to zero whenever $H(k_x, k_y)$ falls below a predetermined 'dangerously low' value. Technically, this is an example of a *band-pass filter* because it allows certain spatial frequency pairs to be passed into the reconstruction and suppresses others. This is, in fact, how the restoration was achieved in Figure 6.2 (the Matlab® code used to generate this figure is given in Example 6.1).

However, whenever *noise is present*, the use of an inverse filter will have unpredictable and often disastrous effects. Consider applying the inverse filter in Equation (6.5) to a situation in which noise is present. We obtain

$$\hat{F}(k_x, k_y) = Y(k_x, k_y)G(k_x, k_y) = \frac{G(k_x, k_y)}{H(k_x, k_y)} + \frac{N(k_x, k_y)}{H(k_x, k_y)}$$
$$= F(k_x, k_y) + \frac{N(k_x, k_y)}{H(k_x, k_y)} \tag{6.6}$$

where we use the common 'hat' notation (i.e. $\hat{F}(k_x, k_y)$) to denote an estimated quantity. Equation (6.6) shows that our recovered frequency spectrum has an additional term: the noise spectrum $N(k_x, k_y)$ divided by the system OTF $H(k_x, k_y)$. Clearly, we would like this additional term to be as small as possible, since the estimated spectrum will then approach the true input spectrum. The noise spectrum, however, is an unknown and random addition

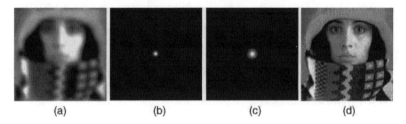

Figure 6.2 Restoration of a blurred but noiseless image through *inverse* filtering. From left to right (a) blurred original; (b) Gaussian PSF; (c) corresponding MTF; (d) recovered original

---

Example 6.1

Matlab code	What is happening?
A=imread('trui.png'); B=fft2(A); B=fftshift(B);	%Read in image and take FT
[x y]=size(A); [X Y]=meshgrid(1:x,1:y);	%Construct Gaussian PSF
h=exp(-(X-x/2).^2./48).*exp(-(Y-y/2).^2./48);	%extending over entire array
H=psf2otf(h,size(h)); H=fftshift(H);	%Get OTF corresponding to PSF
g=ifft2(B.*H); g=abs(g);	%Generate blurred image via
	%Fourier domain
G=fft2(g); G=fftshift(G);	%Take FT of image
indices=find(H>1e-6);	%Do inverse filtering AVOIDING
F=zeros(size(G)); F(indices)=G(indices)./	%small values in OTF !!
H(indices);	
f=ifft2(F); f=abs(f);	%Inverse FT to get filtered image
subplot(1,4,1), imshow(g,[min(min(g))	%Display "original" blurred image
max(max(g))]);	
subplot(1,4,2), imagesc(h); axis square; axis off;	%Display PSF
subplot(1,4,3), imagesc(abs(H)); axis square; axis off;	%Display MTF
subplot(1,4,4), imagesc(f); axis square; axis tight;	%Display filtered image
axis off;	

Comments
New Matlab functions: *find, psf2otf.*
   The first section of code generates a Gaussian blurred image. The second section estimates the original image using an inverse filter.

---

to the data to which we 'do not have access' and which is actually quite inseparable from the output image spectrum. Moreover, it is characteristic of many noise processes that they have *significant high-frequency content*; in other words, there is a spatial frequency regime for which $|N(k_x, k_y)| \gg |G(k_x, k_y)|$. In this case, it is clear that the first term on the right-hand side in Equation (6.6) (the true spatial frequency content) will be completely dominated by the second term (the noise). This is the case shown in the Figure 6.3 (the Matlab code used to generate this figure is given in Example 6.2).

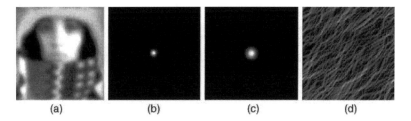

(a)   (b)   (c)   (d)

Figure 6.3 Restoration of a blurred and noisy image through *inverse* filtering. From left to right: (a) blurred original with Gaussian white noise (zero mean, 0.2% variance); (b) system PSF; (c) corresponding MTF; (d) recovered image

---

Example 6.2

Matlab code	What is happening?
A=imread('trui.png'); B=fft2(A); B=fftshift(B);	
[x y]=size(A); [X Y]=meshgrid(1:x,1:y);	
h=exp(-(X-x/2).^2./48).*exp(-(Y-y/2).^2./48);	%CODE SAME AS EXAMPLE 6.1
H=psf2otf(h,size(h)); H=fftshift(H);	%Get OTF
g=ifft2(B.*H); g=mat2gray(abs(g));	%Blur image in Fourier domain
g=imnoise(g,'gaussian',0,0.002);	%Add noise to image
	%CODE HEREAFTER SAME AS 6.1

Comments
Matlab functions: *imnoise*, *mat2gray*.
    Similar to Example 6.1 except that white noise is added to the image.

6.4 The Wiener–Helstrom filter

As Example 6.2 and the corresponding Figure 6.3 show, whenever noise is present in an image we require a more sophisticated approach than a simple inverse filter. Inspection of the LSI imaging equation in the frequency domain

$$\hat{F}(k_x, k_y) = Y(k_x, k_y)G(k_x, k_y) = Y(k_x, k_y)[H(k_x, k_y)F(k_x, k_y) + N(k_x, k_y)] \qquad (6.7)$$

suggests that we would ideally like our frequency-domain filter $Y(k_x, k_y)$ to have the following qualitative properties:

- At those spatial frequency pairs for which the noise component $|N(k_x, k_y)|$ is much smaller than the image component $|G(k_x, k_y)|$, our filter should approach the inverse filter. Thus, we want

$$Y(k_x, k_y) \approx \frac{1}{H(k_x, k_y)} \text{ when } |N(k_x, k_y)| \ll |G(k_x, k_y)|$$

    This ensures accurate recovery of these frequency components in the restored image.

- At those spatial frequency pairs for which the noise component $|N(k_x, k_y)|$ is much larger than the image component $|G(k_x, k_y)|$, our filter should approach *zero*. Thus:

$$Y(k_x, k_y) \approx 0 \text{ when } |N(k_x, k_y)| \gg |G(k_x, k_y)|$$

This ensures that we do not attempt to restore spatial frequency pairs which are dominated by the noise component.

- At those spatial frequency pairs for which the noise component $|N(k_x, k_y)|$ and the image component $|G(k_x, k_y)|$ are comparable, our filter should 'damp' these frequencies, effecting an appropriate compromise between complete acceptance (inverse filter) and total suppression.

These three properties are broadly achieved by the *Wiener–Helstrom* (often abbreviated simply to *Wiener*) *filter*, defined as

$$Y(k_x, k_y) = \frac{H^*(k_x, k_y) W_F(k_x, k_y)}{|H(k_x, k_y)|^2 W_F(k_x, k_y) + W_N(k_x, k_y)} \tag{6.8}$$

In Equation (6.8) H^* denotes the complex conjugate of the OTF and the quantities $W_F(k_x, k_y)$ and $W_N(k_x, k_y)$ are respectively the *input* and *noise power spectra*:

$$W_F(k_x, k_y) = \langle |F(k_x, k_y)|^2 \rangle \quad \text{and} \quad W_N(k_x, k_y) = \langle |N(k_x, k_y)|^2 \rangle \tag{6.9}$$

Division of the right-hand side of Equation (6.8) by the input power spectrum enables us to express the *Wiener–Helstrom filter* in an alternative and more transparent form:

$$Y(k_x, k_y) = \frac{H^*(k_x, k_y)}{|H(k_x, k_y)|^2 + \text{NSR}(k_x, k_y)} \tag{6.10}$$

where the quantity $\text{NSR}(k_x, k_y) = W_N(k_x, k_y)/W_F(k_x, k_y)$ gives the noise/signal power ratio. Equation (6.10) thus shows clearly that the Wiener–Helstrom filter approximates an inverse filter for those frequencies at which the signal/noise power ratio is large, but becomes increasingly small for spatial frequencies at which the signal-noise power ratio is small.

6.5   Origin of the Wiener–Helstrom filter

The Wiener–Helstrom filter defined by Equation (6.8) (or equivalently by Equation (6.10)) is not simply an ad hoc empirical filter but has a firm theoretical basis. It is important because it is, in a clearly defined sense, an *optimal linear restoration filter*. Let us restate the LSI imaging equation presented in Equation (6.1):

$$g(x, y) = \iint f(x', y') h(x-x', y-y') \, dx' \, dy' + n(x, y)$$

which, via the convolution theorem, has the corresponding frequency domain equivalent presented in Equation (6.2):

$$G(k_x, k_y) = F(k_x, k_y)H(k_x, k_y) + N(k_x, k_y)$$

A well-defined optimal criterion is to seek an estimate of the input distribution $\hat{f}(x, y)$, which varies minimally from the true input $f(x, y)$ in the *mean-square* sense. Namely, we will seek an estimate $\hat{f}(x, y)$ such that the following quantity is minimized:

$$Q = \left\langle \int_{-\infty}^{\infty} \int_{-\infty}^{\infty} [\hat{F}(x, y) - F(x, y)]^2 \, dx \, dy \right\rangle \tag{6.11}$$

This optimization criterion is called the minimum mean-square error (MMSE) or minimum variance criterion. It is important to be aware that the angle brackets in Equation (6.11) denote ensemble averaging; that is, we seek an estimator $\hat{f}(x, y)$ that is optimal with respect to statistical averaging over many realizations of the random noise process $n(x, y)$ and, indeed, to the PSF in those special situations where it may be treated as a stochastic quantity.[3]

Application of Parseval's theorem to Equation (6.11) allows us to express an entirely equivalent criterion in the Fourier domain:

$$Q = \left\langle \int_{-\infty}^{\infty} \int_{-\infty}^{\infty} [\hat{F}(x, y) - F(x, y)]^2 \, dx \, dy \right\rangle \tag{6.12}$$

The basic approach to deriving the Wiener–Helstrom filter is to postulate a linear filter $Y(k_x, k_y)$ which acts on the image data $G(k_x, k_y) = F(k_x, k_y)H(k_x, k_y) + N(k_x, k_y)$ and then to take variations in the operator $Y(k_x, k_y)$ such that the quantity Q is *stationary*. The formal derivation is quite lengthy and requires methods from statistical estimation theory and variational calculus, and so is not repeated here.[4]

It is further possible to show that the resulting mean-square error after application of the Wiener–Helstrom filter is given by

$$Q = \int_{-\infty}^{\infty} \int_{-\infty}^{\infty} \frac{W_F(k_x, k_y) W_N(k_x, k_y)}{|H(k_x, k_y)|^2 W_F(k_x, k_y) + W_N(k_x, k_y)} \, dx \, dy \tag{6.13}$$

In summary, the Wiener–Helstrom filter works by accepting low-noise frequency components and rejecting high-noise frequency components. Further, it does so in a

---

[3] A good example of a randomly varying and thus stochastic PSF is the PSF of a ground-based telescope observing through atmospheric turbulence.

[4] Interested readers can study a detailed derivation on the book's website at http://www.fundipbook.com/materials/.

well-defined 'optimal' way. This is very reasonable behaviour and it is tempting to think that image restoration in the frequency domain should begin and end here.

However, despite its practical and historical importance the Wiener–Helstrom filter is no panacea, and two points are essential to make:

(1)   The exact implementation of Equation (6.10) requires knowledge of the *input* power spectrum. Since the input distribution is precisely the quantity we are trying to estimate, we cannot generally expect to know its power spectrum![5] Thus, in the vast majority of cases, the Wiener–Helstrom filter cannot be precisely implemented. However, there are several ways in which we can attempt to get round this problem to produce a practical filter:

(a)   When the input distribution belongs to a well-defined class in which the input is constrained, reasonably accurate estimates of the power spectrum may be available. A typical instance might be a standard medical imaging procedure, such as taking a chest radiograph, in which the input structure can essentially be considered as a modest variation on a basic prototype (i.e. the skeletal structure is very similar across the population). In this case, the prototype may be used to provide the approximation to the input power spectrum.

(b)   When the input is not constrained in this way, the Wiener–Helstrom filter can be approximated by substituting the *output* power spectrum $W_G(k_x, k_y)$ (which we can easily calculate) or preferably some filtered version of it *in place* of the input power spectrum.

(c)   Another simple, but less accurate, approach is to approximate the noise/signal ratio (the NSR function in Equation (6.10)) by an appropriate scalar constant. By removing all spatial frequency dependence from this term, one typically trades off some high-frequency content in the restoration for increased stability.

(2)   Restorations produced by the Wiener–Helstrom filter often produce restorations which are rather too blurred from the perceptual viewpoint – suggesting that the MMSE criterion, although a perfectly valid mathematical optimisation criterion, is almost certainly not the perceptually optimal from a human standpoint. For this reason, many other filters have been developed which can produce better results for specific applications.

Figure 6.4 (for which the Matlab code is given in Example 6.3) shows the results of two different approximations to the ideal Wiener filter on an image exhibiting a significant level of noise. The restored image is far from perfect but represents a significant improvement on the original.

---

[5] This point is strangely neglected in many discussions of the Wiener filter.

Example 6.3

Matlab code	What is happening?
I = imread('trui.png');I=double(I);	%Read in image
noise =15.*randn(size(I));	%Generate noise
PSF = fspecial('motion',21,11);	%Generate motion PSF
Blurred = imfilter(I,PSF,'circular');	%Blur image
BlurredNoisy = Blurred + noise;	%Add noise to blurred image
NSR = sum(noise(:).^2)/sum(I(:).^2);	% Calculate SCALAR noise-to-power ratio
NP = abs(fftn(noise)).^2;	%Calculate noise power spectrum
NPOW = sum(NP(:))/prod(size(noise));	%Calculate average power in noise spectrum
NCORR = fftshift(real(ifftn(NP)));	%Get autocorrelation function of the noise, %centred using fftshift
IP = abs(fftn(I)).^2;	%Calculate image power spectrum
IPOW = sum(IP(:))/prod(size(I));	%Calculate average power in image spectrum
ICORR = fftshift(real(ifftn(IP)));	%Get autocorrelation function of the image, %centred using fftshift
NSR = NPOW./IPOW;	%SCALAR noise-to-signal power ratio

subplot(131);imshow(BlurredNoisy,[min(min(BlurredNoisy)) max(max(BlurredNoisy))]);
　　　　　　　　　　　　　　　%Display blurred and noisy image';
subplot(132);imshow(deconvwnr(BlurredNoisy,PSF,NSR),[]);
　　　　　　　　　　　　　　　%Wiener filtered – PSF and scalar noise/
　　　　　　　　　　　　　　　%signal power ratio
subplot(133);imshow(deconvwnr(BlurredNoisy,PSF,NCORR,ICORR),[]);
　　　　　　　　　　　　　　　%Wiener filtered – PSF and noise and signal
　　　　　　　　　　　　　　　%autocorrelations

Comments

Matlab functions: *fftn*, *ifftn*, *prod*, *deconvwnr*.

　　This example generates a noisy, blurred image and then demonstrates restoration using two variations of the Wiener filter. The first approximates the noise/signal power ratio by a constant and the second assumes knowledge of the noise and signal autocorrelation functions.

　　The Image Processing Toolbox has a function specifically for carrying out Wiener filtering, *deconvwnr*. Essentially, it can be used in three different ways:

- The user supplies the PSF responsible for the blurring only. In this case, it is assumed that there is no noise and the filter reduces to the inverse filter Equation (6.5).

- The user supplies the PSF responsible for the blurring and a *scalar* estimate of the noise/ signal power ratio NSR. In other words, only the total amounts of power in the noise and in the input are provided and their frequency dependence is not supplied.

- The user supplies the PSF responsible for the blurring and a *frequency-dependent* estimate of the noise/signal power ratio, via their respective autocorrelation functions. (Recall that the autocorrelation theorem allows the calculation of the autocorrelation from the power spectrum.)

　　Type *doc deconvwnr* for full details.

(a) (b) (c)

Figure 6.4 Restoration of a blurred and noisy image through *Wiener* filtering. From left to right: (a) blurred original with Gaussian white noise (zero mean, standard deviation ~10 % of mean signal); (b) Wiener filtered with scalar estimate of total power in noise and signal; (c) Wiener filtered with autocorrelation of noise and signal provided

6.6 Acceptable solutions to the imaging equation

A solution to the imaging equation, Equation (6.1), is deemed *acceptable* if it is consistent with all the information that we have about the system PSF and the noise process. Explicitly, this means that if we take a solution for the input distribution $\hat{f}(x, y)$, pass it through our system by convolution with the system PSF and subtract the result from the output, the residual (i.e. $\hat{n} = g - h * \hat{f}$) should be a noise distribution possessing the same statistical properties as that of the noise model. Any spatial structure in the residual which is not consistent with the noise model indicates that the solution is not consistent with the imaging equation. A completely equivalent view can be taken in the frequency domain, where subtraction of the estimated output spectrum $\hat{G}(k_x, k_y)$ from the measured output spectrum $G(k_x, k_y)$ should be consistent with our statistical knowledge of the noise spectrum $N(k_x, k_y)$.

6.7 Constrained deconvolution

The Wiener filter is designed to minimize the sum of the mean-square errors between the actual input distribution and our estimate of it. This is a perfectly sensible criterion for a restoration filter, but we must recognize that it is not the only criterion that we may be interested in. Speaking loosely, the Wiener filter criterion considers *all parts of the image to have equal importance*, as its only goal is to minimize the sum of the squared errors over the entire image. Important visual criteria, such as the preservation of smoothness (typically for noise suppression) or enhancing sharpness (to preserve edges), are not explicitly encapsulated by the Wiener filter. Given that edge structure or overall smoothness, for example, are so important in the perception of images, it stands to reason that we may wish to ensure that the restoration attempts to recover certain characteristics of the image particularly accurately.

In constrained least-squares deconvolution, the task is to restrict ourselves to solutions which minimize some desired quantity in the restored image *but which impose constraints on the solution space*. In many instances, we may know (or at least be able to make an educated estimate) the overall noise power $\|n(x, y)\|^2 = \int n^2(x, y) \, dx \, dy$ in the output distribution. It stands to reason that convolution of any restored image $\hat{f}(x, y)$ with the PSF of our imaging system (this is our predicted output distribution $\hat{g}(x, y) = h(x, y) * *\hat{f}(x, y)$ if you will) should not differ from the *actual* distribution $g(x, y)$ by an amount that would exceed the known noise power. For this reason, one common and sensible constraint is to demand that the noise power in our restored image be similar to its known value. Thus, this constraint requires that:

$$\|g(x, y) - h(x, y) * *\hat{f}(x, y)\|^2 = \|n(x, y)\|^2. \tag{6.14}$$

We stress that there are *many* possible solutions $\hat{f}(x, y)$ which satisfy the particular constraint expressed by Equation (6.14). This follows from the random nature of the noise and ill-conditioned nature of the system. The specific mathematical approach taken to constrained restoration is to specify a *cost function* which comprises two basic parts. The first part consists of some linear operator **L** (often referred to as the *regularization* operator) acting on the output distribution $\mathbf{L}f(x, y)$. The second part consists of one or more Lagrange multiplier terms which specify the constraints. The aim is to minimize the size of $\mathbf{L}f(x, y)$ subject to the chosen constraints. The precise form of the linear operator **L** will naturally depend on what property of the restored image we wish to minimize and will be discussed shortly.

Formally then, we seek a solution $\hat{f}(x, y)$ which will minimize the cost function:

$$Q = \|\mathbf{L}f(x, y)\|^2 + \lambda\{\|g(x, y) - h(x, y) * *f(x, y)\|^2 - \|n(x, y)\|^2\} \tag{6.15}$$

where **L** is some general linear operator and λ is an unknown Lagrange multiplier.

This problem can just as easily be formulated in the frequency domain by considering Equation (6.2), the Fourier equivalent of the imaging equation. Eq. 6.15 then becomes

$$Q = \|\mathbf{L}F(k_x, k_y)\|^2 + \lambda\{\|G(k_x, k_y) - H(k_x, k_y)F(k_x, k_y)\|^2 - \|N(k_x, k_y)\|^2\} \tag{6.16}$$

By taking vector derivatives of Equation (6.16) with respect to $F(k_x, k_y)$ and setting to zero, we can derive the constrained restoration filter[6] $Y(k_x, k_y)$ such that the restored spectrum is given by

$$\hat{F}(k_x, k_y) = Y(k_x, k_y)G(k_x, k_y)$$

where

$$Y(k_x, k_y) = \frac{H^*(k_x, k_y)}{|H(k_x, k_y)|^2 + \alpha|L(k_x, k_y)|^2} \tag{6.17}$$

---

[6] A derivation of constrained least-squares restoration is available on the book website at http://www.fundipbook.com/materials/.

Thus, $Y(k_x, k_y)$ simply multiplies the output frequency spectrum to produce our estimate of the input spectrum. In Equation (6.17), $H(k_x, k_y)$ denotes the system OTF, $L(k_x, k_y)$ is the frequency-domain representation of the selected linear operator and the parameter $\alpha = 1/\lambda$ (the inverse Lagrange multiplier) is treated as a regularization parameter whose value is chosen to ensure that the least-squares constraint is satisfied. Although analytical methods have been developed to estimate α, it is often treated as a 'user tuneable' parameter.

We stress that the generality of the linear operator **L** in Equations (6.15) and (6.16) allows a variety of restoration criteria to be specified. For example, choosing **L** = **I**, the identity operator effectively results in the so-called parametric Wiener filter which produces a minimum norm solution – this is the minimum energy solution which satisfies our constraint. Another criterion is to seek overall maximum 'smoothness' to the image. A good measure of the sharpness in an image is provided by the Laplacian function $\nabla^2 f(x, y)$. By choosing **L** = ∇^2 and minimizing $\nabla^2 f(x, y)$, we effectively ensure a smooth solution.

Figure 6.5 was produced by Matlab Example 6.4 and shows examples of constrained least-squares solutions using a Laplacian operator on a blurred and noisy image. Note the increased smoothness which results from fixing the value of λ to be high.

Figure 6.5 Example of constrained deconvolution. Left: original image; centre: restored by minimizing Laplacian of image subject to least-squares noise constraint; right: similar to centre, except Lagrange multiplier fixed at 10 times the previous value

Example 6.4

Matlab code	What is happening?
I = imread('trui.png');	%Read in image
PSF = fspecial('gaussian',7,10);	%Define PSF
V = .01;	%Specify noise level
BlurredNoisy = imnoise(imfilter(I,PSF), 'gaussian',0,V);	%Produce noisy blurred image
NP = V.*prod(size(I));	%Calculate noise power
[J LAGRA_J] = deconvreg(BlurredNoisy, PSF,NP);	%Constrained deconvolution
	%default Laplacian operator, Lagrange % multiplier optimised

```
[K LAGRA_K] = deconvreg(BlurredNoisy,PSF,
    [],LAGRA_J*10);
```
                                                %Lagrange multiplier fixed
                                                %(10 times larger)
```
subplot(131);imshow(BlurredNoisy);             %Display original
subplot(132);imshow(J);                        %Display 1st deconvolution result
subplot(133);imshow(K);                        %Display 2nd deconvolution result
```

Comments

Matlab functions: *imfilter, imnoise, fspecial, deconvreg*.

This example generates a degraded image and then shows the results of constrained deconvolution using two different regularization parameters.

The Image Processing Toolbox has a function specifically for carrying out constrained or regularized filtering, *deconvreg*. Several points are worthy of mention:

- Unless the user *explicitly specifies otherwise*, the default linear operator is taken to be the Laplacian. The reconstruction will then aim to optimize smoothness.

- The user must supply the PSF responsible for the blurring.

- The overall noise power should be specified. The default value is 0.

- The user may specify a given range within which to search for an optimum value of the Lagrange multiplier or may specify the value of the multiplier directly.

Type *doc deconvreg* for full details.

6.8 Estimating an unknown point-spread function or optical transfer function

As we have seen, in the frequency domain the 'standard' deconvolution problem may be expressed by Equation (6.2):

$$G(k_x, k_y) = F(k_x, k_y)H(k_x, k_y) + N(k_x, k_y)$$

and our discussion of frequency-domain restoration so far has covered the simple inverse, Wiener and constrained least-squares approaches to recovering the input spectrum $F(k_x, k_y)$. All of these methods have implicitly assumed that we have knowledge of the system OTF $H(k_x, k_y)$. In many practical situations, this information is not readily available. For example, image analysis and restoration undertaken for forensic purposes very often falls into this category: corrupted images taken by cameras long since lost or otherwise inaccessible are presented in the hope of verifying personal identity or for some other matter of legal importance. In such cases, attempts at restoration can only proceed if some means is found

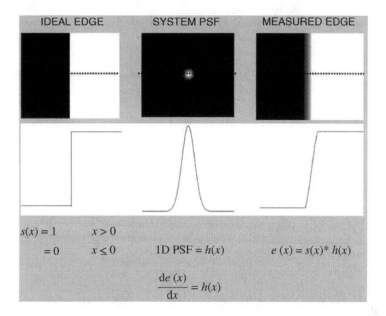

Figure 6.6 The measured 1-D profile through an edge is the result of convolution of the idealized edge with a 1-D PSF $h(x)$. $h(x)$ is calculated as the derivative of the edge $de/dx = h(x)$ and this 1-D PSF is the marginal distribution of the 2-D PSF along the direction of the edge

to sensibly estimate the system OTF (or, equivalently, its PSF) from the image itself. This can be attempted in two basic ways.

- The first is conceptually simple. We can inspect the image for objects which we know *a priori* are point-like in nature (stars are a good example). Their appearance in the image will by definition give an estimate of the PSF. Clearly, such point-like objects are not guaranteed in any image.

- A closely related approach is to look for straight lines and, by implication, *edges*, since the PSF can also be estimated if we can identify these in the image.

To understand this second approach, consider the idealized sharp edge and a blurred version of that edge in Figure 6.6. A profile taken through the edge in a direction perpendicular to the direction of the line is a 1-D signal which is termed the edge-spread function (ESF), which we will denote $e(x)$. Clearly, $e(x)$ is the result of a convolution of the idealized edge with a 1-D system PSF $h(x)$; thus, $e(x) = s(x) * h(x)$. Now, it can be shown formally[7] that the derivative of the ideal edge function is given by a delta function, i.e. $\delta(x-x_0) = ds/dx$ (entirely equivalently, the integral of the delta function yields the ideal edge function). Using this relation, we find that simple differentiation of the ESF yields the system PSF, i.e. $de/dx = h(x)$.

Of course, this expression only yields a *1-D version* of the PSF and actually represents an integration (the marginal distribution) of the true 2-D PSF $h(x, y)$ along the direction of

---

[7] See http://www.fundipbook.com/materials/.

the edge. Now, if there is good reason to believe that $h(x, y)$ is circularly symmetric (and this is reasonable in many cases), then the job is done and our 1-D estimate is simply rotated in the plane through 360 degrees to produce the corresponding 2-D version. However, in the most general case, the situation is rather more involved. Fusing a large number of such marginal distributions of a quite general function $h(x, y)$ can only be done practically by using Fourier techniques (typically employed in computed tomography) which exploit the central slice theorem. The problem becomes much simpler if we can approximate the 2-D PSF $h(x, y)$ by a 2-D Gaussian function. This is very often a good approximation to reality, and the function takes the specific form

$$p(\mathbf{x}) = \frac{1}{2\pi|\mathbf{C}_x|}\exp\left(-\frac{1}{2}\mathbf{x}^T\mathbf{C}_x^{-1}\mathbf{x}\right) \tag{6.18}$$

where the vector $\mathbf{x} = [x\ y]$ and the matrix $\mathbf{C}_x$ can be expressed as

$$\mathbf{C}_x = \mathbf{R}_\theta\begin{bmatrix} \sigma_1^2 & 0 \\ 0 & \sigma_2^2 \end{bmatrix}\mathbf{R}_\theta^T \tag{6.19}$$

Here, the parameters σ_1^2 and σ_2^2 represent the width of the PSF along two orthogonal directions. Note that the matrix $\mathbf{R}_\theta$ is a 2-D rotation matrix (which posesses one degree of freedom, the angle θ) included to allow for the possibility that the principal axes of the PSF may not be aligned with the image axes. When the principal axes of the PSF are aligned with the image axes, the matrix $\mathbf{R}_\theta = \mathbf{I}$, the identity matrix and the 2-D Gaussian is separable in x and y:-

$$p(\mathbf{x}) = \frac{1}{2\pi|\mathbf{C}_x|}\exp\left(-\frac{1}{2}\mathbf{x}^T\mathbf{C}_x^{-1}\mathbf{x}\right)$$
$$\to p(x, y) = \frac{1}{\sqrt{2\pi}\sigma_1}\exp\left(-\frac{x^2}{\sigma_1^2}\right)\frac{1}{\sqrt{2\pi}\sigma_2}\exp\left(-\frac{y^2}{\sigma_2^2}\right) \tag{6.20}$$

Under this assumption of a 2-D Gaussian PSF, the response of the system to an arbitrary edge will be an integrated Gaussian function (the so-called error function familiar from statistics). The error function has a single free parameter and by least-squares fitting the edge response to the error function at a number of edges a robust estimate of the PSF can be formed.

6.9 Blind deconvolution

Consider once more the 'standard' deconvolution problem in the frequency domain given by Equation (6.2):

$$G(k_x, k_y) = F(k_x, k_y)H(k_x, k_y) + N(k_x, k_y)$$

in which we seek to estimate the input spectrum $F(k_x, k_y)$. We have already stressed that we do not know the noise spectrum $N(k_x, k_y)$. **What happens if we also don't know the system OTF $H(k_x, k_y)$?** At first sight, this problem, known as *blind deconvolution*, looks an impossible one to solve. After all, the measured output spectrum $G(k_x, k_y)$ is given by the product of *two unknown quantities* plus a random noise process.

Remarkably, feasible solutions can be found to this problem by iterative procedures (usually carried out in the frequency domain) which enforce only basic constraints on the feasible solutions, namely:

(1) that they have *finite support*, i.e. the sought input distribution is known to be confined to a certain spatial region and is zero outside this region;

(2) any proposed solution must be strictly positive (the input which corresponds to a flux of some nature cannot have negative values).

The blind deconvolution procedure attempts to estimate/restore not only the original input distribution, but also the PSF responsible for the degradation. A detailed account of approaches to blind deconvolution lies outside the scope of this text, but some related references are available on the book website.[8]

Examples of restoration using maximum-likelihood blind deconvolution were generated with Matlab Example 6.5 as shown in Figure 6.7. The PSF of the blurred holiday snap on the far left was unknown and we see that, although the deconvolution procedure has produced some ringing effects, the restored images are indeed sharper.

Example 6.5

Matlab code	What is happening?
A=imread('test_blur.jpg');	%Read in image
A=edgetaper(A,ones(25));	%Smooth edges of image
[J,PSF] = deconvblind(A,ones(10));	%Deconvolve – initial estimate PSF 'flat'
subplot(1,4,1), imshow(A,[]);	%Display original
subplot(1,4,2), imshow(J,[]);	%Display deconvolved
h=fspecial('gaussian',[10 10],3);	
[J,PSF] = deconvblind(A,h);	%Deconvolve – initial estimate PSF normal
subplot(1,4,3), imshow(J,[]);	%Display
J = deconvwnr(A,PSF,0.01);	%Wiener filter with 'blind' recovered PSF
subplot(1,4,4), imshow(J,[]);	%Display Wiener deconvolution

Comments

Matlab functions: ***edgetaper, deconvblind, deconvwnr***.

The example above shows the results of blind deconvolution starting from two different initial estimates of the PSF. In general, the results of blind deconvolution are quite sensitive to the initial PSF estimate. The third restoration shows the results of Wiener filtering but using the PSF recovered from blind deconvolution.

The Image Processing Toolbox has a function specifically for carrying out blind deconvolution, ***deconvblind***. Type *doc deconvblind* for full details on how to use the blind deconvolution algorithm in Matlab.

---

[8] See http://www.fundipbook.com/materials/.

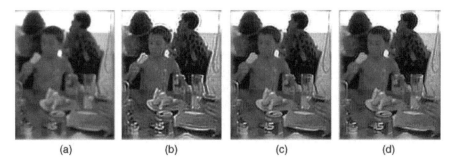

     (a)   (b)   (c)   (d)

Figure 6.7 Maximum likelihood blind deconvolution. (a) Original image obtained with unknown PSF. (b) Restoration by maximum-likelihood blind deconvolution (initial guess for PSF a 10×10 averaging filter). (c) Restoration by maximum-likelihood blind deconvolution (initial guess for PSF a 10×10 Gaussian filter with $\sigma = 3$). (d) Wiener filtered using the final estimated PSF from (c) and an assumed scalar NSR $= 0.01$

6.10 Iterative deconvolution and the Lucy–Richardson algorithm

Image restoration remains a vast field of research. It would be tempting to offer some discussion on the array of Bayesian-related reconstruction techniques and nonlinear image restoration problems. However, this lies beyond our scope, and this chapter aims only to give the reader a flavour of the basic restoration problem and of some common and basic approaches to solutions. Readers interested in pursuing this subject at a deeper level will find a vast amount of literature at their disposal.

However, partly because it has found such popularity amongst the astronomy community and partly because it is an algorithm that is explicitly provided by the Matlab Image Processing Toolbox, we briefly describe one further technique, namely that of the Lucy–Richardson (LR) deconvolution algorithm. The LR algorithm is best understood if we first consider a simple iterative algorithm and then see how the LR method is an extension of this approach.

The linear imaging equation states that $g(x, y) = f(x, y) * * h(x, y) + n(x, y)$, so that the noise $n(x,y)$ is the difference between the output distribution $g(x,y)$ (i.e. the image we actually measured) and the unknown input distribution $f(x,y)$ convolved with the PSF $h(x,y)$:

$$n(x, y) = g(x, y) - f(x, y) * * h(x, y) \tag{6.21}$$

Substitution of a *good* estimate of the input distribution $f(x,y)$ in Equation (6.21) would tend to make $n(x, y)$ small, whereas a poor estimate would make it large and in the limit of negligible noise our ideal estimate would satisfy:

$$g(x, y) - f(x, y) * * h(x, y) = 0 \tag{6.22}$$

If we add the input distribution $f(x,y)$ to both sides of Equation (6.22), then we have:

$$f(x, y) = f(x, y) + [g(x, y) - f(x, y) * * h(x, y)] \tag{6.23}$$

This equation can be viewed as an iterative procedure in which a new estimate of the input (the left-hand side) is given as the sum of the previous estimate (first term of right-hand side) and a *correction term* (in brackets). The correction term is actually the difference between our measured image and our prediction of it using the current estimate of the input. This certainly seems a reasonable correction to add and is certainly an easy one to calculate. Writing this explicitly as an iterative procedure, the $(k + 1)$th estimate of the input is thus given by:

$$f_{k+1}(x,y) = f_k(x,y) + [g(x,y) - f_k(x,y) * *h(x,y)] \qquad (6.24)$$

The procedure described by Equation (6.24) is started by setting $f_0(x,y) = g(x,y)$. In other words, we seed the algorithm by taking our first estimate of the input distribution to be the measured output. Unless the image is very severely degraded, the output is not hugely different from the true input distribution and this simple procedure converges nicely. Note that Equation (6.24) can never be satisfied exactly, unless there is no noise at all, but will reach a point at which the size of the correction term[9] in Equation (6.24) reaches a minimum value.

A variation on Equation (6.24) is the Van Cittert algorithm, which uses a pixel-dependent weighting factor or *relaxation parameter* $w(x,y)$ to control the speed of the convergence:

$$f_{k+1}(x,y) = f_k(x,y) + w(x,y)[g(x,y) - f_k(x,y) * * h(x,y)] \qquad (6.25)$$

The basic assumptions of the LR method are twofold:

- we assume that the PSF is known;

- we assume that the noise in the output is governed by the Poisson density function.

The LR method is an iterative algorithm which attempts to find the *maximum-likelihood* solution given knowledge of the PSF and the assumption of Poisson noise. As is customary in most other discussions of the algorithm, we will consider a discrete form of Equation (6.1) in which the input and output distributions are represented by vectors and the PSF by a 2-D matrix. In other words, let the ith pixel in the input distribution have value f_i. This is related to the observed value of the ith pixel in the output g_i by

$$g_i = \sum_j h_{ij} f_j \qquad (6.26)$$

where the summation over index j provides the contribution of each input pixel, as expressed by the PSF h_{ij}, to the observed output pixel. It is customary to normalize the discrete PSF so that $\sum_i \sum_j h_{ij} = 1$.

---

[9] By *size* we mean here the total sum of the squared pixel values.

The iterative LR formula is given by

$$f_j = f_j \sum_i \left(\frac{h_{ij} g_i}{\sum_k h_{jk} f_k} \right) \tag{6.27}$$

where the kernel in Equation (6.27) approaches unity as the iterations progress. The theoretical basis for Equation (6.27) can be established from Bayes' theorem, and the interested reader is referred to the literature for details.[10] Examples of restorations achieved using the LR deconvolution algorithm are given by the code in Matlab Example 6.6 and shown in Figure 6.8.

Example 6.6

Matlab code	What is happening?
A = imread('trui.png'); A=mat2gray(double(A));	%Read in image and convert to intensity
PSF = fspecial('gaussian',7,10);	%Specify PSF
V = .0001;	%Define variance of noise
J0 = imnoise(imfilter(A,PSF),'gaussian',0,V);	%Create blurred and noisy image
WT = zeros(size(A));WT(5:end-4,5:end-4) = 1;	%Define weighting function
J1 = deconvlucy(J0,PSF,10);	%LR deconvolution 10 iterations
J2 = deconvlucy(J0,PSF,20,sqrt(V));	%20 iterations, deviation of noise provided
J3 = deconvlucy(J0,PSF,20,sqrt(V),WT);	%weight function to suppress ringing
subplot(141);imshow(J0);	%Display various results
subplot(142);imshow(J1);	
subplot(143);imshow(J2);	
subplot(144);imshow(J3);	

Comments

Matlab functions: *imnoise, mat2gray, deconvlucy*.

    The example above shows the results of LR deconvolution under three different conditions. In the first, only the PSF is supplied and 10 iterations are applied. In the second, an estimate of the overall standard deviation of the noise is supplied. In the third example, a weighting function (which assigns zero weight to pixels near the border of the image) is applied to suppress the ringing effects evident in the previous reconstructions.

    The Image Processing Toolbox has a function specifically for carrying out LR deconvolution, *deconvlucy*. Type *doc deconvlucy* for full details on how to use the LR deconvolution algorithm in Matlab.

---

[10] Richardson WH. Bayesian-based iterative method of image restoration, *J. Opt. Soc. Am.* **62** (1972) 55. Lucy, LB, An iterative technique for the rectification of observed distributions, *Astron. J.* **79** (1974) 745.

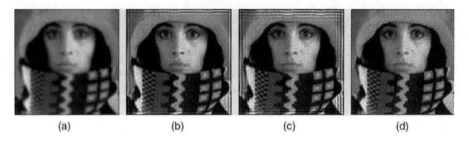

Figure 6.8   LR deconvolution. From left to right: (a) original image; (b) 10 iterations, PSF only supplied; (c) 20 iterations, variance of noise also supplied; (d) 20 iterations, band of five pixels around edge zero-weighted to suppress ringing effects

6.11   Matrix formulation of image restoration

We will end our introductory discussion on image restoration by briefly examining restoration from an alternative perspective. Specifically, we will discuss the *linear algebraic or matrix formulation* of the image restoration problem. This presentation will be fairly mathematical and might fairly be glossed over on a first reading. However, the formulation is a powerful one because it effectively places many seemingly different techniques within an identical mathematical framework. For this reason, we choose to devote the remaining sections to it.

Our starting point is the linear superposition integral introduced in Chapter 5, but with an additional noise term added:

$$g(x, y) = \iint f(x', y')h(x, y; x', y') \, dx' \, dy' + n(x, y) \tag{6.28}$$

Note that this equation expresses a linear mapping between a distribution in some input domain specified by coordinates (x', y') to a distribution in some output domain (x, y). With the noise term added, this constitutes our most general expression for a continuous, 2-D, linear imaging model. If the PSF $h(x, y; x', y')$ (which is, in general, four-dimensional) is shift invariant, then this reduces to the standard convolution imaging given by Equation (6.2). However, we will stick with the more general case here, since it is equally simple to treat.

Digital imaging systems produce a *discretized* approximation to the linear superposition integral in which the input domain (x', y') and the output domain (x, y) are sampled into a finite number of pixels. It is straightforward to show[11] that the discretized version of Equation (6.1) results in the matrix equation

$$\mathbf{g} = \mathbf{Hf} + \mathbf{n} \tag{6.29}$$

---

[11] See http://www.fundipbook.com/materials/.

where **g** is the output or image vector, which contains all the pixels in the image suitably arranged as a single column vector;[12] **f** is the input vector, similarly containing all the pixel values in the input distribution arranged as a single column vector; **n** is the noise vector, where the noise contributions to each pixel are arranged as a single column vector; **H** is the discrete form of the system PSF, describing the transfer of each pixel in the input to each and every pixel in the output, and is sometimes called the system or transfer matrix.

This formulation does not, of course, change anything fundamental about the image restoration problem, which remains essentially the same – namely, how to recover **f** given **g** and various degrees of knowledge of **H** and **n**. However, we do have a new mathematical perspective. Equation (6.29) actually *comprises a large system of linear equations*. This linear model has been the subject of much study within the field of statistical estimation theory and is very well understood. The key advantage of expressing the restoration problem in these terms is that it brings the problem under a simple and unifying framework which is particularly attractive from a theoretical perspective.

The solution of the linear system Equation (6.29) can take a variety of forms. Our discussion must necessarily be modest in scope and will concentrate on just two aspects:

(1)   to explain the essence of the estimation theoretic approach to solving this linear problem;

(2)   to outline some of the more important and commonly used solutions.

Solutions to the linear model of Equation (6.29) are called *estimators* and the goal is to use our knowledge of the specific situation to derive an estimator $\hat{\mathbf{f}}$ which is as close as possible to the actual input vector **f**.

Generally, different estimators are appropriate, depending on the following three criteria:

• which of the quantities in Equation (6.29) we can treat as stochastic (i.e. random) and which as fixed or deterministic;

• the specific minimization criterion used;

• whether we wish to impose any constraints on the permissible solutions.

We look first at the simplest solution: the standard least-squares estimator.

6.12   The standard least-squares solution

Consider our system of linear equations described by Equation (6.29). The simplest solution from a conceptual point of view is to choose that value of **f** that minimizes the squared

---

[12] The term stacking operation is sometimes used to refer to this representation of a 2-D image, as each of the original columns of pixels is stacked one on top of another in an ordered fashion.

length of the error vector **n**. This least-squares criterion effectively says 'find an estimate $\hat{\mathbf{f}}$ which minimizes the sum of the squared errors between the *actual measurement vector* **g** and the predicted value $\hat{\mathbf{g}} = \mathbf{H}\hat{\mathbf{f}}$'. Note that such a criterion does not make any explicit assumptions about the statistics of the noise or the *a priori* probability of specific solution vectors.

Accordingly, we define the scalar cost function

$$Q = \mathbf{n}^{\mathrm{T}}\mathbf{n} = \|\mathbf{H}\mathbf{f}-\mathbf{g}\|^{2} \tag{6.30}$$

and seek a solution vector which minimises Q. Q is a scalar quantity that depends on many free parameters (all the elements of the solution vector $\hat{\mathbf{f}}$). Taking the vector derivative of Q with respect to **f** in Equation (6.30) and demanding that this be zero at a minimum imposes the requirement that

$$\frac{\partial Q}{\partial \mathbf{f}} = \frac{\partial}{\partial \mathbf{f}}\left\{[\mathbf{H}\mathbf{f}-\mathbf{g}]^{\mathrm{T}}[\mathbf{H}\mathbf{f}-\mathbf{g}]\right\} = 0$$

$$\Rightarrow \quad \{2\mathbf{H}^{\mathrm{T}}\mathbf{H}\mathbf{f} -\mathbf{H}^{\mathrm{T}}\mathbf{g}-\mathbf{g}^{\mathrm{T}}\mathbf{H}\} = 2\mathbf{H}^{\mathrm{T}}\mathbf{H}\mathbf{f}-2H^{\mathrm{T}}\mathbf{g} = 0 \tag{6.31}$$

$$\hat{\mathbf{f}} = (\mathbf{H}^{\mathrm{T}}\mathbf{H})^{-1}\mathbf{H}^{\mathrm{T}}\mathbf{g}$$

Note that we have here assumed a zero mean noise process ($\langle \mathbf{n}\rangle = 0$) for simplicity, but a similar result can easily be obtained for a nonzero mean noise process.

The least-squares solution given by Equation (6.31) is conceptually simple and we obtain a neat closed form for the optimal vector $\hat{\mathbf{f}}$.[13] It does, however, have several weaknesses.

(1)   The least-squares solution is 'image blind'. By this we mean that its goal is simply to minimize the squared length of the error vector and, as such, it 'treats everything the same', i.e. it does not take into account any image properties or feature specifics which may be of perceptual importance and which we may wish to preserve in our restored image.

(2)   There is no guarantee that a true inverse to the term $(\mathbf{H}^{\mathrm{T}}\mathbf{H})^{-1}$ exists. However, this is typically accommodated by calculating a pseudo-inverse using a singular value decomposition of $\mathbf{H}^{\mathrm{T}}\mathbf{H}$.

(3)   It makes no assumptions about the *relative likelihood* of particular solutions. In many cases, it is reasonable to treat the input vector as stochastic in which certain input distributions *are a priori more probable*. This particular issue is explained shortly.

6.13 Constrained least-squares restoration

We offered a discussion of constrained deconvolution in Section 6.7. Here, we re-emphasize that several useful estimators can be derived by minimizing some specific linear operator **L**

---

[13] See http://www.fundipbook.com/materials/ for details of the solution.

on the input distribution **f** *subject* to the squared noise constraint $\|\mathbf{Hf}-\mathbf{g}\|^2-\|\mathbf{n}\|^2 = 0$. Note that this constraint was mentioned in Section 6.7 and is entirely equivalent, albeit in matrix formulation, to Equation (6.14). Using the method of Lagrange multipliers, we thus form the scalar cost function

$$Q = \|\mathbf{Lf}\|^2-\lambda\{\|\mathbf{Hf}\|^2-\|\mathbf{n}\|^2\} \tag{6.32}$$

and take vector derivatives of Q with respect to **f**. The general solution is

$$\hat{\mathbf{f}} = [\mathbf{H}^{\mathrm{T}}\mathbf{H}-\alpha\mathbf{L}^{\mathrm{T}}\mathbf{L}]^{-1}\mathbf{H}^{\mathrm{T}}\mathbf{g} \tag{6.33}$$

where $\alpha = 1/\lambda$, the inverse of the Lagrange multiplier. In principle, Equation (6.33) may be used to replace **f** in the noise constraint term $\|\mathbf{Hf}-\mathbf{g}\|^2-\|\mathbf{n}\|^2 = 0$ and the value of α adjusted to ensure that the noise constraint is indeed satisfied. In practice, the value of α is often treated as an adjustable parameter.

We note that some interesting specific solutions can be obtained through a specific choice for the linear operator **L**:

$\mathbf{L} = \mathbf{I}$: parametric Wiener filter

$$\hat{\mathbf{f}} = [\mathbf{H}^{\mathrm{T}}\mathbf{H}-\alpha\mathbf{I}]^{-1}\mathbf{H}^{\mathrm{T}}\mathbf{g} \tag{6.34}$$

The similarity to Equation (6.10) is apparent. α acts like a noise/signal ratio and clearly, in the limit that $\alpha \rightarrow 0$, we tend towards the direct inverse or least-squares filter.

$\mathbf{L} = \nabla^2$: maximum smoothness filter

$$\hat{\mathbf{f}} = [\mathbf{H}^{\mathrm{T}}\mathbf{H}-\Gamma^{\mathrm{T}}\Gamma\mathbf{I}]^{-1}\mathbf{H}^{\mathrm{T}}\mathbf{g} \tag{6.35}$$

where the matrix Γ is the discrete representation of the Laplacian operator.

The two forms given by Equations (6.34) and (6.35) are very common, but we stress that, in principle, *any linear operator* which can be implemented in discrete form can be substituted into Equation (6.33).

To end this brief discussion, we note that the computational implementation of the various solutions such as Equation (6.31) and Equation (6.33) cannot be sensibly attempted naively. This is because of the sheer size of the system matrix **H**. For example, even a digital image of size 512^2 (very modest by today's standards) would result in a system matrix comprising $\sim$69 000 million elements, the direct inversion of which presents a huge calculation. The implementation of Equation (6.33) and similar forms discussed in this section is made possible in practice by the *sparse* nature of **H** (most of its elements are zero) and the fact that it exhibits cyclic patterns which are a consequence of the shift invariance of the PSF of the imaging system. **H** is said to possess a block circulant structure, and algorithms for efficiently computing quantities, such as $\mathbf{H}^{\mathrm{T}}\mathbf{H}$ and its inverse, have been extensively developed.[14]

---

[14] H. C. Andrews and B. R. Hunt (1977) *Digital Image Restoration*, Prentice-Hall (ISBN 0-13-214213-9). This is now an old book, but it contains an excellent and comprehensive discussion of such techniques.

6.14   Stochastic input distributions and Bayesian estimators

A key assumption in deriving the least-squares estimators given by Equations (6.22) and (6.24) is that both the system matrix **H** and the input distribution **f** are treated as *deterministic* quantities. Superficially, it would seem that they could not really be otherwise. After all, an imaging system whose properties are not changing with time is clearly deterministic. Further, treating the input distribution as a deterministic (i.e. non-random) quantity seems very reasonable.

   Taking the input distribution to be deterministic is certainly acceptable in many instances. However, sometimes we know a priori that *certain input distributions are more likely to occur than others are.* In this case, we may treat the input as stochastic. To illustrate this point, consider a simple but topical example. If we know a priori that we are imaging a human face (but do not know who the actual individual is) it stands to reason that only certain configurations of the vector **f** are at all feasible as solutions. In such a case we treat the input distribution as belonging to a specific pattern class – a concept we explore further in Chapter 11 on classification. Briefly, a pattern class is a conceptual or actual group of patterns which shares certain features in common because (typically) they are sampled from the same underlying probability density function. The specification of an individual pattern in the given class is usually made through an N-dimensional vector, each element of which corresponds to a variable used to describe the pattern. Accordingly, if we have knowledge of the input pattern probability density function (or even just some of its moments), we can try to include this knowledge in our estimation procedure. This is the fundamental motivation behind the whole class of Bayesian estimators. Speaking in loose and rather general terms, Bayesian methods effectively attempt to strike the right balance between an estimate based on the data alone (the recorded image, the PSF and the noise characteristics) and *our prior knowledge of how likely certain solutions are in the first place.* We will finish this chapter with a powerful but simple derivation from 'classical' estimation theory which incorporates the idea of prior knowledge about likely solutions without explicitly invoking Bayes' theorem.[15] This is the famous generalized Gauss–Markov estimator.

6.15   The generalized Gauss–Markov estimator

Our starting point is the general linear system written in matrix form in Equation (6.29):

$$\mathbf{Hf} + \mathbf{n} = \mathbf{g}$$

We make the following important assumptions on our linear model:

- The system matrix **H** is a deterministic quantity.

---

[15] It is worth noting that many statisticians are adherents of either classical estimation theory (and do not use Bayes' theorem) or 'die-hard Bayesians' who enthusiastically apply it at more or less every opportunity. Both approaches yield ultimately similar results and the difference really lies in the philosophical perspective.

- The error (noise) vector is a stochastic quantity. We will assume that it has zero mean (this is not restrictive, but it simplifies the algebra somewhat) and a known error covariance function $\mathbf{V} = \langle \mathbf{nn}^T \rangle$. The input distribution $\mathbf{f}$ is a stochastic quantity. We will also assume without restricting our analysis that it has zero mean and a known covariance function given by $\mathbf{C}_f = \langle \mathbf{ff}^T \rangle$.

- We will further assume that the noise and signal are *statistically uncorrelated* (a condition that is usually satisfied in most instances). Thus, we have $\langle \mathbf{nf}^T \rangle = \langle \mathbf{n}^T\mathbf{f} \rangle = 0$.

The angle brackets denote expectation averaging over an ensemble of many observations of the random process. In principle, the covariance matrices $\mathbf{C}_f$ and $\mathbf{V}$ are derived from an infinite number of their respective random realizations. In practice, they must usually be estimated from a finite sample of observations.

We seek an estimator for the input distribution which is given by some linear operator on the observed data -

$$\hat{\mathbf{f}} = \mathbf{Lg} = \mathbf{L}(\mathbf{Hf} + \mathbf{n}) \qquad (6.36)$$

and which will minimize the error covariance matrix of the input distribution given by

$$\mathbf{E} = \langle (\mathbf{f} - \hat{\mathbf{f}})(\mathbf{f} - \hat{\mathbf{f}})^T \rangle \qquad (6.37)$$

where the matrix operator $\mathbf{L}$ is to be determined. Inserting Equation (6.36) into Equation (6.37) and using the definition of $\mathbf{C}_f = \langle \mathbf{ff}^T \rangle$, the quantity to be minimized is

$$\mathbf{E} = [\mathbf{I} - \mathbf{LA}]\mathbf{C}_f[\mathbf{I} - \mathbf{LA}]^T + \mathbf{LVL}^T \qquad (6.38)$$

As a momentary aside, we note that whilst the reader will almost certainly be familiar with the notion of minimizing a scalar function of many variables, the notion of 'minimizing' a multivalued matrix quantity as required by Equation (6.38) may, however, be novel.[16] The required approach is to take first variations in $\mathbf{E}$ with respect to the linear operator $\mathbf{L}$ and set these to zero. This yields

$$\delta\mathbf{E} = \delta\mathbf{L}(\mathbf{HC}_f[\mathbf{I} - \mathbf{LH}]^T + \mathbf{VL}^T) + ([\mathbf{I} - \mathbf{LH}]\mathbf{C}_f\mathbf{H}^T + \mathbf{LV})\delta\mathbf{L}^T = \mathbf{0} \qquad (6.39)$$

The optimal matrix is thus given by

$$\mathbf{L}_{OPT} = \mathbf{C}_f\mathbf{H}^T(\mathbf{HC}_f\mathbf{H}^T + \mathbf{V})^{-1} \qquad (6.40)$$

Matrix algebraic manipulation shows that this may also be written as

$$\mathbf{L}_{OPT} = (\mathbf{H}^T\mathbf{V}^{-1}\mathbf{H} + \mathbf{C}_f^{-1})^{-1}\mathbf{H}^T\mathbf{V}^{-1} \qquad (6.41)$$

---

[16] This has a close relation to the calculus of variations. See the book's website http://www.fundipbook.com/materials/ for a brief discussion of this point.

Substitution of the optimal estimator given by Equation (6.41) into the definition of the error covariance matrix on the input distribution given by Equation (6.37) yields

$$\mathbf{E}_{\text{OPT}} = (\mathbf{H}^{\text{T}}\mathbf{V}^{-1}\mathbf{H} + \mathbf{C}_{\mathbf{f}}^{-1})^{-1} \tag{6.42}$$

Equation (6.42) is a useful expression because it provides an actual measure of the closeness of the reconstructed or estimated input to the actual input.

We note two limiting forms of interest for Equation (6.41):

- If we assume that we have no prior knowledge of the input distribution, this is equivalent to treating it as deterministic quantity and we effectively allow $\mathbf{C_f} \to \infty$ (i.e. infinite variance/covariance is possible), the Gauss–Markov estimator then reduces to

$$\mathbf{L}_{\text{OPT}} = (\mathbf{H}^{\text{T}}\mathbf{V}^{-1}\mathbf{H})^{-1}\mathbf{H}^{\text{T}}\mathbf{V}^{-1} \tag{6.43}$$

  This is known as the BLUE (best linear unbiased estimator).

- If further we may assume that the noise on the pixels has equal variance σ^2 and is uncorrelated (a reasonable assumption in many cases) then we may write $\mathbf{V} = \sigma^2\mathbf{I}$ and we obtain a standard least-squares solution:

$$\mathbf{L}_{\text{OPT}} = [\mathbf{H}^{\text{T}}\mathbf{H}]^{-1}\mathbf{H}^{\text{T}} \tag{6.44}$$

We stress that the solution presented here is a quite general solution to the linear problem. However, computational implementation of the solutions when the underlying signals are images can generally only be attempted by exploiting the sparse and cyclic structure of the system transfer matrix.

For further examples and exercises see http://www.fundipbook.com

7

Geometry

The need for geometric manipulation of images appears in many image processing applications. The removal of optical distortions introduced by a camera, the geometric 'warping' of an image to conform to some standard reference shape and the accurate registration of two or more images are just a few examples requiring a mathematical treatment involving geometric concepts. Inextricably related to our discussion of geometry in images is the notion of shape. In this chapter we will introduce a preliminary definition of shape which provides the necessary basis for subsequent discussion of the geometric techniques that form the bulk of the chapter.

We note in passing that the definition of useful *shape descriptors* and methods for their automated calculation play an important role in *feature extraction*, a key step in pattern recognition techniques. Typically, shape descriptors use or reduce the coordinate pairs in the shape vector representation to produce some compact or even *single-parameter* measure of the approximate shape. A discussion of these techniques is deferred until Chapter 9.

7.1 The description of shape

It is clear that our simple, everyday concept of shape *implies the existence of some boundary.* If we talk of the shape of objects such as a cigar, a football or an egg, we are implicitly referring to the boundary of the object. These objects are simple and the specification of the coordinates constituting the boundary is considered a *sufficient* mathematical description of the shape. For more complex objects, the boundary coordinates are usually not mathematically sufficient. A good example of the latter is the human face. If we are to describe face-shape accurately enough to discriminate between faces then we need to specify not just the overall outline of the head, but also all the internal features, such as eyes, nose, mouth, etc. Thus, to remain general, our basic description of shape is taken to be some ordered set of coordinate pairs (or tuples for higher dimensional spaces) which we deem sufficient for the particular purpose we have in mind. One of the simplest and most direct ways to describe a shape mathematically is to locate a finite number N of points along the boundary and concatenate them to constitute a *shape vector* which, for a 2-D object, we simply denote as:

$$\mathbf{x} = \begin{bmatrix} x_1 & y_1 & x_2 & \cdots & x_N & y_N \end{bmatrix}$$

Fundamentals of Digital Image Processing – A Practical Approach with Examples in Matlab
Chris Solomon and Toby Breckon
© 2011 John Wiley & Sons, Ltd

In general, the points defining the shape may be selected and annotated manually or they may be the result of an automated procedure.

There are two main aspects to the treatment of boundaries and shape in image processing:

- How do we identify and locate a defining feature boundary in an image?

- How, then, can we label boundaries and features to provide a meaningful and useful definition of the shape?

The first aspect is a very important one and belongs largely within the domain of *image segmentation*. Typically, image segmentation is concerned with *automated* methods for identifying object boundaries and regions of interest in images and is a subject that we will address in Chapter 10. In the remainder of this chapter we will be primarily concerned with the second aspect, namely how we can define meaningful descriptions of feature shapes in images and manipulate them to achieve our goals. Thus, we will implicitly assume either that *manual segmentation* (in which an operator/observer can define the region of interest) is appropriate to the problem in question or that an automated method to delineate the boundary is available.

7.2 Shape-preserving transformations

Translating an object from one place to another, *rotating* it or *scaling* (magnifying) it are all operations which change the object's shape vector coordinates but *do not* change its essential shape (Figure 7.1). In other words, the shape of an object is something which is basically defined by its boundary but which is *invariant to the translation, rotation and scaling of the coordinates* that define that boundary. Accordingly, we will adopt the following definition of shape:

> - *Shape* is all the geometrical information that remains after location, scale and rotation effects have been filtered out from an object.

It follows that objects which actually have identical shape may have shape vectors which are quite different and the simple vector description of a shape as given in Section 7.1 is *partially redundant*. If we wish to describe and compare shapes in a compact and meaningful way, therefore, we must seek a *minimalist* representation which removes this redundancy. Such a representation can be achieved by the process known as *Procrustes alignment*. This procedure effectively applies an appropriate sequence of the three shape-preserving transformations of translation, scaling and rotation to two (or more) shape vectors to match them as closely as possible to each other, thereby filtering out all the non-essential differences between their shapes. This important procedure is discussed in Section 7.6. First, however, we consider a simple framework which permits the mathematical treatment of shape within the framework of matrix algebra.

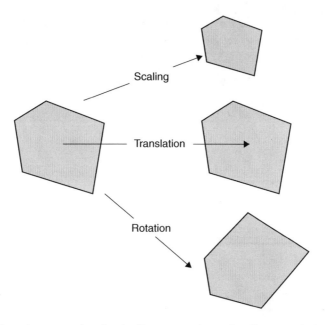

Figure 7.1   Shape is preserved under the linear operations of scaling, translation and rotation

7.3   Shape transformation and homogeneous coordinates

Shape can best be represented as ordered arrays in which the N Euclidean coordinate pairs, $\{(x_1, y_1) \; \cdots \; (x_N, y_N)\}$, are written as the columns of a matrix $\mathbf{S}$:

$$\mathbf{S} = \begin{bmatrix} x_1 & x_2 & x_3 & \cdots & x_N \\ y_1 & y_2 & y_3 & \cdots & y_N \end{bmatrix} \qquad (7.1)$$

The representation $\mathbf{S}$ is often referred to as a *point distribution matrix* (PDM), since each column gives the coordinates of one point in the overall distribution. The advantage of this arrangement is that linear transformations of the shape can be achieved by simple matrix multiplication.

In general, the result of applying a 2×2 transforming matrix $\mathbf{T}$ to the PDM $\mathbf{S}$ is to produce a new set of shape coordinates $\mathbf{S}'$ given by

$$\mathbf{S}' = \mathbf{TS} \qquad (7.2)$$

with the individual coordinate pairs transforming as

$$\begin{bmatrix} x' \\ y' \end{bmatrix} = \mathbf{T} \begin{bmatrix} x \\ y \end{bmatrix} \qquad \text{or} \qquad \mathbf{x}' = \mathbf{Tx} = \begin{bmatrix} \alpha_{11} & \alpha_{12} \\ \alpha_{21} & \alpha_{22} \end{bmatrix} \begin{bmatrix} x \\ y \end{bmatrix} \qquad (7.3)$$

and with the parameters α_{ij} controlling the specific form of the transformation.

For example, operations such as overall scaling in two dimensions, shearing in the x and y directions and rotation about the origin through angle θ can all be expressed in the 2×2 matrix form described by Equation (7.3). Explicitly, these matrices are respectively given by:

$$\mathbf{T}_{sc} = \begin{bmatrix} s & 0 \\ 0 & s \end{bmatrix} \quad \mathbf{T}_x = \begin{bmatrix} 1 & 0 \\ \alpha & 1 \end{bmatrix} \quad \mathbf{T}_y = \begin{bmatrix} 1 & \alpha \\ 0 & 1 \end{bmatrix} \quad \mathbf{T}_\theta = \begin{bmatrix} \cos\theta & \sin\theta \\ -\sin\theta & \cos\theta \end{bmatrix} \quad (7.4)$$

where the scalars s and α are scaling and shear factors respectively.

Note, however, that the simple operation of point translation:

$$\begin{bmatrix} x' \\ y' \end{bmatrix} = \begin{bmatrix} x \\ y \end{bmatrix} + \begin{bmatrix} \alpha_x \\ \alpha_y \end{bmatrix} \qquad \mathbf{x}' = \mathbf{x} + \mathbf{d} \qquad (7.5)$$

cannot be accomplished through a single matrix multiplication as per Equation (7.3), but rather only with a further *vector addition*.

The operation of translation is very common. Most often, the need is to refer a set of coordinates to the origin of our chosen coordinate system, apply one or more transformations to the data and then translate it back to the original system. It is highly desirable, for practical reasons, to be able to express all these linear operations (scaling, rotation, shearing *and* translation) through the same mathematical operation of matrix multiplication. This can be achieved through the use of *homogeneous coordinates*.

In homogeneous coordinates, we express 2-D shape vectors

$$\begin{bmatrix} x \\ y \end{bmatrix}$$

in a space of one higher dimension as

$$\begin{bmatrix} wx \\ wy \\ w \end{bmatrix}$$

with w an arbitrary constant. For our purposes, we select $w = 1$ so that 2-D shape vectors in homogeneous coordinates will be given by the general form

$$\begin{bmatrix} x \\ y \\ 1 \end{bmatrix}$$

Similarly, the general 2-D transformation matrix $\mathbf{T}$ previously described by Equation (7.3), will be expressed as a 3×3 matrix (one additional dimension $\Rightarrow$ one extra row and one extra column) in the form

$$\mathbf{T} = \begin{bmatrix} \alpha_{11} & \alpha_{12} & \alpha_{13} \\ \alpha_{21} & \alpha_{22} & \alpha_{23} \\ \alpha_{31} & \alpha_{32} & \alpha_{33} \end{bmatrix} \qquad (7.6)$$

The reader may well ask at this point why it is reasonable to introduce homogeneous coordinates in this seemingly arbitrary way.[1] On one level, we can think of homogeneous coordinates simply as a computational artifice that enables us to include the operation of translation within an overall matrix expressing the linear transformation. For practical purposes, this is adequate and the *utility* of this will become apparent in the following sections. There is, however, a deeper underlying meaning to homogeneous coordinates which relates to the use of projective spaces as a means of removing the special role of the origin in a Cartesian system. Unfortunately, a thorough discussion of this topic lies outside the scope of this text.

7.4 The general 2-D affine transformation

Consider a coordinate pair (x, y) which is subjected to a linear transformation of the form

$$x' = T_x(x, y) = ax + by + c$$
$$y' = T_y(x, y) = dx + ey + f \tag{7.7}$$

where a, b, c, d, e and f are arbitrary coefficients. Such a transformation is the 2-D (linear) affine transformation.

As is evident from its form, the affine transformation in 2-D space has six free parameters. Two of these (c and f) define the vector corresponding to translation of the shape vector, the other four free parameters (a, b, d and e) permit a combination of rotation, scaling and shearing.

From a geometrical perspective, we may summarize the effect of the general affine transformation on a shape as follows:

The affine transformation: translation, rotation, scaling, stretching and shearing may be included. Straight lines remain straight and parallel lines remain parallel but rectangles may become parallelograms.

It is important to be aware that the *general* affine transformation (in which no constraints are placed on the coefficients) permits shearing and, thus, *does not preserve shape*.

The 2-D affine transformation in regular Cartesian coordinates (Equation (7.7)) can easily be expressed in matrix form:

$$\begin{bmatrix} x' \\ y' \end{bmatrix} = \begin{bmatrix} a & b \\ d & e \end{bmatrix} \begin{bmatrix} x \\ y \end{bmatrix} + \begin{bmatrix} c \\ f \end{bmatrix} \qquad \mathbf{x}' = \mathbf{T}\mathbf{x} + \mathbf{d} \tag{7.8}$$

Typically, the affine transformation is applied not just on a single pair of coordinates but on the many pairs which constitute the PDM. The matrix-vector form of Equation (7.8)

---

[1] This addition of one dimension to express coordinate vectors in homogeneous coordinates extends to 3-D (i.e. $[x, y, z] \rightarrow [x, y, z, 1]$) and higher.

is thus inconvenient when we wish to consider a *sequence of transformations*. We would ideally like to find a single operator (i.e. a single matrix) which expresses the overall result of that sequence and simply apply it to the input coordinates. The need to add the translation vector **d** in Equation (7.8) after each transformation prevents us from doing this. We can overcome this problem by expressing the affine transformation in *homogenous coordinates*.

7.5 Affine transformation in homogeneous coordinates

In the homogeneous system, an extra dimension is added so that the PDM **S** is defined by an augmented matrix in which a row of 1s are placed beneath the 2-D coordinates:

$$\mathbf{S} = \begin{bmatrix} x_1 & x_2 & x_3 & \cdots & x_N \\ y_1 & y_2 & y_3 & \cdots & y_N \\ 1 & 1 & 1 & \cdots & 1 \end{bmatrix} \tag{7.9}$$

The affine transformation in homogeneous coordinates takes the general form

$$\mathbf{T} = \begin{bmatrix} \alpha_{11} & \alpha_{12} & \alpha_{13} \\ \alpha_{21} & \alpha_{22} & \alpha_{23} \\ 0 & 0 & 1 \end{bmatrix} \tag{7.10}$$

where the parameters α_{13} and α_{23} correspond to the translation parameters (c and f) in Equation (7.7) and the other block of four

$$\begin{bmatrix} \alpha_{11} & \alpha_{12} \\ \alpha_{21} & \alpha_{22} \end{bmatrix}$$

correspond directly to *a*, *b*, *d* and *e*.

Table 7.1 summarizes how the parameters in the affine transform are chosen to effect the operations of translation, rotation, scaling and shearing. Figure 7.2 demonstrates the effect of each transformation on a square whose centre is at the origin and has a side of length equal to 2 units.

Table 7.1 Coefficient values needed to effect the linear transformations of translation, rotation, scaling and shearing in homogeneous coordinates

Transformation	α_{11}	α_{12}	α_{13}	α_{21}	α_{22}	α_{23}
Translation by (x, y)	1	0	x	0	1	y
Rotation by θ	$\cos\theta$	$-\sin\theta$	0	$\sin\theta$	$\cos\theta$	0
Uniform scaling by s	s	0	0	0	s	0
Vertical shear by s	1	s	0	0	1	0
Horizontal shear by s	s	0	0	s	1	0

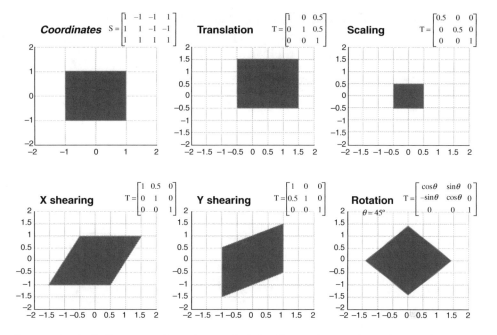

Figure 7.2   Basic linear transformations and their effects. The transformation matrices **T** and the original point distribution matrix **S** are expressed in homogeneous coordinates. (See http://www.fundipbook.com/materials/ for the Matlab code).

Using the homogeneous forms for the transformation matrix and the shape matrix, we can then express the transformed shape coordinates $\widehat{S}$ as

$$\widehat{S} = TS \tag{7.11}$$

Note the following important property of the affine transformation:

- Any sequence of affine transformations reduces to a single affine transformation. A sequence of affine transformations represented by matrices $(T_1, T_2, \ldots, T_N)$ applied to an input point distribution **S** can be represented by a *single* matrix $T = T_1 T_2 \ldots T_N$ which operates on **S**.

Unless explicitly stated otherwise, we will henceforth assume the use of homogeneous coordinates in this chapter.

7.6   The procrustes transformation

The Procrustes transformation matrix is a *special* case of the general affine transformation described by Equation (7.10) and is perhaps the most important form of the affine transform in the field of shape analysis. The key property of the Procrustes transformation is that multiplication of any point distribution model by the Procrustes matrix will *preserve the shape*. For this reason, it is pivotal in many tasks of shape and image alignment. The general 2-D form in homogeneous coordinates is given by

$$\mathbf{T} = \begin{bmatrix} \alpha & \gamma & \lambda_1 \\ -\gamma & \alpha & \lambda_2 \\ 0 & 0 & 1 \end{bmatrix} \quad (7.12)$$

The Procrustes matrix has only four free parameters (unlike the *general* affine form, which has six). This is because Procrustes transformation is a combination of the three shape-preserving operations of *translation*, *rotation* and *scaling* and we may decompose the Procrustes transformation into three successive, primitive operations of translation ($\mathbf{X}$), rotation ($\mathbf{R}$) and scaling ($\mathbf{S}$). Namely, $\mathbf{T} = \mathbf{SRX}$, where

$$\mathbf{X} = \begin{bmatrix} 1 & 0 & \beta_x \\ 0 & 1 & \beta_y \\ 0 & 0 & 1 \end{bmatrix} \quad \mathbf{R} = \begin{bmatrix} \cos\theta & \sin\theta & 0 \\ -\sin\theta & \cos\theta & 0 \\ 0 & 0 & 1 \end{bmatrix} \quad \mathbf{S} = \begin{bmatrix} S & 0 & 0 \\ 0 & S & 0 \\ 0 & 0 & 1 \end{bmatrix} \quad (7.13)$$

For this sequence to define the Procrustes transformation, translation must be applied first,[2] but the order of the rotation and scaling is unimportant. Multiplying the matrices in Equation (7.13) together and equating to the general form defined by Equation (7.12), we have

$$\begin{aligned} \alpha &= S\cos\theta & \beta &= S\sin\theta \\ \lambda_1 &= S(\beta_x\cos\theta + \beta_y\sin\theta) & \lambda_2 &= S(-\beta_x\sin\theta + \beta_y\cos\theta) \end{aligned} \quad (7.14)$$

The most common use of the Procrustes transformation is in the procedure known as *Procrustes alignment*. This involves the alignment of one or more shapes to a particular *reference shape*. The criterion for alignment is to find that combination of translation, rotation and scaling (i.e. the four free parameters β_x, β_y, θ and S) which minimize the sum of the mean-square distances between corresponding points on the boundary of the given shape and the reference.

7.7 Procrustes alignment

The Procrustes alignment procedure is conceptually straightforward and summarized diagrammatically in Figure 7.3. The mathematical derivation presented below yields the precise combination of translation, scaling and rotation that is required to achieve the minimization.

Consider two corresponding sets of N coordinates ordered as the columns of the point distribution matrices $\mathbf{X}$ and $\mathbf{Y}$. In a 2-D space, these sets are written as:

---

[2] This is essential because the subsequent rotation is defined about the origin and rotation on a point model whose centroid is not at the origin will *change the shape* of the point model.

Figure 1.2   Example of grayscale (left) and false colour (right) image display

Original Image (8-bit RGB)
= 1024 x 768 x 3
= 2304Kb ~= 2.3Mb

Lossy Compression : JPEG
Lossless Compression : PNG

JPEG (Quality : 0) = 16k   JPEG (Quality : 20) = 40k

JPEG (Quality : 75) = 168k   PNG (max. compression) = 1.4Mb

Figure 1.5   Example image compressed using lossless and varying levels of lossy compression

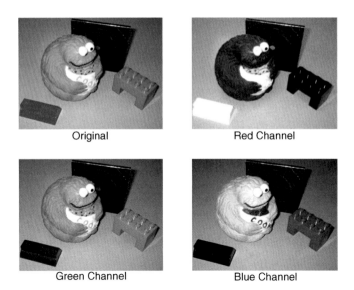

Original Red Channel

Green Channel Blue Channel

Figure 1.6 Colour RGB image separated into its red (R), green (G) and blue (B) colour channels

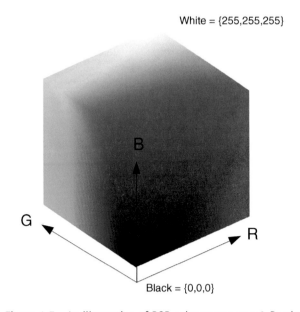

White = {255,255,255}

B

G R

Black = {0,0,0}

Figure 1.7 An illustration of RGB colour space as a 3-D cube

Figure 1.8 An example of RGB colour image (left) to grey-scale image (right) conversion

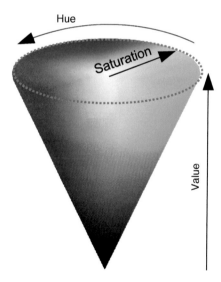

Figure 1.9 HSV colour space as a 3-D cone

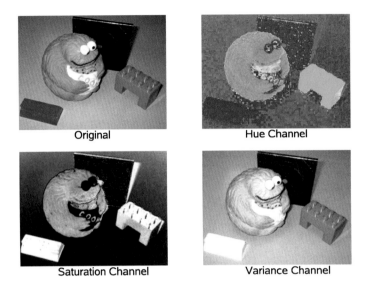

Original   Hue Channel

Saturation Channel   Variance Channel

Figure 1.10   Image transformed and displayed in HSV colour space

Figure 3.21   Adaptive histogram equalization applied to a sample colour image

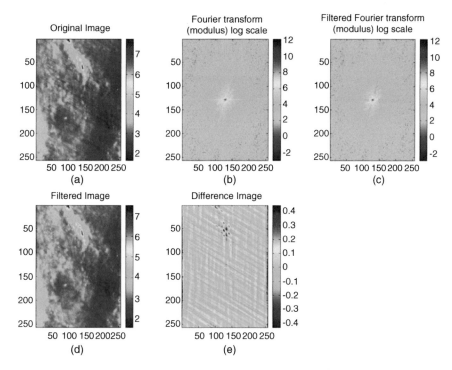

Figure 5.8 An application of frequency-domain filtering. Proceeding from left to right and top to bottom, we have: (a) the original image with striping effect apparent; (b) the Fourier modulus of the image (displayed on log scale); (c) the Fourier modulus of the image after filtering (displayed on log scale); (d) the filtered image resulting from recombination with the original phase; (e) the difference between the original and filtered images

Figure 8.18 Application of the hit-or-miss transformation to detect a target shape in a string of text. Note that the target includes the background and hit-or-miss is strictly sensitive to both the scale and the orientation of the target shape

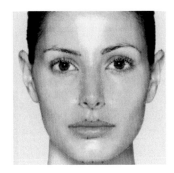

Figure 9.1 Anatomical landmarks (indicated in red) are located at points which can be easily identified visually. Mathematical landmarks (black crosses) are identified at points of zero gradient and maximum corner content. The pseudo-landmarks are indicated by green circles

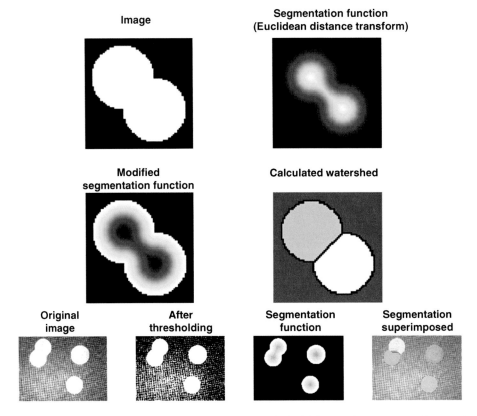

Figure 10.9 In the first, idealized example, the watershed yields a perfect segmentation. In the second image of overlapping coins, morphological opening is required first on the thresholded image prior to calculation of the watershed

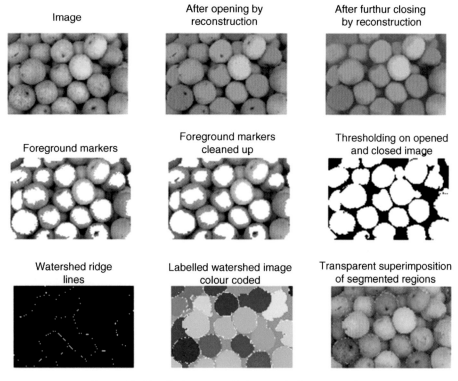

| Image | After opening by reconstruction | After furthur closing by reconstruction |

| Foreground markers | Foreground markers cleaned up | Thresholding on opened and closed image |

| Watershed ridge lines | Labelled watershed image colour coded | Transparent superimposition of segmented regions |

Figure 10.11   Marker-controlled watershed segmentation

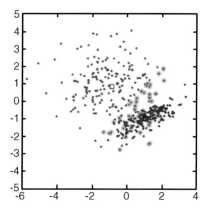

● Training data - Class 1
● Training data - Class 2
○ Misclassified by minimum distance classifier

Figure 11.7   Feature vectors which are statistically more likely to have originated from one class may lie closer to the prototype of another class. This is a major weakness of a simple Euclidean minimum distance classifier

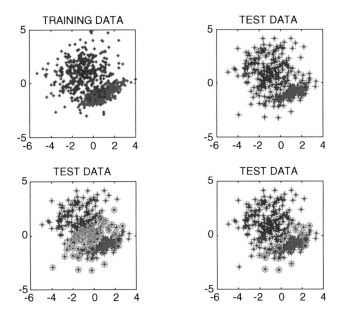

Figure 11.8 A comparison of Bayesian classifiers under different assumptions about the covariance matrix and prior distributions. The class-conditional densities of the training data are both normal. (a) Training data: 200 samples from each class (class 1: blue; class 2: red). (b) Test data: 50 originate from class 1, 100 originate from class 2 (indicated in red). (c) Classification using a Bayesian classifier with a diagonal estimate of the covariance and an erroneous prior distribution of $p(\omega_1) = 0.9$ and $p(\omega_2) = 0.1$. Misclassified test points are circled in green. (d) Classification using a Bayesian classifier with a diagonal estimate of the covariance and correct prior distribution of $p(\omega_1) = 1/3$ and $p(\omega_2) = 2/3$. Misclassified test points are circled in green

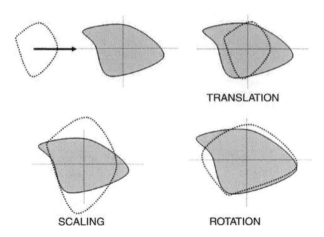

TRANSLATION

SCALING ROTATION

Figure 7.3 Procrustes alignment translates, scales and rotates a shape so as to minimise the sum of the squared distances between the coordinates of the two shapes

$$\mathbf{X} = \begin{bmatrix} x_1 & x_2 & \cdots & \cdots & x_N \\ y_1 & y_2 & \cdots & \cdots & y_N \end{bmatrix} \quad \text{and} \quad \mathbf{XY} = \begin{bmatrix} x_1' & x_2' & \cdots & \cdots & x_N' \\ y_1' & y_2' & \cdots & \cdots & y_N' \end{bmatrix} \quad (7.15)$$

where $\vec{x}_i$ and $\vec{x}_i'$ are the ith points (and thus ith column vectors) of matrices $\mathbf{X}$ and $\mathbf{Y}$ respectively.

The aim is to transform (i.e. align) the *input* coordinates $\mathbf{X}$ to the *reference* coordinates $\mathbf{Y}$ so as to minimize *the total sum of the squared Euclidean distances between the corresponding points*. The three transformations of translation, scaling and rotation are *successively* applied, each independently satisfying this least-squares criterion.

Step 1: translation The aim is to find that global translation $\vec{t}$ of the coordinates in matrix $\mathbf{X}$ which minimizes the total sum of the squared Euclidean distances between the translated coordinates and their corresponding values in the reference matrix $\mathbf{Y}$. In other words:

$$\vec{x}_i \rightarrow \vec{x}_i + \vec{t}$$

and we seek to minimize the least-squares cost function Q, defined as

$$Q = \sum_{i=1}^{N} \left[\vec{x}_i + \vec{t} - \vec{x}_i'\right]^{\mathrm{T}} \left[\vec{x}_i + \vec{t} - \vec{x}_i'\right] \quad (7.16)$$

by solving for $\vec{t}$, the translation vector, which achieves this. It is easy to show that

$$\vec{t} = \langle \vec{y} \rangle - \langle \vec{x} \rangle \quad (7.17)$$

where $\langle \vec{x} \rangle$ and $\langle \vec{y} \rangle$ are the average (centroid) coordinate vectors of the respective sets of N points.

Note that it is common in the alignment procedure to first refer the reference coordinates (matrix **Y**) to the origin, in which case $\langle \vec{\mathbf{y}} \rangle = 0$. In this case, the first step in the Procrustes alignment simply subtracts the sample mean value from each of the coordinates and the required translation is described by

$$\text{Step 1:} \quad \vec{\mathbf{x}}_i \rightarrow \vec{\mathbf{x}}_i - \langle \vec{\mathbf{x}} \rangle \tag{7.18}$$

Step 2: scaling A uniform scaling of all the coordinates in **X** can be achieved by a *diagonal* matrix $\mathbf{S} = s\mathbf{I}$, where s is the scaling parameter and **I** is the identity matrix. Minimization of the least-squares cost function

$$Q = \sum_{i=1}^{N} [\mathbf{S}\vec{\mathbf{x}}_i - \vec{\mathbf{x}}_i']^{\mathrm{T}} [\mathbf{S}\vec{\mathbf{x}}_i - \vec{\mathbf{x}}_i'] \tag{7.19}$$

with respect to the free parameter s yields the solution

$$s = \frac{\displaystyle\sum_{i=1}^{N} \vec{\mathbf{x}}_i'^{\mathrm{T}} \vec{\mathbf{x}}_i}{\displaystyle\sum_{i=1}^{N} \vec{\mathbf{x}}_i^{\mathrm{T}} \vec{\mathbf{x}}_i} \tag{7.20}$$

$$\text{Step 2:} \quad \mathbf{X} \rightarrow \mathbf{SX} \text{ with } \mathbf{S} = s\mathbf{I} \text{ and } s \text{ given by Equation (7.20)} \tag{7.21}$$

Step 3: rotation The final stage of the alignment procedure is to identify the appropriate orthonormal rotation matrix **R**. We define an error matrix $\mathbf{E} = \mathbf{Y} - \mathbf{RX}$ and seek that matrix **R** which minimizes the least-squares cost Q given by

$$Q = \mathrm{Tr}\{\mathbf{E}^{\mathrm{T}}\mathbf{E}\} = \mathrm{Tr}\{(\mathbf{Y} - \mathbf{RX})^{\mathrm{T}}(\mathbf{Y} - \mathbf{RX})\} \tag{7.22}$$

where Tr denotes the trace operator (the sum of the diagonal elements of a square matrix).

In this final step, the minimization of Equation (7.22) with respect to a variable matrix **R** is rather more involved than the translation and scaling stages. It is possible to show that our solution is

$$\mathbf{R} = \mathbf{V}\mathbf{U}^{\mathrm{T}} \tag{7.23}$$

where **U** and **V** are eigenvector matrices obtained from the singular value decomposition (SVD) of the matrix $\mathbf{XY}^{\mathrm{T}}$. (Computationally, the SVD algorithm decomposes the matrix $\mathbf{XY}^{\mathrm{T}}$ into the product of three orthogonal matrices **U**, **S** and **V** such that $\mathbf{XY}^{\mathrm{T}} = \mathbf{USV}^{\mathrm{T}}$.)[3]

---

[3] A detailed and comprehensive derivation of the Procrustes alignment equations for the general N-dimensional case is available at http://www.fundipbook.com/materials/.

Step 3: $\mathbf{X} \rightarrow \mathbf{RX}$ with $\mathbf{R}$ given by Equation(7.23)   (7.24)

In summary, the solution steps for Procrustes alignment are as follows.

- *Translation:* place the origin at the centroid of your reference coordinates (that set of points to which you wish to align) and translate the input coordinates (the points you wish to align) to this origin by subtracting the centroids as per Equation (7.18).

- *Scaling* then scale these coordinates as per Equation (7.21).

- *Rotation*:

    – form the product of the coordinate matrices $\mathbf{XY}^{\mathrm{T}}$

    – calculate its SVD as $\mathbf{XY}^{\mathrm{T}} = \mathbf{USV}^{\mathrm{T}}$

    – calculate the matrix $\mathbf{R} = \mathbf{VU}^{\mathrm{T}}$

    – then multiply the translated and scaled coordinates by $\mathbf{R}$ as per Equation (7.24).

The first plot in Figure 7.4 shows two configurations of points, both approximately in the shape of five-pointed stars but at different scale, rotation and location. The shape vector to be aligned is referred to as the *input*, whereas the shape vector to which the inputs are aligned is called the *base*. The second plot shows them after one has been Procrustes aligned to the other. The code that produced Figure 7.4 is given in Example 7.1.

The Procrustes transformation preserves shape and, thus, does not permit stretching or shearing – it is consequently a special case of the *affine* transformation. Stretching and shearing have an important role in image processing too, particularly in the process of *image registration*, but we will address this later. The other major class of linear transformation is the *projective* transform, to which we now turn.

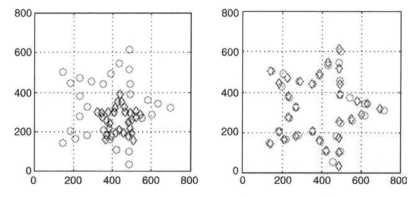

Figure 7.4 Procrustes alignment of two shapes: rotation, scaling and translation are applied to minimize the sum of the squared differences between corresponding pairs of coordinates

---

Example 7.1

Matlab code	What is happening?
load procrustes_star.mat;	%load coordinates of two shapes
whos	%input_points and base_points
subplot(1,2,1),	
plot(base_points(:,1),base_points(:,2),'kd'); hold on;	%Plot the shape coordinates
plot(input_points(:,1),input_points(:,2),'ro'); axis square; grid on	
[D,Z,transform] = procrustes(input_points, base_points);	%Procrustes align input to base
subplot(1,2,2),	
plot(input_points(:,1),input_points(:,2),'kd'); hold on;	
plot(Z(:,1),Z(:,2),'ro'); axis square; grid on; hold off;	%Plot aligned coordinates

Comments

Matlab functions: ***procrustes***.

The input to the Procrustes function is the two point distributions, i.e. sets of shape coordinates that need to be aligned.

The output variables comprise:

D	the sum of the squared differences between corresponding coordinate pairs. This provides some measure of the similarity between the shapes.
Z	the coordinates of the points after alignment to the reference.
transform	a Matlab structure containing the explicit forms for the rotation, scaling and translation components that effected the alignment.

Type ≫ *doc procrustes* at the Matlab prompt for full details on the procrustes function.

7.8 The projective transform

If a camera captures an image of a 3-D object, then, in general, there will be a *perspective* mapping of points on the object to corresponding points in the image. Pairs of points that are the *same distance* apart on the object will *be nearer or further apart* in the image depending on their distance from the camera and their orientation with respect to the image plane. In a perspective transformation we consider 3-D *world* coordinates (X, Y, Z) which are arbitrarily distributed in 3-D space and mapped (typically) to the image plane of a camera. However, in those situations in which the points of interest in the object can be considered as confined (either actually or approximately) to a plane and we are interested in the mapping of those object points into an arbitrarily oriented image plane, the required transformation is termed *projective*.

The form of the projective transform is completely determined by considering how one arbitrary quadrilateral in the object plane maps into another quadrilateral in the image plane. Thus, the task reduces to the following:

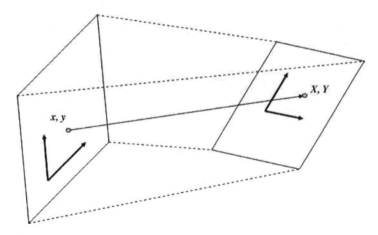

Figure 7.5   The projective transform is defined by the mapping of one arbitrary quadrilateral in the object plane into another in the image plane where the relative orientation of the quadrilaterals is unconstrained. The coordinates (x, y) are the image plane coordinates whereas (X, Y) are referred to as the *world* coordinates of the object point

> Given the coordinates of the four corners of both quadrilaterals, compute the projective transform which maps an arbitrary point within one quadrilateral to its corresponding point in the other (see Figure 7.5).

Since this mapping is constrained at four 2-D points, there are eight coordinates and thus eight degrees of freedom in a projective transform. It is possible to show that the general form of the 2-D projective transformation matrix in a homogeneous coordinate system is given by:

$$\mathbf{T} = \begin{bmatrix} \alpha_{11} & \alpha_{12} & \alpha_{13} \\ \alpha_{21} & \alpha_{22} & \alpha_{23} \\ \alpha_{31} & \alpha_{32} & \alpha_{33} \end{bmatrix}$$

and the transformed (projected) coordinates in the image plane are related to the world points through matrix multiplication as:

$$\widehat{\mathbf{S}} = \mathbf{TS} \rightarrow \begin{bmatrix} x'_1 & \cdots & x'_N \\ y'_1 & \cdots & y'_N \\ 1 & 1 & 1 \end{bmatrix} = \begin{bmatrix} \alpha_{11} & \alpha_{12} & \alpha_{13} \\ \alpha_{21} & \alpha_{22} & \alpha_{23} \\ \alpha_{31} & \alpha_{32} & \alpha_{33} \end{bmatrix} \begin{bmatrix} x_1 & \cdots & x_N \\ y_1 & \cdots & y_N \\ 1 & 1 & 1 \end{bmatrix} \qquad (7.25)$$

Note that there are eight degrees of freedom in the projective matrix (not nine), since the parameters are constrained by the relation

$$\alpha_{31}x + \alpha_{32}y + \alpha_{33} = 1 \qquad (7.26)$$

(a)   (b)   (c)

Figure 7.6   Examples of projective transformation. The original image on the left is projectively transformed to examples (a), (b) and (c)

Figure 7.6 shows the visual effect of some projective transforms. In this case, we have a 2-D (and thus planar) object. We define a square quadrilateral in the original object with coordinates $\{0, 0; 1, 0; 1, 1; 0, 1\}$ and demand that these coordinates transform linearly to the corresponding quadrilaterals defined in Table 7.2. The required transform matrix is calculated by inverting Equation (7.25) subject to the constraint expressed by Equation (7.26).

Projective transforms are useful for registering or aligning images or more generally 'scenes' which can be approximated as flat obtained from different viewpoints. Figure 7.6 illustrates the use of projective transformation as a means to register an image with respect to a different viewpoint. The Matlab® code which produced Figure 7.7 is given in Example 7.2.

Example 7.2

Matlab code	**What is happening?**
A = imread('plate_side.jpg');	%Read image to be registered
figure, imshow(A);	%Display
[x,y] = ginput(4); input_points = [x y];	%Interactively define coords of input quadrilateral
figure, imshow('plate_reference.jpg')	%Read base reference)image.
[x,y] = ginput(4); base_points = [x y];	%Interactively define coords of base quadrilateral
t_carplate = cp2tform(input_points, base_points, 'projective');	%Create projective transformation structure
registered = imtransform(A,t_carplate);	%Apply projective transform
B = imcrop(registered);	%Interactively crop result
figure, imshow(B)	%Display corrected image

Comments

Matlab functions: ***ginput, cp2tform, imtransform, imcrop***.

   Key functions in this example are ***cp2tform*** and ***imtransform***. ***cp2tform*** is a general purpose function for geometric transformation which in this case requires specification of the input and base coordinates. ***imtransform*** applies the transform structure to input image and copies the texture to the mapped locations. See Matlab documentation for further details.

Table 7.2 Coordinate mappings for the projective transform in Figure 7.5: a square in the object plane is mapped to the given quadrilateral in the projected plane

Original coordinates	$\{0, 0; 1, 0; 1, 1; 0, 1\}$
Transformed coordinates	
(a)	$\{0, 0.4; 1, 0; 1, 1; 0, 0.6\}$
(b)	$\{0, 1; 1, 0; 0.7, 1; 0.3, 1\}$
(c)	$\{0, 0.4; 1, 0; 1.2, 1; -0.2, 0.6\}$

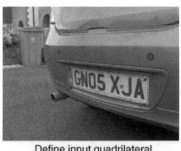

Define input quadrilateral

Define base quadrilateral

After projective transformation

Cropped image

Figure 7.7 Projective transformation. In this example, the coordinates of the corners of the number plate and the reference image form respectively the input and base coordinates required to define the projective transform. Application of the transform results in registration of the plate with respect to the reference (base)

The projective transform is the most general linear transformation and only preserves straight lines in the input. A summary of the different linear transforms in order of increasing generality and the quantities which are preserved under their action is given in Table 7.3.

The geometric transformations described in this chapter so far are all linear transformations, in as much that the coordinates of any point in the transformed image are expressed as some linear combination of the coordinates in the input image. They are particularly appropriate to situations in which we have one or more examples of a shape or image and we wish to align them according to a set rule or method. In the remainder of this chapter, we will take a look at nonlinear transformations and then at piecewise transformation. In this latter technique, an input and a reference image are divided into corresponding triangular regions

Table 7.3 A summary detailing preserved quantities and degrees of freedom of various linear transformations. Each of the transformations is a special case of the successively more general case detailed beneath it

Transformation	Permissible operations	Preserved quantities	Degrees of freedom	
			in 2-D	in 3-D
Euclidean	Rotation and translation	Distances Angles Parallelism	3	6
Similarity	Rotation, scaling and translation	Angles Parallelism	4	7
Affine	Rotation, scaling, shearing and translation	Parallelism	6	12
Projective	Most general linear form	Straight lines	8	15

and this may be used very effectively to achieve image transformations which are locally linear but globally nonlinear.

7.9 Nonlinear transformations

The linear transformations of shape coordinates we have discussed so far are very important. However, many situations occur in which we need to apply a *nonlinear* transformation on the set of input coordinates which constitute the shape vector. In general, we have:

$$x' = T_x(x, y) \qquad y' = T_y(x, y) \tag{7.27}$$

where T_x and T_y are any nonlinear function of the input coordinates (x, y). Whenever T_x and T_y are nonlinear in the input coordinates (x, y), then, in general, straight lines will *not remain straight* after the transformation. In other words, the transformation will introduce a degree of *warping* in the data. One form which can be made to model many nonlinear distortions accurately is to represent T_x and T_y as second-order polynomials:

$$
\begin{aligned}
x' &= T_x(x, y) = a_0 x^2 + a_1 xy + a_2 y^2 + a_3 x + a_4 y + a_5 \\
y' &= T_y(x, y) = b_0 x^2 + b_1 xy + b_2 y^2 + b_3 x + b_4 y + b_5
\end{aligned}
\tag{7.28}
$$

A common form of nonlinear distortion is the pincushion or barrel effect often produced by lower quality, wide-angle camera lenses. This is a radially symmetric aberration, which is most simply represented in polar coordinates as:

$$
\begin{aligned}
r' &= T_r(r, \theta) = r + ar^3 \\
\theta' &= T_\theta(r, \theta) = \theta
\end{aligned}
\tag{7.29}
$$

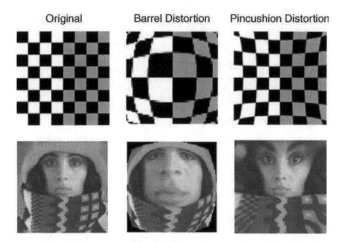

Figure 7.8 Barrel and pincushion distortion: These nonlinear transformations (greatly exaggerated in the examples above) are typical of the distortion that can be introduced by the use of poor-quality wide-angle lenses. The chessboard image of black, white and grey squares in the top row illustrates directly how the reference grid is distorted under the effect of this transformation

where $r = 0$ corresponds to the centre of the image frame. The pincushion ($a < 0$) and barrel ($a > 0$) distortion effects are effectively controlled by the magnitude of the coefficient a in Equation (7.29) and result in the 'hall of mirrors'-type[4] of distortions illustrated in Figure 7.8 (the Matlab code for which is given in Example 7.3).

Example 7.3

Matlab code	What is happening?
I = checkerboard(20,4);	%Read in image
%I = imread('trui.png');	%Read in image
[nrows,ncols] = size(I);	%Extract no. of cols and rows
[xi,yi] = meshgrid(1:ncols,1:nrows);	%Define grid
imid = round(size(I,2)/2);	%Find index of middle element
xt = xi(:) - imid;	%Subtract off and thus
yt = yi(:) - imid;	%shift origin to centre
[theta,r] = cart2pol(xt,yt);	%Convert from cartesian to polar
a = .0005;	%Set the amplitude of the cubic term
s = r + a.*r.^3;	%Calculate BARREL distortion
[ut,vt] = pol2cart(theta,s);	%Return the (distorted) Cartesian coordinates
u = reshape(ut,size(xi)) + imid;	%Reshape the coordinates to original 2-D grid
v = reshape(vt,size(yi)) + imid;	%Reshape the coordinates into original 2-D grid
tmap_B = cat(3,u,v);	%Assign u and v grids as the 2 planes of a 3-D array
resamp = makeresampler('linear', 'fill');	

---

[4] Some readers may recall visiting a room at a funfair or seaside pier in which curved mirrors are used to produce these effects. In my boyhood times on Brighton pier, this was called 'The hall of mirrors'.

```
I_barrel = tformarray(I,[],resamp,
     [2 1],[1 2],[],tmap_B,.3);
[theta,r] = cart2pol(xt,yt);          %Convert from cartesian to polar
a = −0.000015;                        %Set amplitude of cubic term
s = r + a.*r.^3;                      %Calculate PINCUSHION distortion
[ut,vt] = pol2cart(theta,s);          %Return the (distorted) Cartesian coordinates
u = reshape(ut,size(xi)) + imid;      %Reshape the coordinates into original 2-D grid
v = reshape(vt,size(yi)) + imid;      %Reshape the coordinates into original 2-D grid
tmap_B = cat(3,u,v);                  %Assign u and v grids as the 2 planes of a 3-D array
resamp = makeresampler('linear',      %Define resampling structure for use with
     'fill');                              tformarray
I_pin = tformarray(I,[],resamp,       %Transform image to conform to grid in tmap_B
     [2 1],[1 2],[],tmap_B,.3);
subplot(131); imshow(I);
subplot(1,3,2); imshow(I_barrel);
subplot(1,3,3), imshow(I_pin);
```

Comments
Matlab functions: *meshgrid*, *reshape*, *cart2pol* and *pol2cart*.
   Key functions in this example are *makeresampler* and *tformarray*. tformarray is a general purpose function for geometric transformation of a multidimensional array and **makeresampler** produces a structure that specifies how tformarray will interpolate the data onto the transformed coordinate array. See Matlab documentation for further details.

7.10   Warping: the spatial transformation of an image

There are certain special applications where we are concerned only with the object shape and, therefore, it is satisfactory to treat the object simply as a set of geometric landmark points. However, the majority of real images also have intensity values which are associated with the spatial coordinates and it is the spatial transformation of these intensity values over the entire image that is the real goal in most practical applications. This mapping of intensity values from one spatial distribution into another spatial distribution is commonly known as *image warping*. Thus, spatially transforming or warping an image consists of both a geometrical/spatial transformation of landmark coordinates *and* the subsequent transfer of pixel values from locations in one image to corresponding locations in the warped image.

   Typical warping applications require us to transform an image so that its shape (as defined by the landmarks) conforms to some corrected, reference form. The image which is to be transformed is termed the *input*. The normal situation is that we wish to transform the input image so that it conforms in some well-defined sense to a reference geometry. This geometry is defined by identifying a set of landmarks on a reference image which clearly correspond to those in the input. The reference image is termed the *base* image and the base landmarks effectively define the *target* spatial coordinates to which the input landmarks must be transformed. This is depicted in Figure 7.9. Note that a pair of corresponding landmarks in the input and base images are referred to as *tie-points* (as they are in a certain sense tied to each other).

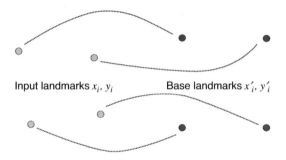

Input landmarks x_i, y_i Base landmarks x'_i, y'_i

The first step in the spatial transformation of an image is to
carry out a geometric transformation of input landmarks. The aim
is to transform the input coordinates such that their new values will
match the reference (base) landmarks.

Figure 7.9 The central concept in geometric transformation or warping is to define a mapping between input and base landmarks

There are theoretically an infinite number of spatial transformations which we might apply to effect a warp, although, in practice, a relatively limited number are considered. Whatever the specific transformation we choose, the basic steps in the overall procedure are the same:

(1)   Locate an equal number of matching landmarks on the image pair (x'_i, y'_i for the base image and x_i, y_i for the input). Note that matching landmarks are often called *tie-points* and we will use these terms interchangeably.

(2)   Define a chosen functional form to map coordinates from the input to the base. Thus, $x' = f(x, y)$, $y' = g(x, y)$.

(3)   Write a system of equations each of which constrains the required mapping from a tie-point in the input to its corresponding tie-point in the base. Thus, for N tie-points, we have

$$x'_1 = f(x_1, y_1) \qquad y'_1 = g(x_1, y_1)$$
$$x'_2 = f(x_2, y_2) \qquad y'_2 = g(x_2, y_2)$$
$$\vdots \qquad\qquad \vdots$$
$$x'_N = f(x_N, y_N) \qquad y'_N = g(x_N, y_N)$$

(4)   The chosen functional forms $f(x, y)$ and $g(x, y)$ will be characterized by one or more free (fit) parameters. We solve the system of equations for the free parameters.

(5)   For every valid point in the base image, calculate the corresponding location in the input using the transforms f and g and copy the intensity value at this point to the base location.

This basic process is illustrated schematically in Figure 7.10.

Consider the simple example depicted in Figure 7.11 (produced using the Matlab code in Example 7.4). The aim is to transform the *input* image containing 10 landmarks shown top

1. Place landmarks

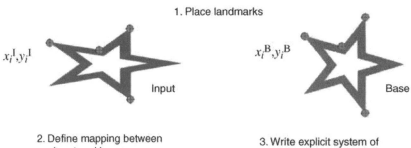

x_i^I, y_i^I Input x_i^B, y_i^B Base

2. Define mapping between   3. Write explicit system of
   input and base - *e.g*   equations for landmarks

$$x^B = a_1 x^I + a_2 y^I + a_3 x^I y^I + a_4$$
$$y^B = a_5 x^I + a_6 y^I + a_7 x^I y^I + a_8$$

4. Solve system for   5. Calculate all corresponding locations in
   mapping parameters   input and base and copy intensity values

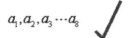

Figure 7.10 The basic steps in the warping transformation

Figure 7.11 Top: the 10 input landmarks on the image on the left need to be transformed to conform to the 10 base landmarks on the right. An affine transformation (as defined by Equation (7.8) or equivalently by Equation (7.10) in homogeneous coordinates) can approximate this quite well. Bottom: the original and transformed images are displayed

---

Example 7.4

Matlab code	**What is happening?**
I = imread('cameraman.tif');	%Read in image
cpstruct=cpselect(I,I);	%Mark the tie points and save within gui
	%Gives input_points and output_points
tform = cp2tform(input_points,	%Infer affine transformation from tie points
base_points,'affine');	
B = imtransform(I,tform);	%Transform input
subplot(1,2,1), imshow(I),subplot(1,2,2),	%Display
imshow(B)	

Comments

Matlab functions: ***cpselect*, *cp2tform*, *imtransform*.**

cpselect is a graphical tool which forms part of the image processing toolbox and which is designed to make the identification, placement and storage of landmarks on image pairs easier.

cp2tform is a function that calculates the transform parameters which best achieves the warp from the input to the base coordinates in a least-squares sense. The user must supply the input and base coordinates and specify the type of transform from a number of permissible options. The output is a Matlab structure which contains the parameters of the transform and associated quantities of use.

imtransform is a function that carries out the warping transformation. The user must supply the input image and the transform structure (typically provided by use of cp2tform).

See the Matlab documentation for more details of these functions and their usage.

---

left to conform to the *base* shape depicted by the corresponding landmarks on the top right. The input shape is characterized by two approximately horizontal lines and the base by two rotated and scaled but also approximately parallel lines. We might expect that a general affine transformation (which permits a combination of translation, shearing, scaling and rotation) will be effective in ensuring a smooth warp. The image shown bottom right, in which the input image has been mapped to its new location, shows this to be the case.

7.11 Overdetermined spatial transformations

The first step in image warping is to transform the input landmarks such that their new coordinate values will match the corresponding base landmarks. In general, it is rarely possible to achieve *exact* matching using a global spatial transform. The fundamental reason for this is that, whatever transformation function we choose to act on the input coordinates, it must always be characterized by *a fixed number of free parameters*. For example, in a bilinear spatial transform, the coordinates of the input are transformed to the base according to

$$x' = f(x, y) = a_1 x + a_2 y + a_3 xy + a_4$$
$$y' = g(x, y) = a_5 x + a_6 y + a_7 xy + a_8$$

(7.30)

The bilinear transform is thus characterized by eight free parameters a_1–a_8. It follows that any attempt to warp the input coordinates (x, y) to precisely conform to the base (x', y') at more than four landmark pairs using Equation (7.30) cannot generally be achieved, since the resulting system has more equations than unknown parameters, i.e. it is *overdetermined*. In general, whenever the number of equations exceeds the number of unknowns, we cannot expect an exact match and must settle for a least-squares solution. A least-squares solution will select those values for the transform parameters which minimize the sum of the squared distances between the transformed input landmarks and the base landmarks. This effectively means that the input shape is made to conform *as closely as possible* to that of the base but cannot match it exactly.

Consider Figure 7.12, in which two different faces are displayed. Each face image has been marked with a total of 77 corresponding landmarks. The aim is to spatially transform image A on the left such that it assumes the shape of the base image on the right (B). Image C shows the original face and image D the result of a global spatial transformation using a second-order, 2-D polynomial form. Although the result is a smooth warp, careful comparison with image E (the base) indicates that the warp has been far from successful in transforming the shape of the features to match the base.

This is the direct result of our highly overdetermined system (77 landmark pairs, which is far greater than the number of free parameters in the global transformation function). The Matlab code which produces Figure 7.12 is given in Example 7.5. Speaking generally, global transformation functions (of which Equations (7.29) and (7.30) are just two possible examples), are a suitable choice when the distortion introduced into the imaging process has a relatively simple form and can be well approximated by a specific function. However, when the requirement is to warp an image to conform exactly to some new shape at a number of points (particularly if this is a large number of points), they are not suitable. The solution to this particular problem is through the *piecewise* method described in the next section.

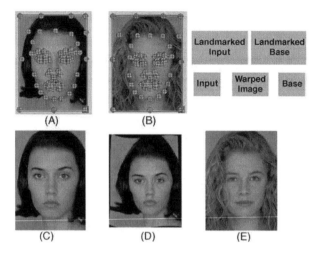

Figure 7.12 Warping using a global polynomial transformation. In this case, the warped image only approximates the tie-points in the base image

---

Example 7.5

Matlab code	What is happening?
load ('ripollfaces.mat')	%Load images and control point %structure
cpselect(rgb2gray(A),rgb2gray(B),cpstruct);	%Load up the tie points
tform = cp2tform(input_points,base_points, 'polynomial');	
	%Infer affine transformation from %tie-points
C = imtransform(A,tform);	%Transform input
subplot(1,2,1), imshow(I),subplot(1,2,2), imshow(C)	%Display

Comments

Matlab functions: **cpselect**, **cp2tform**, **imtransform**.

7.12 The piecewise warp

In general, only the polynomial class of functions can be easily extended to higher order to allow exact matching at a larger number of tie-points (an nth-order polynomial has $n + 1$ free parameters). However, the fitting of a limited set of data points by a high-order polynomial generally results in increasingly wild and unpredictable behaviour in the regions between the tie-points. The inability to define reasonably behaved functions which can *exactly* transform the entire input image to the base at more than a modest number of landmarks but still maintain reasonable behaviour in between the landmark points is the major weakness of the global spatial transformations outlined in the previous section. A means of successfully overcoming this limitation of global, spatial transforms is the *piecewise* warp.

In the *piecewise* approach, the input and base coordinates are divided into a similar number of piecewise, contiguous regions in *one-to-one* correspondence. Typically, each such region is defined by a small number of landmarks; the major advantage is that it allows us to define a simple transformation function whose domain of validity is restricted to each small segment of the image. In this way, we can enforce an exact transformation of the input landmarks to the base values over the entire image plane.

7.13 The piecewise affine warp

The simplest and most popular piecewise warp implicitly assumes that the mapping is locally linear. Such an assumption becomes increasingly reasonable the larger the density of reliable landmark points which are available. In many cases, a linear, piecewise affine transformation produces excellent results, and its computational and conceptual simplicity are further advantages. The key to the piecewise approach is that the landmarks are used to divide the input and base image into an equal number of contiguous regions which

Figure 7.13 Delaunay triangulation divides the image piecewise into contiguous triangular regions. This is a common first stage in piecewise warping

correspond on a one-to-one basis. The simplest and most popular approach to this is to use Delaunay triangulation (see Figure 7.13), which connects an irregular point set by a mesh of triangles each satisfying the Delaunay property.[5] A division of the image into nonoverlapping quadrilaterals is also possible, though not as popular.

Once the triangulation has been carried out, each and every pixel lying within a specified triangle of the base image (i.e. every valid pair of base coordinates) is mapped to its corresponding coordinates in the input image triangle and the input intensities (or colour values as appropriate) are copied to the base coordinates.

Let us now consider the mapping procedure explicitly. Figure 7.14 depicts two corresponding triangles in the input and base images. Beginning (arbitrarily) at the vertex with position vector $\mathbf{x}_1$, we can identify any point within the triangle by adding suitable multiples of the shift vectors $(\mathbf{x}_2-\mathbf{x}_1)$ and $(\mathbf{x}_3-\mathbf{x}_1)$. In other words, any point $\mathbf{x}$ within the input region t can be expressed as a linear combination of the vertices of the triangle:

$$\begin{aligned} \mathbf{x} &= \mathbf{x}_1 + \beta(\mathbf{x}_2-\mathbf{x}_1)+\gamma(\mathbf{x}_3-\mathbf{x}_1) \\ &= \alpha\mathbf{x}_1 + \beta\mathbf{x}_2 + \gamma\mathbf{x}_3 \end{aligned} \tag{7.31}$$

where we define

$$\alpha + \beta + \gamma = 1 \tag{7.32}$$

---

[5] This property means that no triangle has any points inside its circumcircle (the unique circle that contains all three points (vertices) of the triangle).

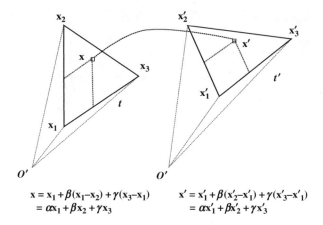

$$x = x_1 + \beta(x_1 - x_2) + \gamma(x_3 - x_1) \qquad x' = x'_1 + \beta(x'_2 - x'_1) + \gamma(x'_3 - x'_1)$$
$$= \alpha x_1 + \beta x_2 + \gamma x_3 \qquad\qquad = \alpha x'_1 + \beta x'_2 + \gamma x'_3$$

Figure 7.14   The linear, piecewise affine warp. Each point in the input image (vector coordinate **x**) is mapped to a corresponding position in the base (warped) image by assuming the texture variation from corresponding triangular regions is locally linear.

The required values of the parameters α, β and γ to reach an arbitrary point $\mathbf{x} = \begin{bmatrix} x & y \end{bmatrix}$ within the input triangle can be obtained by writing Equations (7.31) and (7.32) in matrix form:

$$\begin{bmatrix} x_1 & x_2 & x_3 \\ y_1 & y_2 & y_3 \\ 1 & 1 & 1 \end{bmatrix} \begin{bmatrix} \alpha \\ \beta \\ \gamma \end{bmatrix} = \begin{bmatrix} x \\ y \\ 1 \end{bmatrix} \qquad \mathbf{T}\underline{\alpha} = \mathbf{x} \qquad\qquad (7.33)$$

and solving the resulting system of linear equations as

$$\underline{\alpha} = \mathbf{T}^{-1}\mathbf{x} \qquad\qquad (7.34)$$

The corresponding point in the base (warped) image is now given by *the same linear combination* of the shift vectors in the partner triangle t'. Thus, we have

$$\begin{bmatrix} x'_1 & x'_2 & x'_3 \\ y'_1 & y'_2 & y'_3 \\ 1 & 1 & 1 \end{bmatrix} \begin{bmatrix} \alpha \\ \beta \\ \gamma \end{bmatrix} = \begin{bmatrix} x' \\ y' \\ 1 \end{bmatrix} \qquad \mathbf{S}\underline{\alpha} = \mathbf{x}' \qquad\qquad (7.35)$$

But in this case the coordinate matrix is $\mathbf{S}$ containing the vertices of the triangle t'.

Combining Equations (7.34) and (7.35), it thus follows that coordinates map from the input to the base as

$$\mathbf{x}' = \mathbf{S}\underline{\alpha} = \mathbf{S}\mathbf{T}^{-1}\mathbf{x} = \mathbf{M}\mathbf{x} \qquad\qquad (7.36)$$

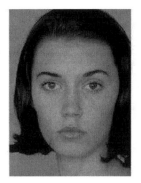

Figure 7.15   A linear piecewise affine warp (right) of the input (centre) to the base (left) using a total of 89 tie-points

---

Piecewise affine warp

(1)   Divide the input and base images into piecewise, corresponding triangular regions using Delaunay triangulation.

(2)   Calculate the mapping matrix **M** as defined by Equations (7.34)–(7.36) for each pair of corresponding triangles in the input and base images.

(3)   Find all pixels lying within the base image triangle (all valid coordinates x_i).

(4)   Apply Equation (6.7) to all valid x_i in the base image to get the corresponding locations in the input image triangle (x_i').

(5)   Set the pixel values in the base at coordinates x_i to the values in the input at coordinates x_i'.

Figure 7.15 shows the result (right) of applying a linear piecewise affine warp of the input image (centre) to the base (left). The tie-points used for the warp are similar to those displayed in Figure 7.12. The vastly improved result is evident.

7.14   Warping: forward and reverse mapping

As we have seen, image warping is a geometric operation that defines a coordinate mapping between an input image and a base image and assigns the intensity values from corresponding locations accordingly. There are, however, *two distinct ways* of carrying out this process. If we take each coordinate location in the input and calculate its corresponding location in the base image, we are implicitly considering a forward mapping (i.e. from 'source' to 'destination'). Conversely, we may consider a reverse mapping ('destination' to 'source') in which we successively consider each spatial location in the base image and calculate its corresponding location in the source. Initially, forward mapping might appear the most logical approach. However, the piecewise affine procedure described in the previous section employs *reverse* mapping. Digital images are discrete by nature and, thus, forward mapping procedures can give rise to problems:

(1) Depending on the specific transformation defined, pixels in the source may be mapped to positions beyond the boundary of the destination image.

(2) Some pixels in the destination may be assigned more than one value, whereas others may not have any value assigned to them at all.

The pixels with no values assigned to them are particularly problematic because they appear as 'holes' in the destination image that are aesthetically unpleasing (and must be filled through interpolation). In general, the greater the difference between the shape of the input and base triangles, the more pronounced this problem becomes. Reverse mapping, however, guarantees a single value for each and every pixel in the destination – there are no holes in the output image and no mapping out-of-bounds. This is the main advantage of reverse mapping.

For further examples and exercises see http://www.fundipbook.com

8

Morphological processing

8.1 Introduction

The word morphology signifies the study of *form* or *structure*. In image processing we use mathematical morphology as a means to identify and extract meaningful image descriptors based on properties of form or shape within the image. Key areas of application are segmentation together with automated counting and inspection. Morphology encompasses a powerful and important body of methods which can be precisely treated mathematically within the framework of *set theory*. While this set-theoretic framework does offer the advantages associated with mathematical rigour, it is not readily accessible to the less mathematically trained reader and the central ideas and uses of morphology can be much more easily grasped through a practical and intuitive discussion. We will take such a pragmatic approach in this chapter.

Morphological operations can be applied to images of all types, but the primary use for morphology (or, at least, the context in which most people will *first* use it) is for processing *binary* images and the key morphological operators are the relatively simple ones called *dilation* and *erosion*. It is in fact possible to show that many more sophisticated morphological procedures can be reduced to a sequence of dilations and erosions. We will, however, begin our discussion of morphological processing on binary images with some preliminary but important definitions.

8.2 Binary images: foreground, background and connectedness

A binary image is an image in which each pixel assumes one of only two possible discrete, logical values: 1 or 0 (see Section 3.2.3). The logical value 1 is variously described as 'high', 'true' or 'on', whilst logical value 0 is described as 'low', 'false' or 'off'. In image processing, we refer to the pixels in a binary image having logical value 1 as the image *foreground* pixels, whilst those pixels having logical value 0 are called the image *background* pixels. An *object* in a binary image consists of any group of *connected* pixels. Two definitions of connection between pixels are commonly used. If we require that a given foreground pixel must have at least one neighbouring foreground pixel to the north, south, east or west of

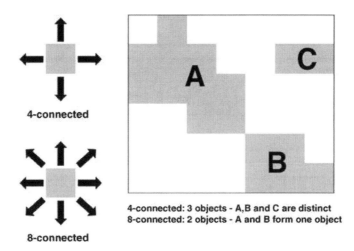

Figure 8.1 The binary image depicted above contains two objects (groups of connected pixels) under 8-connectedness but three objects under 4-connectedness

itself to be considered as part of the same object, then we are using 4-connection. If, however, a neighbouring foreground pixel to the north-east, north-west, south-east or south-west is sufficient for it to be considered as part of the same object, then we are using 8-connection. These simple concepts are illustrated in Figure 8.1.

By their very nature, binary images have no textural (i.e. grey-scale or colour) content; thus, the only properties of interest in binary images are the *shape*, *size* and *location* of the objects in the image. Morphological operations can be extended to grey-scale and colour images, but it is easier, at least initially, to think of morphological operations as *operating on a binary image input to produce a modified binary image output*. From this perspective, the effect of any morphological processing reduces simply to the *determination of which foreground pixels become background and which background pixels become foreground*.

Speaking quite generally, whether or not a given foreground or background pixel has its value changed depends on three things. Two of them are the *image* and the *type of morphological operation* we are carrying out. These two are rather obvious. The third factor is called the structuring element and is a key element in any morphological operation. The structuring element is the entity which determines exactly which image pixels surrounding the given foregound/background pixel must be considered in order to make the decision to change its value or not. The particular choice of structuring element is central to morphological processing.

8.3   Structuring elements and neighbourhoods

A structuring element is a rectangular array of pixels containing the values either 1 or 0 (akin to a small binary image). A number of example structuring elements are depicted in Figure 8.2. Structuring elements have a designated *centre pixel*. This is located at the true

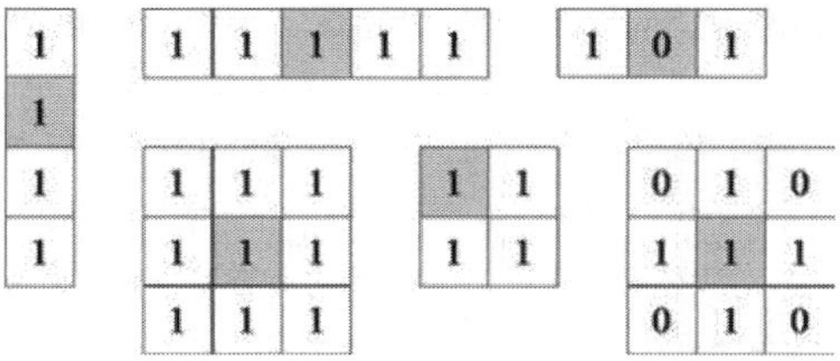

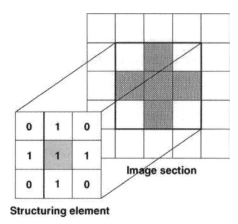

Figure 8.2 Some examples of morphological structuring elements. The centre pixel of each structuring element is shaded

Figure 8.3 The local neighbourhood defined by a structuring element. This is given by those shaded pixels in the image which lie beneath the pixels of value 1 in the structuring element

centre pixel when both dimensions are odd (e.g. in 3×3 or 5×5 structuring elements). When either dimension is even, the centre pixel is chosen to be that pixel north, north-west or west (i.e. *above and/or to the left*) of the geometric centre (thus, a 4×3, 3×4 and a 4×4 structuring element would all have centre pixels at location [2,2]).[1] If we visualize the centre pixel of the structuring element being placed directly above the pixel under consideration in the image, then the *neighbourhood* of that pixel is determined by those pixels which lie underneath those pixels having value 1 in the structuring element. This is illustrated in Figure 8.3.

In general, structuring elements may consist of ones and zeros so as to define any neighbourhood we please, but the practicalities of digital image processing mean that they must be padded with zeros in an appropriate fashion to make *them rectangular in shape* overall. As we shall see in the examples and discussion that follow, much of the art in morphological processing is to *choose the structuring element so as to suit the particular application or aim* we have in mind.

---

[1] In Matlab, the coordinates of the structuring element centre pixel are defined by the expression *floor((size (nhood) + 1)/2)*, where *nhood* is the structuring element.

8.4   Dilation and erosion

The two most important morphological operators are dilation and erosion. All other morphological operations can be defined in terms of these primitive operators. We denote a general image by A and an arbitrary structuring element by B and speak of the erosion/dilation of A by B.

Erosion   To perform erosion of a binary image, we successively place the centre pixel of the structuring element on each foreground pixel (value 1). If *any* of the neighbourhood pixels are background pixels (value 0), then the foreground pixel is switched to background. Formally, the erosion of image A by structuring element B is denoted $A \ominus B$.

Dilation   To perform dilation of a binary image, we successively place the centre pixel of the structuring element on each background pixel. If *any* of the neighbourhood pixels are foreground pixels (value 1), then the background pixel is switched to foreground. Formally, the dilation of image A by structuring element B is denoted $A \oplus B$.

The mechanics of dilation and erosion operate in a very similar way to the convolution kernels employed in spatial filtering. The structuring element slides over the image so that its centre pixel successively lies on top of each foreground or background pixel as appropriate. The new value of each image pixel then depends on the values of the pixels in the neighbourhood defined by the structuring element. Figure 8.4 shows the results of dilation

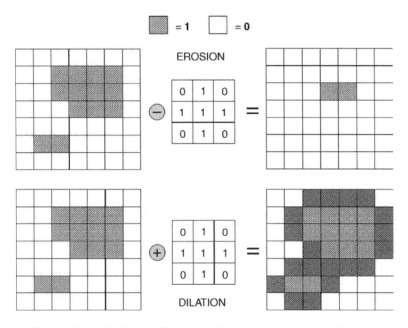

Figure 8.4   The erosion and dilation of a simple binary image. *Erosion*: a foreground pixel only remains a foreground pixel if the 1s in the structuring element (in this example, a cross) are *wholly contained* within the image foreground. If not, it becomes a background pixel. *Dilation*: a background pixel only remains a background pixel if the 1s in the structuring element are wholly contained within the image background. If not, it becomes a foreground pixel. The foreground pixels are shaded and the background pixels are clear. In the diagram demonstrating dilation, the newly created foreground pixels are shaded *darker* to differentiate them from the original foreground pixels

and erosion on a simple binary image. The foreground pixels are shaded and the background pixels are clear. In the diagram demonstrating dilation, the newly created foreground pixels are shaded darker to differentiate them from the original foreground pixels. Note that whenever the structuring element goes over the boundary of the image, we only consider that part of the neighbourhood that lies within the boundary of the image.

There is a powerful and simple way of visualizing the outcome of applying erosion or dilation to an image. For erosion, consider all those locations within the image at which the structuring element can be placed and still remain wholly contained within the image foreground. Only the pixels at these positions will survive the erosion and thus constitute the eroded image. We can consider dilation in an entirely analogous way; namely, we consider those locations at which the structuring element can be placed so as to remain entirely within the image background. Only these pixels will remain background pixels after the dilation and form the dilated image.

8.5 Dilation, erosion and structuring elements within Matlab

In Matlab, we can carry out image erosion and dilation using the Image Processing Toolbox functions *imerode* and *imdilate*. The following simple examples, Examples 8.1 and 8.2, illustrate their use.

Example 8.1

Matlab code	What is happening?
>>bw = imread('text.png');	%Read in binary image
>>se=[0 1 0; 1 1 1; 0 1 0];	%Define structuring element
>>bw_out=imdilate(bw,se);	%Erode image
>>subplot(1,2,1), imshow(bw);	%Display original
>>subplot(1,2,2), imshow(bw_out);	%Display dilated image

Comments
Matlab functions: *imdilate*.
In this example, the structuring element is defined explicitly as a 3×3 array.
The basic syntax requires the image and structuring element as input to the function and the dilated image is returned as output.

Example 8.2

Matlab code	What is happening?
>>bw = imread('text.png');	%Read in binary image
>>se=ones(6,1);	%Define structuring element
>>bw_out=imerode(bw,se);	%Erode image
>>subplot(1,2,1), imshow(bw);	%Display original
>>subplot(1,2,2), imshow(bw_out);	%Display eroded image

Comments
Matlab functions: *imerode*.
In this example, the structuring element is defined explicitly as a 6×1 array.

Explicitly defining the structuring element as in Examples 8.1 and 8.2 is a perfectly acceptable way to operate. However, the best way to define the structuring element is to use the Image Processing Toolbox function **strel**. The basic syntax is

$$\gg \text{se} = \text{strel}(\text{'shape'}, \text{'parameters'})$$

where *shape* is a string that defines the required shape of the structuring element and *parameters* is a list of parameters that determine various properties, including the size. Example 8.3 shows how structuring elements generated using **strel** are then supplied along with the target image as inputs to **imdilate** and **imerode**.

Example 8.3

Matlab code	What is happening?
bw = imread('text.png');	%Read in binary image
se1 = strel('square',4)	%4-by-4 square
se2 = strel('line',5,45)	%line, length 5, angle 45 degrees
bw_1=imdilate(bw,se1);	%Dilate image
bw_2=imerode(bw,se2);	%Erode image
subplot(1,2,1), imshow(bw_1);	%Display dilated image
subplot(1,2,2), imshow(bw_2);	%Display eroded image

Comments
Matlab functions: **strel**.
   This example illustrates the use of the Image Processing Toolbox function **strel** to define two different types of structuring element and demonstrates their effects by dilating and eroding an image.
   Type *doc strel* at the Matlab prompt for full details of the structuring elements that can be generated.

Use of the image processing toolbox function **strel** is recommended for two reasons. First, all the most common types of structuring element can be specified directly as input and it is rare that one will encounter an application requiring a structuring element that cannot be specified directly in this way. Second, it also ensures that the actual computation of the dilation or erosion is carried out in the *most efficient way*. This optimal efficiency relies on the principle that dilation by large structuring elements can often actually be achieved by successive dilation with a sequence of smaller structuring elements, a process known as *structuring element decomposition*. Structuring element decomposition results in a computationally more efficient process and is performed automatically whenever function **strel** is called.

8.6 Structuring element decomposition and Matlab

It is important to note from these examples that the function **strel** *does not return a normal Matlab array*, but rather an entity known as a **strel** *object*. The **strel** object is used to enable the *decomposition* of the structuring element to be stored together with the desired structuring element.

The example below illustrates how Matlab displays when a ***strel*** object is created:

```
>> se3 = strel('disk,5');     % A disk of radius 5

se3 =

Flat STREL object containing 69 neighbors.

Decomposition: 6 STREL objects containing a total of 18
  neighbors

Neighborhood:

0  0  1  1  1  1  1  0  0
0  1  1  1  1  1  1  1  0
1  1  1  1  1  1  1  1  1
1  1  1  1  1  1  1  1  1
1  1  1  1  1  1  1  1  1
1  1  1  1  1  1  1  1  1
1  1  1  1  1  1  1  1  1
0  1  1  1  1  1  1  1  0
0  0  1  1  1  1  1  0  0
```

This is basically telling us two things:

- The neighbourhood is the structuring element that you have created and will effectively be applied; it has 69 nonzero elements (neighbours).

- The decomposition of this neighbourhood (to improve the computational efficiency of dilation) results in six smaller structuring elements with a total of only 18 nonzero elements. *Thus, successive dilation by each of these smaller structuring elements will achieve the same result with less computation.* Since the computational load is proportional to the number of nonzero elements in the structuring element, we expect the execution time to be approximately 18/69 of that using the original form. This is the whole purpose of structuring element decomposition.

If required, the function ***getsequence*** can be applied to find out what precisely these smaller structuring elements are. Try

```
>> decomp=getsequence(se3); whos decomp
```

The structuring elements of the decomposition can be accessed by indexing into decomp; e.g.:

```
>> decomp(1)
```

produces an output

```
ans =

Flat STREL object containing 3 neighbors.

Neighborhood:

1

1

1

   >> decomp(2)
```

produces an output

```
ans =

Flat STREL object containing 3 neighbors.

Neighborhood:

1  0  0

0  1  0
```

and so on.

When you supply the strel object returned by the function ***strel*** to the functions ***imdilate*** and ***imopen***, the decomposition of the structuring element is automatically performed. Thus, in the majority of cases, there is actually no practical need to concern oneself with the details of the structuring element decomposition.

8.7 Effects and uses of erosion and dilation

It is apparent that erosion reduces the size of a binary object, whereas dilation increases it. Erosion has the effect of removing small isolated features, of breaking apart thin, joining regions in a feature and of reducing the size of solid objects by 'eroding' them at the boundaries. Dilation has an approximately reverse effect, broadening and thickening narrow regions and growing the feature around its edges. Dilation and erosion are the reverse of each other, although they are not inverses in the strict mathematical sense. This is because we cannot restore by dilation an object which has previously been completely removed by erosion. Whether it is dilation or erosion (or both) that is of interest, we stress that the appropriate choice of structuring element is often crucial and will depend strongly

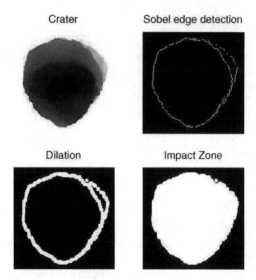

Figure 8.5 Illustrating a simple use of dilation to join small breaks in a defining contour (image courtesy of C.J. Solomon, M. Seeger, L. Kay and J. Curtis, 'Automated compact parametric representation of impact craters' *Int. J. Impact Eng.*, vol. 21, no. 10, 895–904 (1998))

on the application. In particular, *we generally seek to contrive structuring elements which are sensitive to specific shapes or structures* and, therefore, identify, enhance or delete them.

One of the simplest applications of dilation is to join together broken lines which form a contour delineating a region of interest. For example, the effects of noise, uneven illumination and other uncontrollable effects often limit the effectiveness of basic segmentation techniques (e.g. edge detection). In our chosen example, taken from the field of space debris studies, a 3-D depth map of an impact crater caused by a high-velocity, micro-sized aluminium particle has been obtained with a scanning electron microscope. This is displayed at the top left in Figure 8.5. The binary image (top right in Figure 8.5) resulting from edge detection using a Sobel kernel (Section 4.5.2) is reasonably good, but there are a number of breaks in the contour defining the impact region. In this particular example, the aim was to define the impact region of such craters *automatically* in order to express the 3-D depth map as an expansion over a chosen set of 2-D orthogonal functions. We can achieve this task in three simple steps.

(1)   Dilate the edge map until the contour is closed.

(2)   Fill in the background pixels enclosed by the contour. This is achieved by a related morphological method called *region filling* which is explained in Section 8.11.

(3)   Erode the image (the same number of times as we originally dilated) to maintain the overall size of the delineated region.

Our second example is illustrated in Figure 8.6 (which is produced by the Matlab code produced in Example 8.4) and demonstrates the importance of choosing an appropriate structuring element for the given task at hand. Figure 8.6a depicts the layout of a printed

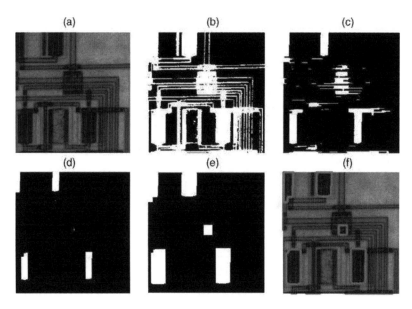

Figure 8.6 Using dilation and erosion to identify features based on shape: (a) original; (b) result after thresholding; (c) After erosion with horizontal line. (d) after erosion with vertical line; (e) after dilation with same vertical and horizontal lines; (f) boundary of remaining objects superimposed on original

Example 8.4

Matlab code	What is happening?
>>length=18; tlevel=0.2;	%Define SE and percent threshold level
>>A=imread('circuit.tif'); subplot(2,3,1), imshow(A)	%Read image and display
>>B=im2bw(A,tlevel); subplot(2,3,2), imshow(~B);	%Threshold image and display
>>SE=ones(3,length); bw1=imerode(~B,SE);	%Erode vertical lines
>>subplot(2,3,3), imshow(bw1);	%Display result
>>bw2=imerode(bw1,SE'); subplot(2,3,4), imshow(bw2);	%Erode horizontal lines
>>bw3=imdilate(bw2,SE');bw4=imdilate(bw3,SE);	%Dilate back
>>subplot(2,3,5), imshow(bw4);	%Display
>>boundary=bwperim(bw4);[i,j]=find(boundary);	%Superimpose boundaries
>> subplot(2,3,6), imshow(A); hold on; plot(j,i,'r.');	

Comments

Matlab functions: ***bwperim***.

This function produces a binary image consisting of the boundary pixels in the input binary image.

circuit in which there are two types of object: the rectangular chips and the horizontal and vertical conduction tracks. The aim is to identify (i.e. segment) the chips automatically. The microprocessor chips and the tracks are darker than the background and can be identified reasonably well by simple intensity thresholding (see Figure 8.6b). The thin *vertical* tracks can first be removed by eroding with a suitable long *horizontal* structuring element. As is evident (see Figure 8.6c), the horizontal structuring element used here (a 3 × 18 array of 1s) tends to preserve horizontal lines but remove thin vertical ones. Analogously, we can remove horizontal lines (see Figure 8.6d) using an appropriate vertical structuring element (an 18 × 3 array of 1s). These two erosions tend to remove most of the thin horizontal and vertical lines. They leave intact the rectangular chips but have significantly reduced their size. We can remedy this by now dilating twice with the same structuring elements (Figure 8.6e). The boundaries of the structures identified in this way are finally super-imposed on the original image for comparison (Figure 8.6f). The result is quite good considering the simplicity of the approach (note the chip in the centre was not located properly due to poor segmentation from thresholding).

8.7.1   Application of erosion to particle sizing

Another simple but effective use of erosion is in granulometry - the counting and sizing of granules or small particles. This finds use in a number of automated inspection applications. The typical scenario consists in having a large number of particles of different size but approximately similar shape and the requirement is to rapidly estimate the probability distribution of particle sizes. In automated inspection applications, the segmentation step in which the individual objects are identified is often relatively straightforward as the image background, camera view and illumination can all be controlled.

We begin the procedure by first counting the total number of objects present. The image is then repeatedly eroded using the same structuring element until no objects remain. At each step in this procedure, we record the total number of objects F which have been removed from the image as a function of the number of erosions n. Now it is clear that the number of erosions required to make an object vanish is *directly proportional* to its size; thus, if the object disappears at the kth erosion we can say that its size $X \approx \alpha k$, where α is some proportionality constant determined by the structuring element.[2] This process thus allows us to form an estimate of $F(X)$ which is essentially a cumulative distribution function (CDF) of particle size, albeit unnormalized.

Elementary probability theory tells us that the CDF and the probability density function (PDF) are related as:

$$F(X) = \int_{-\infty}^{X} p(x)\,\mathrm{d}x \qquad p(x) = \frac{\mathrm{d}F}{\mathrm{d}X}\bigg|_{x}$$

---

[2] This is not an exact relationship as we have not properly defined 'size'. The astute reader will recognize that its accuracy depends on the shape of the objects and of the structuring element. However, provided the structuring element is of the same approximate shape as the particles (a normal condition in such applications), it is a good approximation.

Thus, by differentiating $F(X)$, we may obtain an estimate of the particle size distribution. As the structuring element is necessarily of finite size in a digital implementation, X is a discrete variable. Therefore, it can be useful in certain instances to form a smooth approximation to $F(X)$ by a fitting technique before differentiating to estimate the PDF.

Morphological operations can be implemented very efficiently on specialist hardware, and so one of the primary attractions of this simple procedure is its speed. Figure 8.7 illustrates the essential approach using a test image. The Matlab code which generates Figure 8.7 is given in Example 8.5 (note that the estimated distribution and density functions remain *unnormalized* in this code)

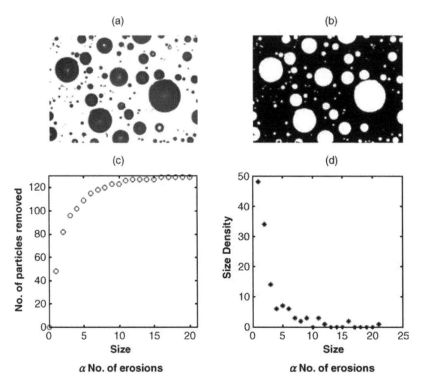

Figure 8.7 Using repeated erosion and object counting to estimate the distribution of particle sizes: (a) original image; (b) binary image resulting from intensity thresholding; (c) estimated cumulative distribution (unnormalized); (d) estimated size density function (unnormalized)

Example 8.5

Matlab code	What is happening?
A=imread('enamel.tif'); subplot(1,3,1), imshow(A);	%Read in image and display
bw=~im2bw(A,0.5); bw = imfill(bw,'holes');	%Threshold and fill in holes
subplot(1,3,2), imshow(bw);	%Display resulting binary image
[L,num_0]=bwlabel(bw);	%Label and count number of objects
	%in binary image

```
se=strel('disk',2);                              %Define structuring element,
                                                    radius=2
count =0;                                         %Set number of erosions = 0
num=num_0;                                        %Initialise number of objects in
                                                    image

while num>0                                       %Begin iterative erosion
count=count + 1
   bw=imerode(bw,se);                             %Erode
   [L,num]=bwlabel(bw);                           %Count and label objects
   P(count)=num_0-num;                            %Build discrete distribution
   figure(2); imshow(bw); drawnow;                %Display eroded binary image
end

   figure(2); subplot(1,2,1), plot(0:count,[0 P],'ro');   %Plot Cumulative distribution
   axis square;axis([0 count 0 max(P)]);          %Force square axis
   xlabel('Size'); ylabel('Particles removed')    %Label axes
   subplot(1,2,2), plot(diff([0 P]),'k*'); axis square;   %Plot estimated size density
                                                    function
```

Comments

Matlab functions: ***bwlabel, imfill.***

The function ***bwlabel*** analyses an input binary image so as to identify all the connected components (i.e. objects) in the image. It returns a so-called labelled image in which all the pixels belonging to the first connected component are assigned a value 1, all the pixels belonging to the second connected component are assigned a value 2 and so on. The function ***imfill*** identifies the pixels constituting holes within connected components and fills them (ie. sets their value to 1).

8.8   Morphological opening and closing

Opening is the name given to the morphological operation of *erosion followed by dilation with the same structuring element.* We denote the opening of A by structuring element B as $A \circ B = (A \ominus B) \oplus B$.

The general effect of opening is to remove small, isolated objects from the foreground of an image, placing them in the background. It tends to smooth the contour of a binary object and breaks narrow joining regions in an object.

Closing is the name given to the morphological operation of *dilation followed by erosion with the same structuring element.* We denote the closing of A by structuring element B as $A \bullet B = (A \oplus B) \ominus B$.

Closing tends to remove small holes in the foreground, changing small regions of background into foreground. It tends to join narrow isthmuses between objects.

On initial consideration of these operators, it is not easy to see how they can be useful or indeed why they differ from one another in their effect on an image. After all, erosion and dilation are logical opposites and superficial consideration would tempt us to conclude that it will make little practical difference which one is used first? However, their different effects stems from two simple facts.

(1)   If erosion eliminates an object, *dilation cannot recover it* – dilation needs at least one
       foreground pixel to operate.

(2)   Structuring elements can be selected to be both *large* and have arbitrary shape if this
       suits our purposes.

8.8.1   The rolling-ball analogy

The best way to illustrate the difference between morphological opening and closing is to use
the 'rolling-ball' analogy. Let us suppose for a moment that the structuring element *B* is a
flat (i.e. 2-D) rolling ball and the object *A* being opened corresponds to a simple binary
object in the shape of a hexagon, as indicated in Figure 8.8. (These choices are quite arbitrary
but illustrate the process well.) Imagine the ball rolling around freely within *A* but
constrained so as to always stay inside its boundary. *The set of all points within object* A
which can be reached by B *as it rolls around in this way belongs in the opening of* A *with* B. For a
'solid' object (with no holes), such as that depicted in Figure 8.8, the opening can be
visualized as the region enclosed by the contour which is generated by rolling the ball all the
way round the inner surface such that the ball always maintains contact with the boundary.

    We can also use the rolling-ball analogy to define the closing of *A* by *B*. The analogy is the
same, except that this time we roll structuring element *B* all the way around the *outer*
boundary of object *A*. The resulting contour defines the boundary of the closed object *A* • *B*.
This concept is illustrated in Figure 8.9.

    The effect of closing and opening on binary images is illustrated in Figure 8.10 (which is
produced using the Matlab code in Example 8.6). Note the different results of opening and

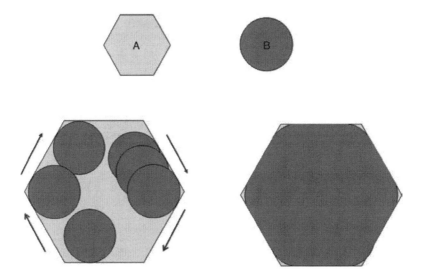

Figure 8.8   The *opening* of object *A* by structuring element *B*, *A* ∘ *B*. This can be visualized as all
possible points within object *A* which can be reached by moving the ball within object *A* without
breaching the boundary. For a solid object *A* (no holes), the boundary of *A* ∘ *B* is simply given by
'rolling' *B* within *A* so as to never lose contact with the boundary. This is the circumference of the area
shaded dark grey

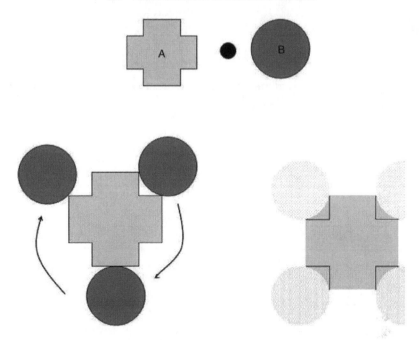

Figure 8.9 The *closing* of object *A* by structuring element *B*, *A•B*. This can be visualized as all possible points contained within the boundary defined by the contour as *B* is rolled around the outer boundary of object *A*. Strictly, this analogy holds only for a 'solid' object *A* (one containing no holes)

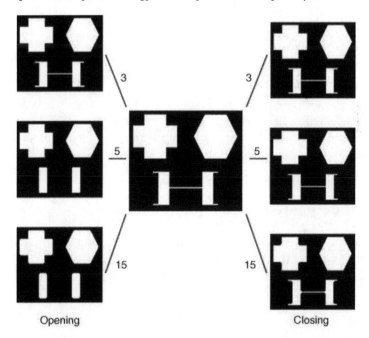

Opening   Closing

Figure 8.10 Illustrating the effects of opening and closing upon some example shapes. The original image shown in the centre was *opened* using disk-shaped structuring elements of radii 3, 5 and 15 pixels, producing the images to the left. The images to the right were similarly produced by *closing* the image using the same structuring elements. The differences become more pronounced the larger the structuring element employed

closing on the test objects and how this becomes more pronounced as the structuring element increases in size.

Example 8.6

Matlab code	**What is happening?**
A=imread('openclose_shapes.png'); A=~logical(A);	%Read in image and convert to binary
se=strel('disk',3); bw1=imopen(A,se); bw2=imclose(A,se);	%Define SEs then open and close
subplot(3,2,1), imshow(bw1); subplot(3,2,2), imshow(bw2);	%Display results
se=strel('disk',6); bw1=imopen(A,se); bw2=imclose(A,se);	%Define SEs then open and close
subplot(3,2,3), imshow(bw1); subplot(3,2,4), imshow(bw2);	%Display results
se=strel('disk',15); bw1=imopen(A,se); bw2=imclose(A,se);	%Define SEs then open and close
subplot(3,2,5), imshow(bw1); subplot(3,2,6), imshow(bw2);	%Display results

Comments

Matlab functions: *imopen, imclose*.

The Image Processing Toolbox provides functions to carry out morphological opening and closing on both binary and grey-scale images.

We have seen how the basic morphological operators of dilation and erosion can be combined to define opening and closing. They can also be used together with simple logical operations to define a number of practical and useful morphological transforms. In the following sections we discuss some of the most important examples.

8.9 Boundary extraction

We can define the boundary of an object by first eroding the object with a suitable small structuring element and then subtracting the result from the original image. Thus, for a binary image A and a structuring element B, the boundary A_P is defined as

$$A_P = A - A \ominus B \tag{8.1}$$

The example in Figure 8.11 (the Matlab code for which is given in Example 8.7) illustrates the process. Note that the thickness of the boundary can be controlled through the specific choice of the structuring element B in Equation (8.1).

 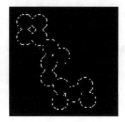 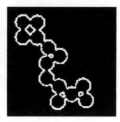

Figure 8.11 Boundary extraction. Left: original.; centre: single-pixel boundary; right: thick boundary extracted through use of larger structuring element

---

Example 8.7

Matlab code	What is happening?
A=imread('circles.png');	%Read in binary image
bw=bwperim(A);	%Calculate perimeter
se=strel('disk',5); bw1=A-imerode(A,se);	%se allows thick perimeter extraction
subplot(1,3,1), imshow(A);	
subplot(1,3,2), imshow(bw);	
subplot(1,3,3), imshow(bw)1;	%Display results

Comments

Matlab functions: ***bwperim, imerode***.

Boundary extraction which is a single-pixel thick is implemented in the Matlab Image Processing Toolbox through the function ***bwperim***. Arbitrary thickness can be obtained by specifying an appropriate structuring element.

8.10 Extracting connected components

Earlier in this chapter we discussed the 4-connection and 8-connection of pixels. The set of all pixels which are connected to a given pixel is called the *connected component* of that pixel. A group of pixels which are all connected to each other in this way is differentiated from others by giving it a unique label. Typically, the process of extracting connected components leads to a new image in which the connected groups of pixels (the objects) are given sequential integer values: the background has value 0, the pixels in the first object have value 1, the pixels in the second object have value 2 and so on. To identify and label 8-connected components in a binary image, we proceed as follows:

- Scan the entire image by moving sequentially along the rows from top to bottom.

- Whenever we arrive at a foreground pixel p we examine the four neighbours of p which have already been encountered thus far in the scan. These are the neighbours (i) to the left of p, (ii) above p, (iii) the upper left diagonal and (iv) the upper right diagonal.

The labelling of p occurs as follows:

- if all four neighbours are background (i.e. have value 0), assign a new label to p; else

- if only one neighbour is foreground (i.e. has value 1), assign its label to p; else

- if more than one of the neighbours is foreground, assign one of the labels to p and resolve the equivalences.

An *alternative* method exists for extracting connected components which is based on a constrained region growing. The iterative procedure is as follows.

Let A denote our image, x_0 be an arbitary foreground pixel identified at the beginning of the procedure, x_k the set of connected pixels existing at the kth iteration and B be a structuring element. Apply the following iterative expression:

$$\begin{array}{ll} \text{Do} & x_k = (x_{k-1} \oplus B) \cap A \quad k = 1, 2, 3, \ldots \\ \text{until} & x_k = x_{k-1} \end{array} \tag{8.2}$$

The algorithm thus starts by dilating from a single pixel within the connected component using the structuring element B. At each iteration, the intersection with A ensures that no pixels are included which do not belong to the connected component. When we reach the point at which a given iteration produces the same connected component as the previous iteration (i.e. $x_k = x_{k-1}$), all pixels belonging to the connected component have been found.

This is repeated for another foreground pixel which is again arbitrary except that it must not belong to the connected components found this far. This procedure continues until all foreground pixels have been assigned to a connected component.

Extracting connected components is a very common operation, particularly in automated inspection applications. Extracting connected components can be achieved with the Matlab Image Processing Toolbox function ***bwlabel***. A simple example is given in Example 8.8 and Figure 8.12.

Example 8.8

Matlab code	What is happening?
bw=imread('basic_shapes.png');	%Read in image
[L,num]=bwlabel(bw);	%Get labelled image and number %of objects
subplot(1,2,1), imagesc(bw); axis image; axis off; colorbar('North'); subplot(1,2,2), imagesc(L); axis image; axis off; colormap(jet); colorbar('North')	%Plot binary input image %Display labelled image

Comments

The function ***bwlabel*** returns a so-called labelled image in which all the pixels belonging to the first connected component are assigned a value 1, the pixels belonging to the second connected component are assigned a value 2 and so on.

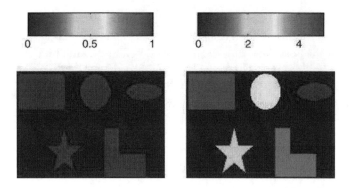

Figure 8.12 Connected components labelling. The image displayed on the left is a binary image containing five connected components or objects. The image displayed on the right is a 'label matrix' in which the first group of connected pixels is assigned a value of 1, the second group a value of 2 and so on. The false-colour bar indicates the numerical values used

8.11 Region filling

Binary images usually arise in image processing applications as the result of thresholding or some other segmentation procedure on an input grey-scale or colour image. These procedures are rarely perfect (often due to uncontrollable variations in illumination intensity) and may leave 'holes', i.e. 'background' pixels within the binary objects that emerge from the segmentation. Filling the holes within these objects is often desirable in order that subsequent morphological processing can be carried out effectively.

Assume that we have a binary object A within which lies one or more background pixels (a hole). Let B be a structuring element and let x_0 be a 'seed' background pixel lying within the hole (for the moment, we gloss over the question of how we find x_0 in the first place). Setting x_0 to be foreground (value 1) to initialize the procedure, the hole may be filled by applying the following iterative expression:

$$\text{Do} \quad x_k = (x_{k-1} \oplus B) \cap \bar{A} \quad \text{where } k = 1, 2, 3, \ldots$$
$$\text{until} \quad x_k = x_{k-1}$$

(8.3)

The filled region is then obtained as $A \cup x_k$.

The algorithm thus works by identifying a point (x_0) within the hole and then growing this region by successive dilations. After each dilation, we take the intersection (logical AND) with the logical complement of the object A. Without this latter step, the filled region would simply grow and eventually occupy the entire image. Taking the intersection with the complement of A only allows the object to grow within the confines of the internal boundary. When the region has grown to the extent that it touches the boundary at all points, the next iteration will grow the region into the boundary itself. However, the intersection with $\bar{A}$ will produce the *same pixel set* as the previous iteration, at which point the algorithm stops.[3] In the final step, the union of A with the grown region gives the entire filled object.

---

[3] The astute reader will note that this is true provided the dilation only extends the growing region into and *not beyond* the boundary. For this reason, the structuring element must not be larger than the boundary thickness.

Now, a hole is defined as a set of connected, background pixels that cannot be reached by filling in the background from the edge of the image. So far, we have glossed over how we *formally* identify the holes (and thus starting pixels x_0 for Equation (8.3)) in the first place. Starting from a background pixel x_0 at the edge of the image, we apply exactly the same iterative expression Equation (8.3):

$$\text{Do} \quad x_k = (x_{k-1} \oplus B) \cap \bar{A} \quad \text{where } k = 1, 2, 3, \ldots$$
$$\text{until} \quad x_k = x_{k-1}$$

The complete set of pixels belonging to all holes is then obtained as $H = (\overline{A \cup x_k})$. The seed pixels x_0 for the hole-filling procedure described by Equation (8.3) can thus be obtained by: (i) sampling arbitrarily from H; (ii) applying Equation (8.3) to fill in the hole; (iii) removing the set of filled pixels resulting from step (ii) from H; (iv) repeating (i)–(iii) until H is empty.

Region filling is achieved in the Matlab Image Processing Toolbox through the function *imfill* (see Example 8.2).

8.12   The hit-or-miss transformation

The hit-or-miss transform indicates the positions where a certain pattern (characterized by a structuring element B) occurs in the input image. As such, it operates as a basic tool for shape detection. This technique is best understood by an illustrative example. Consider Figure 8.13, in which we depict a binary image A alongside a designated 'target' pixel configuration B (foreground pixels are shaded). The aim is to identify all the locations within the binary image A at which the target pixel configuration defined by B can be found. It is important to stress that we are seeking the correct combination of *both* foreground (shaded) and background (white) pixels, not just the foreground, and will refer to this combination of foreground and background as the *target shape*. The reader should note that by this definition the target shape occurs at just one location within image A in Figure 8.13.

Recalling our earlier definitions, it is readily apparent that the erosion of image A by the target shape B_1 will preserve *all those pixels in image A at which the foreground pixels of the target B_1 can be entirely contained within foreground pixels* in the image. These points are indicated in Figure 8.14 as asterisks and are designated as 'hits'. In terms of our goal of finding the precise target shape B_1, the hits thus identify all locations at which the correct

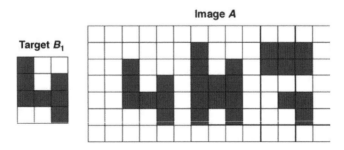

Figure 8.13   The hit-or-miss transformation. The aim is to identify those locations within an image (right) at which the specified target configuration of pixels or *target shape* (left) occurs

Image $A \ominus B_1$

Targer B_1

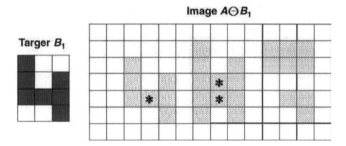

Figure 8.14 The hit-or-miss transformation. The first step is erosion of the image by the target configuration. This produces the *hits* denoted by asterisks

configuration of foreground pixels are found. However, this step does not test for the required configuration of background pixels in B_1.

Consider now Figure 8.15. This depicts the logical complement of image A together with a target pixel configuration B_2 which is the logical complement of B_1 (i.e. $B_2 = \bar{B}_1$).

Consider now applying $\bar{A} \ominus B_2$, namely erosion of the image *complement* $\bar{A}$ by the *complement* of the target shape B_2. By definition, this erosion will preserve all those pixels at which the foreground pixels of the target complement B_2 can be entirely contained within the foreground pixels of the image complement $\bar{A}$. These points are indicated in Figure 8.16 as crosses and are designated as 'misses'. Note, however, that because we have carried out the erosion using the complements of the target and image, this second step logically translates to the *identification of all those locations at which the* background *pixel configuration of the target is entirely contained within the* background *of the image*. The misses thus identify all locations at which the correct configuration of background pixels are found but does not test for the required configuration of foreground pixels.

The first step has identified all those points in the image foreground at which the required configuration of foreground pixels in the target may be found (but has ignored the required background configuration). By contrast, the second step has identified all those points in the image background at which the required configuration of background pixels in the target may be found (but has ignored the required foreground configuration). It follows, therefore, that any point identified by *both* of these steps gives a location for the target configuration. Accordingly, the third and final step in the process is to take a logical intersection (AND) of the two images. Thus, the hit-or-miss transformation of A by $\bar{B}_1$ is

Image $\bar{A}$

Target $\bar{B}_1$

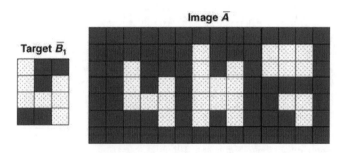

Figure 8.15 The hit-or-miss transformation. For the second step where we locate the misses, we consider the *complement* of the image $\bar{A}$ and the *complement* of the target shape $B_2 = \bar{B}_1$

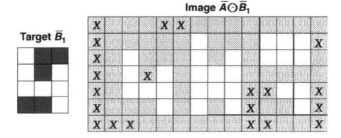

Figure 8.16 The hit-or-miss transformation. The second step consists of eroding the *complement* of the image $\bar{A}$ by the *complement* of the target shape ($B_2 = \bar{B}_1$). This produces the *misses* denoted here by crosses

formally defined as:

$$A \otimes B = (A \ominus B) \cap (\bar{A} \ominus \bar{B}) \tag{8.4}$$

The result obtained for our example is indicated in Figure 8.17. The sole surviving pixel (shaded) gives the only location of the target shape to be found in the image.

   In summary then, the hit-or-miss transformation identifies those locations at which both the defining foreground and background configurations for the target shape are to be simultaneously found – in this sense, the method might more properly be called the hit-*and*-miss, but the former title is the most common. Example 8.9 and Figure 8.18 illustrate the use of the hit-and-miss transform to identify the occurrences of a target letter 'e' in a string of text.

   Note that the two structuring elements B_1 and B_2 described in our account so far of the hit-or-miss transform are logical complements. This means that only *exact* examples of the target shape (including background and foreground pixel configurations) are identified. If even one pixel of the configuration probed in the image differs from the target shape, then the hit-or-miss transform we have used will not identify it (see Figure 8.19).

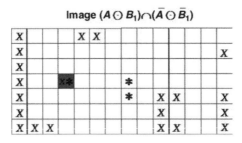

Figure 8.17 The hit-or-miss transformation. The final step takes the intersection (logical AND) of the two eroded images $(A \odot B) \cap (\bar{A} \odot \bar{B})$. Any surviving pixels give the location of the target shape in the original image A

Example 8.9

Matlab code	What is happening?
imread('text.png');	%Read in text
B=imcrop(A);	%Read in target shape interactively
se1=B; se2=~B;	%Define hit and miss structure elements
bw=bwhitmiss(A,se1,se2);	%Perform hit-miss transformation
[i,j]=find(bw==1);	%Get explicit coordinates of locations
subplot(1,3,1), imshow(A);	%Display image
subplot(1,3,2), imagesc(B); axis image; axis off;	%Display target shape
subplot(1,3,3), imshow(A); hold on; plot(j,i,'r*');	%Superimpose locations on image

Comments

Matlab functions: **bwhitmiss**.

The Matlab Image Processing Toolbox has a purpose-made function for carrying out the hit-or-miss transformation: **bwhitmiss**. This example illustrates its use to identify the occurrences of a target letter 'e' in a string of text.

8.12.1 Generalization of hit-or-miss

The hit-or-miss transformation given by Equation (8.4) and demonstrated in Example 8.9 was chosen to be relatively straightforward to understand. However, it is important to stress that it is actually a *special* case. The most general form of the hit-or-miss transformation is given by

$$A \otimes B = (A \ominus B_1) \cap (\bar{A} \ominus B_2) \qquad \text{where } B = (B_1, B_2) \qquad (8.5)$$

$B = (B_1, B_2)$ signifies a pair of structuring elements and B_2 is a general structuring element that is *not necessarily the complement* of B_1 but which is mutually exclusive (i.e. $B_1 \cap B_2 = \varnothing$, the empty set). By allowing the second structuring element B_2 to take on forms other than the complement of B_1, we can effectively relax the constraint that the hit-or-miss operation only identifies the *exact* target shape and widen the scope of the method. This relaxation of the

Figure 8.18 Application of the hit-or-miss transformation to detect a target shape in a string of text. Note that the target includes the background and hit-or-miss is strictly sensitive to both the scale and the orientation of the target shape (*See colour plate section for colour version*)

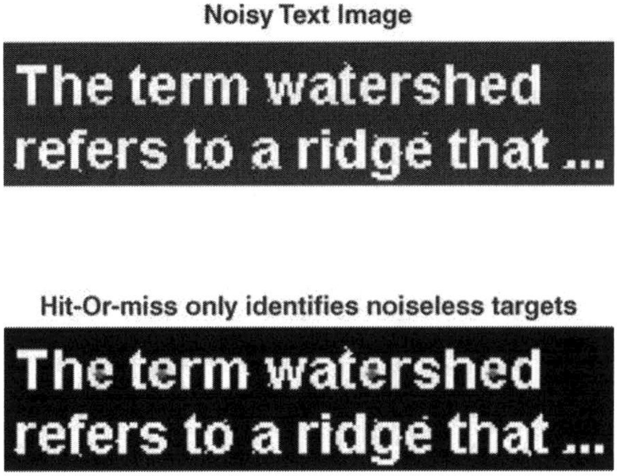

Noisy Text Image

Hit-Or-miss only identifies noiseless targets

Figure 8.19 The fully constrained hit-or-miss transform is sensitive to noise and uncontrolled variations in the target feature. In the example above, three instances of the target letter "e" on the second line were missed due to the presence of a small amount of random noise in the image

method can make it less sensitive to noise and small but insignificant variations in the shape. Note that, in the limiting case where B_2 is an empty structuring element, hit-or-miss transformation reduces to simple erosion by B_1. We discuss this generalization of hit-or-miss next.

8.13 Relaxing constraints in hit-or-miss: 'don't care' pixels

It is apparent that the application of the hit-or-miss transformation as carried out in Figure 8.19 is 'unforgiving', since a discrepancy of even one pixel between the target shape and the probed region in the image will result in the target shape being missed. In many cases, it is desirable to be less constrained and allow a combination of foreground and background pixels that is 'sufficiently close' to the target shape to be considered as a match. We can achieve this relaxation of the exact match criterion through the use of 'don't care' pixels (also known as wild-card pixels). In other words, the target shape now conceptually comprises three types of pixel: strict foreground, strict background and 'don't care'. In the simple example shown in Figure 8.20 (produced using the Matlab code in Example 8.10), we consider the task of automatically locating the 'upper right corners' in a sample of shapes.

Example 8.10

Matlab code	What is happening?
A=imread	%CASE 1
Noisy_Two_Ls.png');	
se1=[0 0 0; 1 1 0; 0 1 0];	%SE1 defines the hits
se2=[1 1 1; 0 0 1; 0 0 1];	%SE2 defines the misses
bw=bwhitmiss(A,se1,se2);	%Apply hit-or-miss transform
subplot(2,2,1), imshow(A,[0 1]);	%Display Image
subplot(2,2,2), imshow(bw,[0 1]);	%Display located pixels
	%NOTE ALTERNATIVE SYNTAX

```
interval=[-1 -1 -1; 1 1 -1; 0 1 -1];      %1s for hits, -1 for misses; 0s for don't
                                            care
bw=bwhitmiss(A,interval);                  %Apply hit-or-miss transform
subplot(2,2,3), imshow(bw,[0 1]);          %Display located pixels

                                           %CASE 2
interval=[0 -1 -1; 0 1 -1; 0 0 0];         %1s for hits, -1 for misses; 0s for don't
                                            care
bw=bwhitmiss(A,interval);                  %Apply hit-or-miss transform
subplot(2,2,4), imshow(bw,[0 1]);          %Display located pixels
```

Comments

Note that bw2 = ***bwhitmiss***(bw,interval) performs the hit–miss operation
defined in terms of a single array, called an 'interval'. An interval is an array
whose elements can be 1, 0 or −1. The 1-valued elements specify the domain
of the first structuring element SE1 (hits); the −1-valued elements specify
the domain of the second structuring element SE2 (misses); the 0-valued
elements, which act as 'don't care' pixels, are ignored.

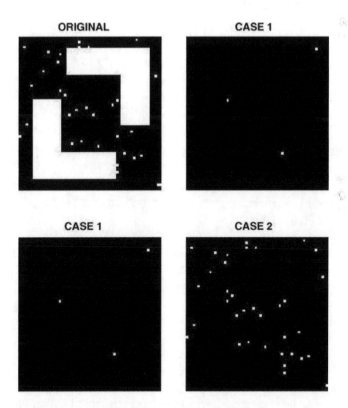

Figure 8.20 Generalizing the hit-or-miss transform. This illustrates the effect of relaxing constraints on the hit-or-miss transform. Top left: original image. Top right and bottom left (which uses an alternative computation): the result of hit-or-miss for strict definition of upper right corner pixels – only the upper right corner pixels of the solid L shapes are identified. Bottom right: in the second (relaxed) case, the noisy pixels are also identified

In the example shown in Figure 8.20, we consider two definitions for upper right corner pixels:

(1) A strict definition which requires that an 'upper right corner pixel' must have *no* neighbours to *north, north-west, north-east or east* (these are the misses) and must have neighbours to the south and west (these are the hits). The south-west neighbour is neither hit nor miss and, thus, a *don't care* pixel.

(2) A looser definition which only requires that an upper right corner pixel must have *no* neighbours to *north, north-west and north-east* (the misses). All other neighbours are 'don't care' pixels.

The differing effects of both definitions are shown in Figure 8.20 using the hit and miss transform.

8.13.1 Morphological thinning

Closely related to the hit-or-miss transformation, the thinning of an image A by a structuring element B is defined as

$$\text{thin}(A, B) = A \cap \overline{A \otimes B}$$

where $A \otimes B$ is the hit-or-miss transformation of A with B. The thinning of A with B is calculated by successively translating the origin of the structuring element to each pixel in the image and comparing it with the underlying image pixels. If both the foreground and background pixels in the structuring element exactly match those in the underlying image, then the image pixel underneath the origin of the structuring element is set to background (zero). If no match occurs, then the pixel is left unchanged. Thinning is a key operation in practical algorithms for calculating the skeleton of an image.

8.14 Skeletonization

Rather as its name implies, the *skeleton* of a binary object is a representation of the basic form of that object which has been reduced down to its minimal level (i.e. a 'bare bones' representation). A very useful conceptualization of the morphological skeleton is provided by the *prairie-fire analogy*: the boundary of an object is set on fire and spreads with uniform velocity in all directions inwards; the skeleton of the object will be defined by the points at which the fire fronts meet and quench (i.e. stop) each other.

Consider an arbitrary, binary object A. A point p within this binary object belongs to the skeleton of A if and only if the two following conditions hold:

(1) A disk D_z may be constructed, with p at its centre, that lies entirely within A and touches the boundary of A at two or more places.

(2) No other larger disk exists that lies entirely within A and yet contains D_z.

An equivalent geometric construction for producing the skeleton is to:

(1) Start at an arbitrary point on the boundary and consider the maximum possible size of disk which can touch the boundary at this point and at least one other point on the boundary and yet remain within the object.

(2) Mark the centre of the disk.

(3) Repeat this process at all points along the entire boundary (moving in infinitesimal steps) until you return to the original starting point. The trace or locus of all the disk centre points is the skeleton of the object.

Figure 8.21 shows a variety of different shapes and their calculated skeletons. Consideration of the calculated skeletons in Figure 8.21 should convince the reader that the circle is actually slightly elliptical and that the pentagon and star are not regular.

The skeleton is a useful representation of the object morphology, as it provides both topological information and numerical metrics which can be used for comparison and categorization. The topology is essentially encapsulated in the number of nodes (where branches meet) and number of end points and the metric information is provided by the lengths of the branches and the angles between them.

Figure 8.21 Some shapes and their corresponding morphological skeletons

A weakness of the skeleton as a representation of a shape is that it is sensitive (sometimes highly) to small changes in the morphology. Slight irregularities in a boundary can lead to spurious 'spurs' in the skeleton which can interfere with recognition processes based on the topological properties of the skeleton. So-called *pruning* can be carried out to remove spurs of less than a certain length, but this is not always effective as small perturbations in the boundary of an image can sometimes lead to large spurs in the skeleton.

Actual computation of a skeleton can proceed in a number of ways. The preferred method is based on an iterative thinning algorithm whose details we do not repeat here. As well its use in a variety of image segmentation tasks, skeletonization has been usefully applied in a number of medical applications, such as unravelling the colon and assessment of laryngotracheal stenosis.

8.15 Opening by reconstruction

One simple but useful effect of morphological opening (erosion followed by dilation) is the removal of small unwanted objects. By choosing a structuring element of a certain size, erosion with this element guarantees the removal of any objects within which that structuring element cannot be contained. The second step of dilation with the same structuring element acts to restore the surviving objects to their original dimensions. However, if we consider an arbitrary shape, opening will not *exactly* maintain the shape of the primary objects except in the simplest and most fortuitous of cases. In general, the larger the size of the structuring element and the greater the difference in shape between structuring element and object, the larger the error in the restored shape will be. The uneven effect on an object of erosion or dilation with a structuring element whose shape is different from the object itself is referred to as the *anisotropy* effect. The examples in Figure 8.22 (generated by Example 8.11) illustrate this effect.

In contrast to morphological opening, *opening by reconstruction* is a morphological transformation which enables the objects which survive an initial erosion to be *exactly* restored to their original shape. The method is conceptually simple and requires two images which are called the *marker* and the *mask*. The mask is the original binary image. The *marker image* is used as the starting point of the procedure and is, in many cases, the image obtained

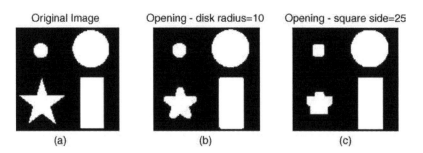

| Original Image | Opening - disk radius=10 | Opening - square side=25 |
| (a) | (b) | (c) |

Figure 8.22 Effects of morphological opening. (a) Original binary image. (b) Result of opening using a circular structuring element of radius 10. Note the rounding on the points of the star and the corners of the rectangle. (c) Result of opening using a square structuring element of side length 25. Only the rectangle is properly restored, the other shapes are significantly distorted

---

Example 8.11

Matlab code	What is happening?
A=imread('open_shapes.png');	%Read in image
se=strel('disk',10); bw=imopen(A,se);	%Open with disk radius 10
subplot(1,3,1), imshow(A);	%Display original
title('Original Image');	
subplot(1,3,2), imshow(bw);	%Display opened image
title('Opening - disk radius=10');	
se=strel('square',25); bw=imopen	%Open with square side 25
(A,se);	
subplot(1,3,3), imshow(bw);	%Display opened image
title('Opening - square side=25');	

---

after an initial erosion of the original image. The marker image is then iteratively dilated using an elementary 3×3 structuring element with the condition that the output image at each iteration is given by the intersection (logical AND) of the marker image with the mask. In this way, *the mask constrains the marker*, never allowing foreground pixels to appear which were not present in the original image. When the output image which results from a given iteration is the same as the image resulting from the previous iteration, this indicates that the dilation is having no further effect and all objects which survived the initial erosion have been restored to their original state. The procedure then terminates. Formally, we can describe this procedure by the following simple algorithm:

- denote the marker image by A and the mask image by M;

- define a 3×3 structuring element of $1s = B$;

- iteratively apply $A_{n+1} = (A_n \oplus B) \cap M$;

- when $A_{n+1} = A_n$, stop.

Example 8.12 and Figure 8.23 illustrate the use of morphological reconstruction. We preserve all alphabetic characters which have a long vertical stroke in a printed text sequence whilst completely removing all others.

Example 8.12

Matlab code	What is happening?
mask=~imread('shakespeare.pbm');	%Read in binary text
mask=imclose(mask,ones(5));	%Close to bridge breaks in letters
se=strel('line',40,90);	%Define vertical se length 40
marker=imerode(mask,se);	%Erode to eliminate characters
im=imreconstruct(marker,mask);	%Reconstruct image

```
subplot(3,1,1), imshow(~mask);
    title('Original mask Image');
subplot(3,1,2), imshow(~marker);
    title('marker image');
subplot(3,1,3), imshow(~im);
    title('Opening by reconstruction');
```

Comments

Opening by reconstruction is implemented in the Matlab Image Processing Toolbox through the function *imreconstruct*.

Note that the "~" operator in this example is the logical NOT (inversion) operator being applied to binary images.

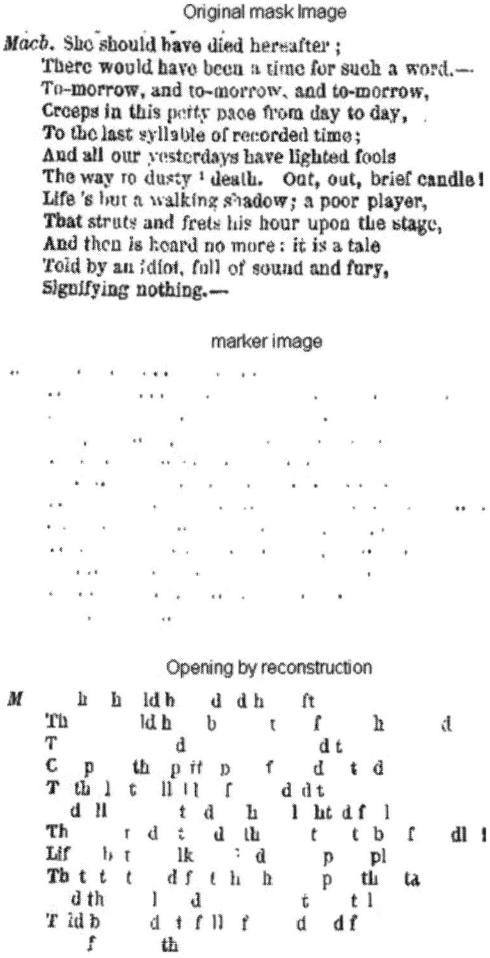

Original mask Image

Macb. She should have died hereafter ;
    There would have been a time for such a word.—
    To-morrow, and to-morrow, and to-morrow,
    Creeps in this petty pace from day to day,
    To the last syllable of recorded time;
    And all our yesterdays have lighted fools
    The way to dusty ¹ death. Out, out, brief candle !
    Life 's but a walking shadow; a poor player,
    That struts and frets his hour upon the stage,
    And then is heard no more: it is a tale
    Told by an idiot, full of sound and fury,
    Signifying nothing.—

marker image

Opening by reconstruction

Figure 8.23 Opening by reconstruction. Top: the original text – the mask image. Middle: the text after erosion by a long, vertical structuring element. This acts as the marker image. Bottom: the image after opening by reconstruction. Only characters with long vertical strokes have been restored

8.16   Grey-scale erosion and dilation

To date, our discussion has only considered the application of morphology to binary images. The same basic principles of morphological processing can, however, be extended to intensity[4] images. Just as in the binary case, the building blocks of grey-scale morphology are the fundamental operations of erosion and dilation. For intensity images, the definitions of these operators are subtly different and can, in principle, be defined in a variety of ways. However, in practice, the following (informal) definitions are more or less universal.

Grey-scale erosion of image A by structuring element B is denoted $A \ominus B$ and the operation may be described as follows:

- successively place the structuring element B over each location in the image A;

- for each location, select the *minimum* value of $A - B$ occurring within the local neighbourhood defined by the structuring element B.

Grey-scale dilation of image A by structuring element B is denoted $A \oplus B$ and the operation may be described as follows:

- successively place the structuring element B over each location in the image;

- for each location, select the *maximum* value of $A + B$ occurring within the local neighbourhood defined by the structuring element B.

8.17   Grey-scale structuring elements: general case

Where grey-scale morphology is being considered, structuring elements have, in the most general case, two parts:

(1)   An array of 0s and 1s, where the 1s indicate the domain or local neighbourhood defined by the structuring element. We denote this by b.

(2)   An array of identical size containing the actual numerical values of the structuring element. We denote this by v_b.

To illustrate this general form of grey-scale erosion and dilation, consider the simple example in Example 8.13.

---

[4] Colour images can also be morphologically processed. Since these simply comprise three 2-D intensity planes (i.e. the three R,G and B colour channels), our discussion of grey-scale morphology will apply to these images too.

Example 8.13

A structuring element comprising a domain

$$b = \begin{array}{ccc} 0 & 1 & 0 \\ 1 & 1 & 1 \\ 0 & 1 & 0 \end{array}$$

and values

$$v_b = \begin{array}{ccc} 0 & 3 & 0 \\ 3 & 1 & 3 \\ 0 & 3 & 0 \end{array}$$

is placed on a section of an image given by

$$A = \begin{array}{ccc} 12 & 13 & 11 \\ 7 & 14 & 8 \\ 10 & 9 & 10 \end{array}$$

Consider positioning the central pixel of the structuring element on the centre pixel of the image segment (value = 14). What is the value after erosion?

Solution
The value of this pixel after grey-scale erosion is the minimum of the values $A - v_b$ over the domain defined by

$$b = \min \left\{ \begin{array}{ccc} - & 13{-}3 & - \\ 7{-}3 & 14{-}1 & 8{-}3 \\ - & 9{-}3 & - \end{array} \right\} = \min \left\{ \begin{array}{ccc} - & 10 & - \\ 4 & 13 & 5 \\ - & 6 & - \end{array} \right\} = 4$$

8.18 Grey-scale erosion and dilation with flat structuring elements

Although grey-scale morphology can use structuring elements with two parts defining the neighbourhood b and height values v_b respectively, it is very common to use *flat* structuring elements. Flat structuring elements have height values which are all zero and are thus specified entirely by their neighbourhood.

When a flat structuring element is assumed, grey-scale erosion and dilation are equivalent to *local minimum and maximum* filters respectively. In other words, erosion with a flat element results in each grey-scale value being replaced by the minimum value in the vicinity defined by the structuring element neighbourhood. Conversely, dilation with a flat element results in each grey-scale value being replaced by the maximum. Figure 8.24 shows how grey-scale erosion and dilation may be used to calculate a so-called

(a) (b) (c)

Figure 8.24 Calculating the morphological gradient: (a) grey-scale dilation with flat 3×3 structuring element; (b) grey-scale erosion with flat 3×3 structuring element; (c) difference of (a) and (b) = morphological gradient

morphological image gradient. The Matlab code in Example 8.14 was used to produce Figure 8.24.

Example 8.14

Matlab code	**What is happening?**
A=imread('cameraman.tif');	%Read in image
se=strel(ones(3));	%Define flat structuring element
Amax=imdilate(A,se);	%Grey-scale dilate image
Amin=imerode(A,se);	%Grey-scale erode image
Mgrad=Amax-Amin;	%subtract the two
subplot(1,3,1), imagesc(Amax); axis image; axis off;	%Display
subplot(1,3,2), imagesc(Amin); axis image; axis off;	
subplot(1,3,3), imagesc(Mgrad); axis image; axis off;	
colormap(gray);	

It is easy to see why the morphological gradient works: replacing a given pixel by the minimum or maximum value in the local neighbourhood defined by the structuring element will effect little or no change in smooth regions of the image. However, when the structuring element spans an edge, the response will be the difference between the maximum and minimum-valued pixels in the defined region and, hence, large. The thickness of the edges can be tailored by adjusting the size of the structuring elements if desired.

8.19 Grey-scale opening and closing

Grey-scale opening and closing are defined in exactly the same way as for binary images and their effect on images is also complementary. Grey-scale opening (erosion followed by dilation) tends to suppress small bright regions in the image whilst leaving the rest of the image relatively unchanged, whereas closing (dilation followed by erosion) tends to suppress small dark regions. We can exploit this property in the following example.

(a)   (b)   (c)   (d)

Figure 8.25 Correction of nonuniform illumination through morphological opening. Left to right: (a) original image; (b) estimate of illumination function by morphological opening of original; (c) original with illumination subtracted; (d) contrast-enhanced version of image (c)

Example 8.15

Matlab code	What is happening?
I = imread('rice.png');	%Read in image
background = imopen(I,strel('disk',15));	%Opening to estimate background
I2 = imsubtract(I,background);	%Subtract background
I3 = imadjust(I2);	%Improve contrast

subplot(1,4,1), imshow(I);subplot(1,4,2), imshow(background);
subplot(1,4,3), imshow(I2);subplot(1,4,4), imshow(I3);

Comments
imerode, imdilate, imclose and *imopen* may be used for both binary and grey-scale images.

A relatively common problem in automated inspection and analysis applications is for the field to be unevenly illuminated. This makes segmentation (the process of separating objects of interest from their background) more difficult. In Figure 8.25 (produced using the Matlab code in Example 8.15), in which the objects/regions of interest are of similar size and separated from one another, opening can be used quite effectively to estimate the illumination function.

8.20 The top-hat transformation

The top-hat transformation is defined as the difference between the image and the image after opening with structuring element b, namely $I - I \ominus b$. Opening has the general effect of removing small light details in the image whilst leaving darker regions undisturbed. The difference of the original and the opened image thus tends to lift out the local details of the image *independently of the intensity* variation of the image as a whole. For this reason, the top-hat transformation is useful for uncovering detail which is rendered invisible by

Figure 8.26 Morphological top-hat filtering to increase local image detail. Left to right: (a) original image; (b) after application of top-hat filter (circular structuring element of diameter approximately equal to dimension of grains); (c) after contrast enhancement

illumination or shading variation over the image as a whole. Figure 8.26 shows an example in which the individual elements are enhanced. The Matlab code corresponding to Figure 8.26 is given in Example 8.16.

Example 8.16

Matlab code	What is happening?
A = imread('rice.png');	%Read in unevenly illuminated image
se = strel('disk',12);	%Define structuring element
Atophat = imtophat(original,se);	%Apply tophat filter
subplot(1,3,1), imshow(A);	%Display original
subplot(1,3,2), imshow(Atophat);	%Display raw filtered image
B = imadjust(tophatFiltered);	%Contrast adjust filtered image
subplot(1,3,3), imshow(B);	%Display filtered and adjusted mage

Comments
imtophat is a general purpose function for top hat filtering and may be used for both binary and grey-scale images.

8.21 Summary

Notwithstanding the length of this chapter, we have presented here a relatively elementary account of morphological processing aimed at conveying some of the core ideas and methods. The interested reader can find a large specialist literature on the use of morphology in image processing. The key to successful application lies largely in the judicious combination of the standard morphological operators and transforms and intelligent choice of structuring elements; detailed knowledge of how specific transforms work is often not needed. To use an old analogy, it is possible to construct a variety of perfectly good houses

Table 8.1 A summary of some important morphological operations and corresponding Matlab functions

Operation	Matlab function	Description	Definition Image A Structuring element B
Erosion	*imerode*	If any foreground pixels in neighbourhood of structuring element are background, set pixel to background	$A \ominus B$
Dilation	*imdilate*	If any background pixels in neighbourhood of structuring element are foreground, set pixel to foreground	$A \oplus B$
Opening	*imopen*	Erode image and then dilate the eroded image using the same structuring element for both operations	$A \circ B = (A \ominus B) \oplus B$
Closing	*imclose*	Dilate image and then erode the dilated image using the same structuring element for both operations	$A \bullet B = (A \oplus B) \ominus B$
Hit-or-miss	*bwhitmiss*	Logical AND between (i) the image eroded with one structuring element and (ii) the image *complement* eroded with a second structuring element	$A \otimes B = (A \ominus B_1) \cap (A \ominus B_2)$
Top Hat	*imtophat*	Subtracts a morphologically opened image from the original image	$A - (A \circ B) \oplus B$
Bottom Hat	*imbothat*	Subtracts the original image from a morphologically closed version of the image	$[(A \oplus B) \ominus B] - A$
Boundary extraction	*bwperim*	Subtracts eroded version of the image from the original	$A - (A \ominus B)$
Connected components labelling	*bwlabel*	Finds all connected components and labels them distinctly	See Section 8.10
Region filling	*imfill*	Fills in the holes in the image	See Section 8.11
Thinning	*bwmorph*	Subtracts the hit-or-miss transform from the original image.	$thin(A, B) = A \cap \overline{A \otimes B}$
Thickening	*bwmorph*	The original image plus additional foreground pixels switched on by the hit-and-miss transform	$thicken(A, B) = A \cup A \otimes B$
Skeletonization	*bwmorph*		See Section 8.14

without knowing how to manufacture bricks. With this in mind, we close this chapter with Table 8.1, which gives a summary of the standard morphological transforms and a reference to the appropriate function in the Matlab Image Processing Toolbox.

Exercises

Exercise 8.1 Consider the original image in Figure 8.4 and the following structuring element:

$$\begin{matrix} 1 & 0 & 0 \\ 0 & 1 & 0 \\ 0 & 0 & 1 \end{matrix}$$

Sketch the result of erosion with this structuring element.
Sketch the result of dilation with this structuring element.

Exercise 8.2 Consider the original image in Figure 8.4 and the following structuring element:

$$\begin{matrix} 1 & 1 & 1 \\ 1 & 1 & 1 \\ 1 & 1 & 1 \end{matrix}$$

What is the result of first eroding the image with this structuring element and then dilating the eroded image with the same structuring element (an operation called *closing*)?
What is the result of first dilating the image with this structuring element and then eroding the dilated image with the same structuring element (an operation called *opening*)?

Exercise 8.3 Consider the following structuring element:

$$\begin{matrix} 1 & 1 & 1 & 1 \\ 1 & 1 & 1 & 1 \\ 1 & 1 & 1 & 1 \\ 1 & 1 & 1 & 1 \end{matrix}$$

- Read in and display the Matlab image 'blobs.png'.

- Carry out erosion using this structuring element directly on the image (do not use the *strel* function). Display the result.

- Find the decomposition of this structuring element using *strel*. Explicitly demonstrate that successive erosions using the decomposition produce the same result as erosion with the original structuring element.

Exercise 8.4 Based on the geometric definitions of the skeleton given, what form do the skeletons of the following geometric figures take:

(1) A perfect circle?

(2) A perfect square?

(3) An equilateral triangle?

Exercise 8.5 Using the same structuring element and image segment as in Example 8.13, show that

$$A \ominus B = \begin{matrix} 4 & 8 & 5 \\ 6 & 4 & 7 \\ 4 & 7 & 5 \end{matrix} \quad \text{and that} \quad A \oplus B = \begin{matrix} 16 & 17 & 16 \\ 17 & 16 & 17 \\ 12 & 17 & 12 \end{matrix}$$

(Remember that, just as with binary images, if the structuring element goes outside the bounds of the image we consider only those image pixels which intersect with the structuring element.)

For further examples and exercises see http://www.fundipbook.com

9

Features

In this chapter we explore some techniques for the representation and extraction of image features. Typically, the aim is to process the image in such a way that the image, or properties of it, can be adequately represented and extracted in a compact form amenable to subsequent recognition and classification. Representations are typically of two basic kinds: internal (which relates to the pixels comprising a region) or external (which relates to the boundary of a region). We will discuss shape-based representations (external), features based on texture and statistical moments (internal) and end the chapter with a detailed look at principal component analysis (PCA), a statistical method which has been widely applied to image processing problems.

9.1 Landmarks and shape vectors

A very basic representation of a shape simply requires the identification of a number N of points along the boundary of the features. In two dimensions, this is simply

$$\mathbf{x} = \begin{bmatrix} x_1 & y_1 & x_2 & \cdots & x_N & y_N \end{bmatrix}$$

Two basic points about the shape vector representation should be immediately stressed:

- In practice, we are restricted to a finite number of points. Therefore, we need to ensure that a sufficient number of points are chosen to represent the shape to the accuracy we require.

- Selection of the coordinate pairs which constitute the shape vector must be done in a way that ensures (as far as possible) good agreement or *correspondence*. By correspondence, we mean that the selected coordinates constituting the boundary/shape must be chosen in such a way as to ensure close agreement across a group of observers. If this were not satisfied, then different observers would define completely different shape vectors for what is actually the same shape. Correspondence is achieved by defining criteria for identifying appropriate and well-defined key points along the boundaries. Such key points are called *landmarks*.

Fundamentals of Digital Image Processing – A Practical Approach with Examples in Matlab
Chris Solomon and Toby Breckon
© 2011 John Wiley & Sons, Ltd

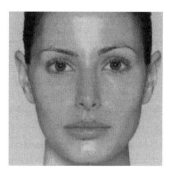

Figure 9.1 Anatomical landmarks (indicated in red) are located at points which can be easily identified visually. Mathematical landmarks (black crosses) are identified at points of zero gradient and maximum corner content. The pseudo-landmarks are indicated by green circles (*See colour plate section for colour version*)

Landmarks essentially divide into three categories:

- *Mathematical landmarks* These correspond to points located on an object according to some mathematical or geometric property (e.g. according to local gradient, curvature or some extremum value). 'Interest' detectors, such as those devised by Harris and Haralick (discussed in Chapter 10) combined with thresholding constitute a good example of a mathematical criterion for the calculation of landmarks.

- *Anatomical or true landmarks* These are points assigned by an expert which are identifiable in (and correspond between) different members of a class of objects. Examples of anatomical landmarks are the corners of the mouth and eyes in human faces.[1]

- *Pseudo-landmarks* These are points which are neither mathematically nor anatomically completely well defined, though they may certainly be intuitively meaningful and/ or require some skilled judgment on the part of an operator. An example of a pseudo-landmark might be the 'centre of the cheek bones' in a human face, because the visual information is not definite enough to allow unerring placement. Despite their 'second-rank' status, pseudo-landmarks are often very important in practice. This is because true landmarks are often quite sparse and accurate geometric transformations (discussed in the geometry chapter) usually require correspondence between two similar (but subtly different) objects at a large and *well-distributed* set of points over the region of interest. In many cases, pseudo-landmarks are constructed as points between mathematical or anatomical landmarks. Figure 9.1 depicts an image of a human face in which examples of all three kinds of landmark have been placed.

Landmarks[2] can also be defined in an analogous way for 3-D shapes or (in principle) even in abstract higher dimensional 'objects'.

---

[1] 'Anatomical' landmarks can be defined for any pattern class of objects. The term is not confined to strictly biological organisms.

[2] Labelling images with anatomical landmarks can become very tedious and is subject to a certain degree of uncontrollable error. Accordingly, one of the big challenges of current image processing/computer vision research is to develop methods which can *automate* the procedure of finding landmarks.

Table 9.1 Some common, single-parameter descriptors for approximate shape in 2-D

Measure	Definition	Circle	Square	Rectangle as $b/a \to \infty$
Form factor	$\dfrac{4\pi \times \text{Area}}{\text{Perimeter}^2}$	1	$\pi/4$	$\to 0$
Roundness	$\dfrac{4 \times \text{Area}}{\pi \times \text{MaxDiameter}^2}$	1	$2/\pi$	$\to 0$
Aspect ratio	$\dfrac{\text{MaxDiameter}}{\text{MinDiameter}}$	1	1	$\to \infty$
Solidity	$\dfrac{\text{Area}}{\text{ConvexArea}}$	1	1	$\to 0$
Extent	$\dfrac{\text{TotalArea}}{\text{Area Bounding Rectangle}}$	$\pi/4$	1	indeterminate
Compactness	$\dfrac{\sqrt{(4 \times \text{Area})/\pi}}{\text{MaxDiameter}}$	1	$\sqrt{2/\pi}$	$\to 0$
Convexity	$\dfrac{\text{Convex Perimeter}}{\text{Perimeter}}$	1	1	1

9.2 Single-parameter shape descriptors

In certain applications, we do not require a shape analysis/recognition technique to provide a precise, regenerative representation of the shape. The aim is simply to characterize a shape as succinctly as possible in order that it can be differentiated from other shapes and classified accordingly. Ideally, a small number of descriptors or even a single parameter can be sufficient to achieve this task. Differentiating between a boomerang and an apple, a cigar and a cigarette can intuitively be achieved by quite simple methods, a fact reflected in our everyday language ('long and curved', 'roughly spherical', 'long, straight and thin', etc).

Table 9.1 gives a number of common, single-parameter measures of approximate shape which can be employed as shape features in basic tasks of discrimination and classification. Note that many of these measures can be derived from knowledge of the perimeter length, the extreme x and y values and the total area enclosed by the boundary – quantities which can be calculated easily in practice. All such measures of shape are a dimensionless combination of size parameters which, in many cases, exploit the variation in perimeter length to surface area as shape changes. Meaningful reference values for such measures are usually provided by the circle or square. Table 9.1 gives these measures for three basic shapes: the circle, the square and the limiting case of a rectangle of length a and side b when $b/a \to \infty$.

Figure 9.2 (produced using the Matlab® code in Example 9.1) shows an image on the left containing a number of metallic objects: a key, a drill bit and a collection of coins. The coins are actually divisible into two basic shapes: those which are circular (1 and 2 pence coins) and those which are heptagonal (the 20 pence and 50 pence coins). The centre image in Figure 9.2 shows the processed binary objects after thresholding and basic morphological processing

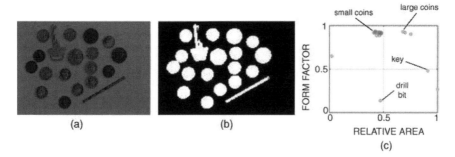

Figure 9.2 From left to right: (a) original image; (b) binary image after thresholding and morphological processing; (c) the normalized area and the form factor of each object in (b) are plotted in a 2-D feature space

and the image on the right plots their associated areas and form factors. The key and drill bit are easily distinguished from the coins by their form-factor values. However, the imperfect result from thresholding means that the form factor is unable to distinguish between the heptagonal coins and circular coins. These are, however, easily separated by area.

Example 9.1

Matlab code	**What is happening?**
A=imread('coins_and_keys.png');	%Read in image and display
subplot(1,2,1), imshow(A);	
bw=~im2bw(rgb2gray(A),0.35); bw=imfill(bw,'holes');	%Threshold and fill in holes
bw=imopen(bw,ones(5)); subplot(1,2,2), imshow(bw,[0 1]);	%Morphological opening
[L,num]=bwlabel(bw);	%Create labelled image
s=regionprops(L,'area','perimeter');	%Calculate region properties
for i=1:num	%Object's area and perimeter
x(i)=s(i).Area;	
y(i)=s(i).Perimeter;	
form(i)=4.*pi.*x(i)./(y(i).^2);	%Calculate form factor
end	
figure; plot(x./max(x),form,'ro');	%Plot area against form factor

Comments

In this example, intensity thresholding produces a binary image containing a variety of objects of different shapes. The ***regionprops*** function is then used to calculate the area and perimeter of each of the objects in the binary object. ***regionprops*** is a powerful in-built function called which can calculate a number of basic and useful shape-related measures. It requires as input a so-called *labelled image* (a matrix in which all pixels belonging to object 1 have value 1, all pixels belonging to object 2 have value 2 and so on). This is generated through use of the Matlab function ***bwlabel***. The form factor is then calculated and plotted against the object area.

   This example also uses the morphological functions ***imfill*** (to fill in the holes in the objects) and ***imopen*** (to remove any small isolated objects), both of which result from imperfect thresholding. These and other functions are discussed more fully in Chapter 8.

9.3 Signatures and the radial fourier expansion

A *signature* is the representation of a 2-D boundary as a 1-D function. Typically, this is achieved by calculating the centroid of the given shape (as defined by the boundary coordinates) and plotting the distance from the centroid to the boundary r as a function of polar angle θ. The signature then of a *closed* boundary is a periodic function, repeating itself on an angular scale of 2π. Thus, one simple and neat way to encode and describe a closed boundary to arbitrary accuracy is through a 1-D (radial) Fourier expansion. The signature can be expressed in real or complex form as follows:

$$r(\theta) = \frac{a_0}{2} + \sum_{n=1}^{\infty} a_n \cos(n\theta) + \sum_{n=1}^{\infty} b_n \sin(n\theta) \quad \text{(real form)}$$

$$r(\theta) = \sum_{n=-\infty}^{\infty} c_n \exp(in\theta) \quad \text{(complex form)}$$

(9.1)

and the shape can be parametrically encoded by the real Fourier expansion coefficients $\{a_n, b_n\}$ or by the complex coefficients $\{c_n\}$.[3] These coefficients can easily be calculated through use of the orthogonality relations for Fourier series (see Chapter 5). The real coefficients for a radial signature are given by

$$a_n = \frac{1}{\pi} \int_{-\pi}^{\pi} r(\theta)\cos(n\theta)\, d\theta$$

$$b_n = \frac{1}{\pi} \int_{-\pi}^{\pi} r(\theta)\sin(n\theta)\, d\theta$$

(9.2)

whereas the complex coefficients $\{c_n\}$ are given by

$$c_n = \frac{1}{2\pi} \int_{-\pi}^{\pi} r(\theta)\exp(-in\theta)\, d\theta$$

(9.3)

Typically, a good approximation to the shape can be encoded using a relatively small number of parameters, and more terms can be included if higher accuracy is required. Figure 9.3 (produced using the Matlab code in Example 9.2) shows how a relatively complex boundary (the Iberian Peninsula, comprising Spain and Portugal) can be increasingly well approximated by a relatively modest number of terms in the Fourier expansion.

Example 9.2

Matlab code	What is happening?
A=imread('spain.png'); iberia=logical(A);	%Read in image convert to binary
ibbig=imdilate(iberia,ones(3));	%Calculate boundary pixels
bound=ibbig-iberia;	
[i,j]=find(bound>0); xcent=mean(j);	%Calculate centroid
ycent=mean(i);	
hold on; plot(xcent,ycent,'ro');	

---

[3] Note that for a digital boundary specified by N points, the order of the expansion n cannot exceed N.

```
subplot(4,2,1),imagesc(bound); axis image;
axis on; colormap(gray);
                                                    %Plot perimeter
[th,r]=cart2pol(j-xcent,i-ycent);                   %Convert to polar coordinates
subplot(4,2,2); plot(th,r,'k.'); grid on; axis off;  %Plot signature

                                                    %Calculate Fourier series boundary
N=0;L=2.*pi;f=r;M=length(f);                        %N = 0 – DC term only
[a,b,dc]=fseries_1D(f,L,N);                         %Calculate expansion coeffs
fapprox=fbuild_1D(a,b,dc,M,L);                      %Build function using the coeffs
[x,y]=pol2cart(th,fapprox);
x=x+xcent; x=x-min(x)+1; y=y+ycent;                 %Convert back to Cartesian coordinates
y=y-min(y)+1;
prm=zeros(round(max(y)),round(max(x)));
i=sub2ind(size(prm),round(y),round(x)); prm(i)=1;
subplot(4,2,3); imagesc(prm); axis image;
axis ij; axis on; colormap(gray);
                                                    %Display 2-D boundary
subplot(4,2,4); plot(th,fapprox,'k.'); axis off;    %Display corresponding signature
s=sprintf('Approximation %d terms',N); title(s);

N=5;L=2.*pi;f=r;M=length(f);                        %Repeat for N = 5 terms
[a,b,dc]=fseries_1D(f,L,N);                         %Calculate expansion coeffs
fapprox=fbuild_1D(a,b,dc,M,L);                      %Build function using the coeffs
[x,y]=pol2cart(th,fapprox);
x=x+xcent; x=x-min(x)+1; y=y+ycent;
y=y-min(y)+1;
prm=zeros(round(max(y)),round(max(x)));
i=sub2ind(size(prm),round(y),round(x)); prm(i)=1;
subplot(4,2,5); imagesc(prm); axis image;
axis ij; axis off; colormap(gray);
                                                    %Display 2-D boundary
subplot(4,2,6); plot(th,fapprox,'k.'); axis off;    %Display corresponding signature
s=sprintf('Approximation %d terms',N); title(s);

N=15;L=2.*pi;f=r;M=length(f);                       %Repeat for N = 15 terms
[a,b,dc]=fseries_1D(f,L,N);                         %Calculate expansion coeffs
fapprox=fbuild_1D(a,b,dc,M,L);                      %Build function using the coeffs
[x,y]=pol2cart(th,fapprox);                         %Convert back to Cartesian coordinates
x=x+xcent; x=x-min(x)+1; y=y+ycent; y=y-min(y)+1;
prm=zeros(round(max(y))+10,round(max(x))+10);
i=sub2ind(size(prm),round(y),round(x)); prm(i)=1;
subplot(4,2,7); imagesc(prm); axis image;
axis ij; axis off; colormap(gray);
                                                    %Display 2-D boundary
subplot(4,2,8); plot(th,fapprox,'k.'); axis off;    %Display corresponding signature
s=sprintf('Approximation %d terms',N); title(s);
```

Comments

Calculates the Fourier expansion of the radial signature of the Iberian landmass using DC only, 5 and 15 terms. This uses two user-defined Matlab functions **fbuild_1D** and **fseries_1D** available on the book's website http://www.fundipbook.com/materials/.

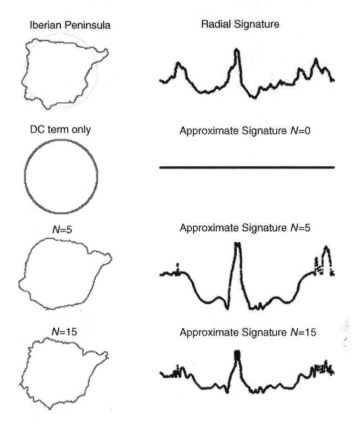

Figure 9.3   An example of the use of the radial Fourier expansion technique to approximate the shape of the Iberian Peninsula using $N = 0$, 5 and 15 terms

The use of radial Fourier expansions can, however, become problematic on complicated boundary shapes, particularly those in which the boundary 'meanders back' on itself (see Figure 9.4). The signature function $r(\theta)$ may not be single valued, there being two or more possible radial values for a given value of θ. In such cases, the choice of which value of $r(\theta)$ to select is somewhat arbitrary and the importance of these unavoidable ambiguities will depend on the specific application. In general, however, strongly meandering boundaries, such as the second example depicted in Figure 9.4, are not suitable for radial Fourier expansion.

As discussed in Chapter 7 on geometry, a robust shape descriptor should be invariant to translation, scaling and rotation. The Fourier descriptors calculated according to Equations (9.2) and (9.3) are certainly translation invariant. This follows because the radial distance in the signature is calculated with respect to an origin defined by the centroid coordinates of the boundary.

Turning to the question of scale invariance, we note the linear nature of the expansion given by Equation (9.1). Multiplication of the signature by an arbitrary scale factor is reflected in the same scale factor multiplying each of the individual Fourier coefficients. A form of scale invariance can thus be achieved most simply by dividing the signature by its maximum value (thus fixing its maximum value as one). Finally, rotational invariance can

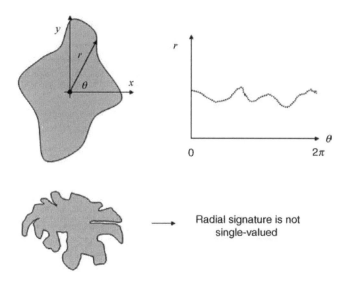

Figure 9.4 The signature is given by the distance from the centroid to the boundary as a function of polar angle. For complex, boundaries (second example) it will not, in general, be single-valued

also be achieved relatively simply as follows. First, note that, for a radial signature, rotational invariance is the same as start-point invariance. In other words, we need the same set of descriptors for a shape irrespective of what point on the boundary we choose to define as zero angle. Consider, then, a radial signature $r(\theta)$ described by a set of complex Fourier coefficients $\{c_n\}$ according to Equation (9.3). Rotation of this shape by some positive, arbitrary angle θ' results in a radial signature $r(\theta - \theta')$. The complex Fourier coefficients for this function, which we distinguish from the unrotated version as $\{c'_n\}$, can be calculated using Equation (9.3) as

$$c'_n = \frac{1}{2\pi} \int_{-\pi}^{\pi} r(\theta - \theta')\exp(-in\,\theta)\,d\theta \tag{9.4}$$

Making the change of variable $\theta'' = \theta - \theta'$ and substituting in Equation (9.4) gives

$$c'_n = \exp(-in\theta')\frac{1}{2\pi} \int_{-\pi}^{\pi} r(\theta'')\exp(-in\theta'')\,d\theta''$$

$$= \exp(-in\theta')c_n \tag{9.5}$$

and we see that the coefficients of the rotated shape differ only by a multiplicative phase factor $\exp(-in\theta')$. A rotation-invariant set of descriptors can thus easily be obtained by taking the modulus squared of the coefficients: $\{|c_n|^2 = |c'_n|^2\}$. Such a parameterization can form an effective way of comparing and distinguishing between different shapes for purposes of recognition and classification. The shape itself, however, must clearly be regenerated using the original complex values $\{c'_n\}$.

9.4 Statistical moments as region descriptors

The concept of moments forms an important part of elementary probability theory. If we have a probability density function $p(x)$ which describes the distribution of the random variable x, then the nth moment is defined as

$$m_n = \int_{-\infty}^{\infty} x^n p(x) dx \qquad (9.6)$$

The zeroth moment $m_0 = \int_{-\infty}^{\infty} p(x) dx$ gives the total area under the function $p(x)$, and is always equal to unity if $p(x)$ is a true probability density function. The first moment, $\mu = m_1 = \int_{-\infty}^{\infty} x p(x) dx$, corresponds to the mean value of the random variable.

The *central* moments of the density function describe the variation about the mean and are defined as

$$M_n = \int_{-\infty}^{\infty} (x - \mu)^n p(x) \, dx \qquad (9.7)$$

The most common central moment, $M_2 = \int_{-\infty}^{\infty} (x - \mu)^2 p(x) dx$, is the well-known *variance* and forms the most basic measure of how 'spread out' the density function is. Higher order moments can yield other information on the shape of the density function, such as the skewness (the tendency to shoot further on one side of the mean than the other). An important theorem of probability theory states that knowledge of all the moments uniquely determines the density function. Thus, we can understand that the moments collectively encode information on the shape of the density function.

Moments extend naturally to 2-D (and higher dimensional) functions. Thus, the pqth moment of a 2-D density function $p(x, y)$ is given by

$$m_{pq} = \int_{-\infty}^{\infty} \int_{-\infty}^{\infty} x^p y^q p(x, y) dx dy \qquad (9.8)$$

We calculate moments of images, however, by replacing the density function with the intensity image $I(x, y)$ or, in some cases, a corresponding binary image $b(x, y)$. The intensity image is thus viewed as an (unnormalized) probability density which provides the likelihood of a particular intensity occurring at location $I(x, y)$.

Let us also note at this juncture that if the image in question is binary then the moments directly encode information about the *shape*.

Thus, in a completely analogous way to the 1-D case, we can define the $(p - q)$th central moment of our 2-D shape $I(x, y)$ as

$$M_{pq} = \int_{-\infty}^{\infty} \int_{-\infty}^{\infty} (x - \mu_x)^p (y - \mu_y)^q I(x, y) \, dx \, dy \qquad (9.9)$$

Since the central moments are measured with respect to the centroid of the shape, it follows that they are necessarily translation invariant. In general, however, we require shape descriptors which will not change when the shape is scaled and/or rotated, i.e. that are also scale and rotation invariant. It is beyond the scope of our immediate discussion to offer a proof, but it is possible to show that *normalized central moments* possess scale invariance.

The $(p - q)$th normalized central moment is defined as

$$\eta_{pq} = \frac{M_{pq}}{M_{00}^{\beta}} \qquad \text{where } \beta = \frac{p+q}{2} + 1 \text{ and } p + q \geq 2 \qquad (9.10)$$

From these normalized central moments, it is possible to calculate seven derived quantities attributed to Hu (also referred to as moments) which are invariant to translation, scale and rotation:

$$\Lambda_1 = \eta_{20} + \eta_{02}$$
$$\Lambda_2 = (\eta_{20} - \eta_{02})^2 + 4\eta_{11}^2$$
$$\Lambda_3 = (\eta_{30} - 3\eta_{12})^2 + (3\eta_{21} - \eta_{03})^2$$
$$\Lambda_4 = (\eta_{30} + \eta_{12})^2 + (\eta_{21} + \eta_{30})^2$$
$$\Lambda_5 = (\eta_{30} - 3\eta_{12})(\eta_{30} + \eta_{12})[(\eta_{30} + \eta_{12})^2 - 3(\eta_{21} - \eta_{03})^2] + (3\eta_{21} - \eta_{03})(\eta_{03} + \eta_{21})$$
$$\times [3(\eta_{30} + \eta_{12})^2 - (\eta_{03} + \eta_{21})^2]$$
$$\Lambda_6 = (\eta_{20} - \eta_{02})[(\eta_{12} + \eta_{30})^2 - (\eta_{21} + \eta_{03})^2] + 4\eta_{11}(\eta_{21} + \eta_{03})(\eta_{12} + \eta_{30})$$
$$\Lambda_7 = (3\eta_{21} - \eta_{03})(\eta_{30} + \eta_{12})[(\eta_{30} + \eta_{12})^2 - 3(\eta_{03} + \eta_{21})^2] + (3\eta_{21} - \eta_{30})(\eta_{21} + \eta_{03})$$
$$\times [3(\eta_{30} + \eta_{12})^2 - (\eta_{03} + \eta_{21})^2]$$

$$(9.11)$$

Figure 9.5 (produced using the Matlab code in Example 9.3) shows an object extracted from an image along with identically shaped objects which have been translated, scaled and rotated with respect to the first. The values of the first three invariant moments are calculated according to Equation (9.11) for all three shapes in code example 9.3. As is evident, they are equal to within the computational precision allowed by a discrete, digital implementation of the equations in Equation (9.11).

Figure 9.5   Example objects for Hu moment calculation in Example 9.3

Example 9.3

Matlab code	What is happening?

```
A=rgb2gray(imread('spanners.png'));      %Read in image, convert to grey
bwin=~im2bw(A,0.5);                       %Threshold and display
[L, num]=bwlabel(bwin);                   %Create labelled image
subplot (2,2,1), imshow(A);               %Display input image

for i=1:num                               %Loop through each labelled object
I=zeros(size(A));                         %array for ith object
ind=find(L==i); I(ind)=1;                 %Find pixels belonging to ith object and set=1
subplot(2, 2, i+1), imshow(I);            %Display identified object

                                          %I=double(bw)./(sum(sum(bw)));
[rows,cols]=size(I); x=1:cols;y=1:rows;   %get indices
[X,Y]=meshgrid(x,y);                      %Set up grid for calculation

                                          %calculate required ordinary moments
M_00=sum(sum(I));
M_10=sum(sum(X.*I)); M_01=sum(sum(Y.*I));
xav=M_10./M_00; yav=M_01./M_00;

X=X-xav; Y=Y-yav;                         %mean subtract the X and Y coordinates

hold on; plot(M_10,M_01,'ko'); drawnow
                                          %calculate required central moments
M_11=sum(sum(X.*Y.*I));
M_20=sum(sum(X.^2.*I)); M_02=sum(sum(Y.^2.*I));
M_21=sum(sum(X.^2.*Y.*I)); M_12=sum(sum(X.*Y.^2.*I));
M_30=sum(sum(X.^3.*I)); M_03=sum(sum(Y.^3.*I));
                                          %calculate normalised central moments
eta_11=M_11./M_00.^2;
eta_20=M_20./M_00.^2;
eta_02=M_02./M_00.^2;

eta_21=M_21./M_00.^(5./2);
eta_12=M_12./M_00.^(5./2);
eta_30=M_30./M_00.^(5./2);
eta_03=M_02./M_00.^(5./2);
                                          %calculate Hu moments
Hu_1=eta_20 + eta_02;
Hu_2=(eta_20 - eta_02).^2 + (2.*eta_11).^2;
Hu_3=(eta_30 - 3.*eta_12).^2 + (3.*eta_21 - eta_03).^2;

s=sprintf('Object number is %d', i)
s=sprintf('Hu invariant moments are %f %f %f',Hu_1,Hu_2,Hu_3)

pause;
end
```

Comments

This example takes an image containing three similar objects at different scale, location and rotation and calculates the first three Hu invariant moments.

9.5 Texture features based on statistical measures

In the simplest terms, texture is loosely used to describe the 'roughness' of something. Accordingly, texture measures the attempt to capture characteristics of the intensity fluctuations between groups of neighbouring pixels, something to which the human eye is very sensitive. Note that texture measures based on statistical measures must generally be defined with respect to a certain neighbourhood Ω which defines the local region over which the calculation is to be made. The simplest such measures are the range defined as

$$\Re_\Omega = \{\max(I(x,y)) - \min(I(x,y))\}_\Omega \qquad (9.12)$$

and the local variance defined as

$$\mathrm{Var}_\Omega = \{\langle I^2(x,y)\rangle - \langle I(x,y)\rangle^2\}_\Omega \qquad (9.13)$$

where the angle brackets denote averaging over the neighbourhood Ω. By first applying a degree of smoothing and varying the size of Ω, we can attempt to extract multiscale measures of texture.

Figure 9.6 (produced using the Matlab code in Example 9.4) shows an image with two contrasting textures and the results of applying the local range, entropy and variance operators.

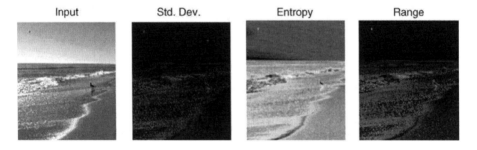

| Input | Std. Dev. | Entropy | Range |

Figure 9.6   Some basic local texture operators: The response to the image is given by applying filters extending over a certain neighbourhood region of the target pixel. Proceeding from left to right: The original input image is followed by the output images resulting from a local operators calculating the standard deviation, entropy and range

---

Example 9.4

Matlab code	What is happening?
A=imread('sunandsea.jpg') ;	%Read image
I=rgb2gray(A);	%Convert to grey-scale
J = stdfilt(I);	%Apply local standard deviaton filter
subplot(1,4,1), imshow(I);	
subplot(1,4,2),imshow(J,[]);	%Display original and processed
J = entropyfilt(I,ones(15));	%Apply entropy filter over 15x15 neighbourhood
subplot(1,4,3),imshow(J,[]);	%Display processed result
J = rangefilt(I,ones(5));	%Apply range filter over 5x5 neighbourhood
subplot(1,4,4),imshow(J,[]);	%Display processed result

Comments
This file shows the effect of three different texture filters based on calculating the variance (standard devn), range and entropy over a locally defined neighbourhood

9.6 Principal component analysis

The reader is probably familiar with the common saying that goes something along the lines of 'Why use a hundred words when ten will do?' This idea of expressing information in its most succinct and compact form embodies very accurately the central idea behind Principal Component Analysis (PCA) a very important and powerful statistical technique. Some of the more significant aspects of digital imaging in which it has found useful application include the classification of photographic film, remote sensing and data compression, and automated facial recognition and facial synthesis, an application we will explore later in this chapter. It is also commonly used as a *low-level* image processing tool for certain tasks, such as determination of the orientation of basic shapes. Mathematically, PCA is closely related to eigenvector/eigenvalue analysis and, as we shall see, can be conveniently understood in terms of the geometry of vectors in high-dimensional spaces.

All the applications to which PCA is put have one very important thing in common: they consider a sequence of images (or speaking more generally, data) which are *correlated*. Whenever we have many examples of data/feature vectors which exhibit a significant degree of correlation with each other, we may consider applying PCA as a means of closely approximating the feature vectors, but using considerably fewer parameters to describe them than the original data. This is termed *dimensionality reduction* and can significantly aid interpretation and analysis of both the sample data and new examples of feature vectors drawn from the same distribution. Performing PCA often allows simple and effective classification and discrimination procedures to be devised which would otherwise be difficult. In certain instances, PCA is also useful as a convenient means to synthesize new data examples obeying the same statistical distribution as an original training sample.

9.7 Principal component analysis: an illustrative example

To grasp the essence of PCA, suppose we have a large sample of M children and that we wish to assess their physical development. For each child, we record exhaustive information on N variables, such as age, height, weight, waist measurement, arm length, neck circumference, finger lengths, etc. It is readily apparent that each of the variables considered here is *not independent* of the others. An older child can generally be expected to be taller, a taller child will generally weigh more, a heavier child will have a larger waist measurement and so on. In other words, we expect that the value of a given variable in the set is, to some degree, *predictive* of the values of the other variables. This example illustrates the primary and essential requirement for PCA to be useful: there *must be some degree of correlation* between the data vectors. In fact, the more strongly correlated the original data, the more effective PCA is likely to be.

To provide some concrete but simple data to work with, consider that we have just $M = 2$ measurements on the height h and the weight w of a sample group of $N = 12$ people as given in Table 9.2. Figure 9.7 plots the mean-subtracted data in the 2-D feature space (w, h) and it is evident that the data shows considerable variance over both

Table 9.2   A sample of 2-D feature vectors (height and weight)

Weight (kg)	65	75	53	54	61	88	70	78	52	95	70	72
Height (cm)	170	176	154	167	171	184	182	190	166	168	176	175

variables and that they are correlated. The sample covariance matrix for these two variables is calculated as:

$$C_x = \frac{1}{N-1}(\mathbf{x} - \bar{\mathbf{x}})(\mathbf{x} - \bar{\mathbf{x}})^{\mathrm{T}} \qquad (9.14)$$

where $\mathbf{x} = (h \quad w)^{\mathrm{T}}$ and $\bar{\mathbf{x}} = (\bar{h} \quad \bar{w})^{\mathrm{T}}$ is the sample mean and confirms this (see Table 9.3).[4]

The basic aim of PCA is to effect a rotation of the coordinate system and thereby express the data in terms of a new set of variables or equivalently axes which are uncorrelated. How are these found? The first principal axis is chosen to satisfy the following criterion:

> The axis passing through the data points which *maximizes the sum of the squared lengths of the perpendicular projection* of the data points onto that axis is the *principal axis*.

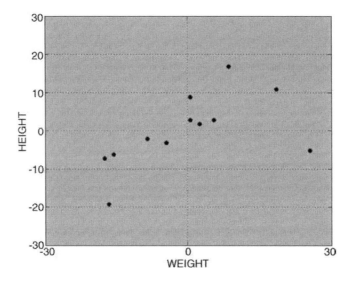

Figure 9.7   Distribution of weight and height values within a 2-D feature space. The mean value of each variable has been subtracted from the data

---

[4] Note that the correlation matrix is effectively a normalized version of the covariance and given by $\rho_{ij} = C_x(i,j)/\sigma_{ij}$, where $\sigma_{ij} = \langle(x_i - \bar{x}_i)(x_j - \bar{x}_j)\rangle$.

Table 9.3 Summary statistics for data in Table 9.2

	Covariance matrix				Correlation matrix	
	Weight	Height	Mean		Weight	Height
Weight	90.57	73.43	$\overline{w} = 69.42$	Weight	1	0.57
Height	73.43	182.99	$\overline{h} = 173.25$	Height	0.57	1

Thus, the principal axis is oriented so as to maximize the overall variance of the data with respect to it (a variance-maximizing transform). We note in passing that an alternative but entirely equivalent criterion for a principal axis is that it *minimizes* the sum of the squared errors (differences) between the actual data points and their *perpendicular projections* onto the said straight line.[5] This concept is illustrated in Figure 9.8.

Let us suppose that the first principal axis has been found. To calculate the second principal axis we proceed as follows:

- Calculate the projections of the data points onto the first principal axis.

- Subtract these projected values from the original data. The modified set of data points is termed the *residual* data.

- The second principal axis is then calculated to satisfy an identical criterion to the first (i.e. the variance of the residual data is maximized along this axis).

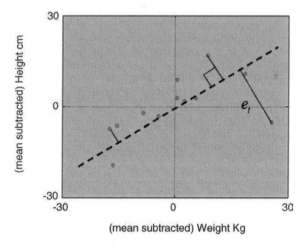

(mean subtracted) Weight Kg

Figure 9.8 The principal axis minimizes the sum of the squared differences between the data and their orthogonal projections onto that axis. Equivalently, this maximizes the variance of the data along the principal axis

---

[5] Note also that this is *distinct* from fitting a straight line by regression. In regression, x is an independent variable and y the dependent variable and we seek to minimize the sum of the squared errors between the actual y values and their predicted values. In PCA, x and y are on an equal footing.

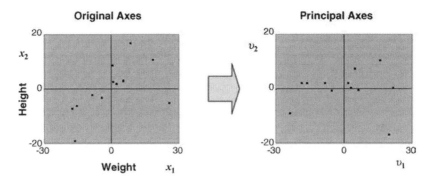

Figure 9.9 The weight–height data of Example 9.5 referred to the original and principal axis coordinate systems. Expressing the data with respect to the principal axes has increased the variance of the data along one axis and (necessarily) reduced it with respect to the other

For a 2-D space, we must stop here (there are no more dimensions left). Figure 9.9 shows our example height–weight data referred to both the original axes and to the principal axes. However, for an N-dimensional feature space (i.e. when N variables exist), this procedure is simply repeated N times to derive N principal axes each of which successively maximizes the variance of the residual data along the given principal axis.

Hopefully, the concept of PCA is clear, but how are these principal axes *actually calculated* ? In the next two sections we will present the basic theory. There are in fact two subtly different viewpoints and thus two different approaches to the calculation of principal components. Both produce the same result, but, as we will see, each offers a slightly different viewpoint on PCA. As this subtlety is of direct relevance to the calculation of principal-component images in image processing, we will present both.

9.8 Theory of principal component analysis: version 1

Assume we have N observations on M variables $(N \geq M)$. Let us denote the mean-subtracted observations on the ith variable by x_i $(i = 1, 2, \ldots, N)$. The aim is to define a set of new, *uncorrelated* variables y_i which are some appropriate linear combination of the x_i. Thus:

$$y_i = \sum_{j=1}^{M} a_{ij} x_j \tag{9.14}$$

with the coefficients a_{ij} to be determined. The set of M linear equations expressed by Equation (9.14) can be written compactly in matrix form as

$$\vec{y} = \mathbf{R}\vec{x} \tag{9.15}$$

where $\vec{y} = [y_1, y_2 \cdots y_M]^T$, $\vec{x} = [x_1, x_2 \cdots x_M]^T$ and $\mathbf{R}$ is a matrix of unknown coefficients. Note that the variables x_i have had their sample means subtracted and are zero-mean variables, so that $\langle \vec{y} \rangle = \mathbf{R} \langle \vec{x} \rangle = 0$. By definition, the covariance matrix $C_{\vec{y}}$ on the new variables represented by $\vec{y}$ is then

$$C_{\vec{y}} = \langle \vec{y}\vec{y}^T \rangle = \mathbf{R} \langle \vec{x}\vec{x}^T \rangle \mathbf{R}^T = \mathbf{R}C_{\vec{x}}\mathbf{R}^T \qquad (9.16)$$

where the angle brackets denote ensemble averaging over the observations on the data vector. We demand that the new variables be uncorrelated. This is equivalent to requiring that the covariance matrix $C_{\vec{y}}$ on the new variables is *diagonal*. Accordingly, we must form the covariance matrix $C_{\vec{x}} = \langle \vec{x}\vec{x}^T \rangle$ on our original data and find a diagonalizing matrix $\mathbf{R}$ such that

$$C_{\vec{y}} = \langle \vec{y}\vec{y}^T \rangle = \mathbf{R}C_{\vec{x}}\mathbf{R}^T = \Lambda \text{ a diagonal matrix} \qquad (9.17)$$

Fortunately, finding such a matrix is easy. Equation (9.16) describes the classic eigenvalue/eigenvector problem,[6] which can be solved by standard numerical procedures. The columns of $\mathbf{R}$ are the set of orthogonal eigenvectors which achieve the diagonalization of $C_{\vec{x}}$ and the diagonal elements of Λ give the corresponding eigenvalues. Computationally, the solution steps are as follows:

- Form the sample covariance matrix $C_{\vec{x}} = \langle \vec{x}\vec{x}^T \rangle$.

- Provide this as the input matrix to an eigenvector/eigenvalue decomposition routine which will return the eigenvectors and eigenvalues in the matrices $\mathbf{R}$ and Λ respectively.

- The new variables (axes) are then readily calculable using Equation (9.15).

From this viewpoint, PCA has effected a rotation of the original coordinate system to define new variables (or, equivalently, new axes) which are linear combinations of the original variables. We have calculated the *principal axes*.

Note that the data itself can be expressed in terms of the new axes by exploiting the orthonormality of the eigenvector matrix (i.e. it consists of unit length, mutually orthogonal vectors such that $\mathbf{R}\mathbf{R}^T = \mathbf{R}^T\mathbf{R} = \mathbf{I}$). Multiplying Equation (9.15) from the left by $\mathbf{R}^T$, we thus have an expression for the data in the new system:

$$\vec{x} = \mathbf{R}^T\vec{y} \qquad (9.18)$$

9.9 Theory of principal component analysis: version 2

In this second approach to the PCA calculation, we take an alternative viewpoint. We consider each sample of N observations on each of the M variables to form our basic data/

---

[6] More commonly written as $\mathbf{R}C_{\vec{x}} = \Lambda\mathbf{R}$ with Λ a diagonal matrix

observation vector. The M data vectors $\{\vec{x}_i\}$ thus exist in an N-dimensional space and each 'points to a particular location' in this space. The more strongly correlated the variables are, it follows that the more similar will be the pointing directions of the data vectors. We propose new vectors $\vec{v}_j$ which are linear combinations of the original vectors $\{\vec{x}_i\}$. Thus, in tableau form, we may write

$$
\begin{bmatrix} \uparrow & \uparrow & & \uparrow \\ \vec{v}_1 & \vec{v}_2 & \cdots & \vec{v}_M \\ \downarrow & \downarrow & & \downarrow \end{bmatrix} = \begin{bmatrix} \uparrow & \uparrow & & \uparrow \\ \vec{x}_1 & \vec{x}_2 & \cdots & \vec{x}_M \\ \downarrow & \downarrow & & \downarrow \end{bmatrix} \begin{bmatrix} a_{11} & a_{12} & \cdots & a_{1M} \\ a_{21} & \ddots & & \vdots \\ \vdots & & \ddots & \vdots \\ a_{M1} & a_{M2} & & a_{MM} \end{bmatrix} \qquad \text{or} \qquad \mathbf{V} = \mathbf{X}\mathbf{R}
$$

$$(9.19)$$

and demand that the $\vec{v}_j$ are mutually orthogonal. The task is to find the matrix of coefficients $\mathbf{R}$ which will accomplish this task and, therefore, we require that the new vectors satisfy

$$
\mathbf{V}^{\mathsf{T}}\mathbf{V} = [\mathbf{X}\mathbf{R}]^{\mathsf{T}}[\mathbf{X}\mathbf{R}] = \Lambda \text{ a diagonal matrix}
$$

Multiplying both sides by the factor $1/(N-1)$ and rearranging we have

$$
\frac{1}{N-1}\mathbf{V}^{\mathsf{T}}\mathbf{V} = \mathbf{R}^{\mathsf{T}}\frac{[\mathbf{X}^{\mathsf{T}}\mathbf{X}]}{N-1}\mathbf{R} = \Lambda \text{ a diagonal matrix} \qquad (9.20)
$$

Note that the insertion on both sides of the factor $1/(N-1)$ is not strictly required but is undertaken as it enables us to identify terms on the left- and right-hand sides which are sample covariance matrices on the data, i.e. $\mathbf{C_X} = (N-1)^{-1}\mathbf{X}^{\mathsf{T}}\mathbf{X}$. Thus, we require a matrix $\mathbf{R}$ such that

$$
\mathbf{R}^{\mathsf{T}}\mathbf{C_X}\mathbf{R} = \frac{1}{N-1}\mathbf{V}^{\mathsf{T}}\mathbf{V} = C_V = \Lambda \text{ a diagonal matrix} \qquad (9.21)
$$

and we again arrive at a solution which requires the eigenvector/eigenvalue decomposition of the data covariance matrix $\mathbf{C_X} = (N-1)^{-1}\mathbf{X}^{\mathsf{T}}\mathbf{X}$.[7] Computationally, things are the same as in the first case:

- Form the data covariance matrix $\mathbf{C_X} = (N-1)^{-1}\mathbf{X}^{\mathsf{T}}\mathbf{X}$.

- Perform an eigenvector/eigenvalue decomposition of $\mathbf{C_X}$, which will return the eigenvectors and eigenvalues in the matrices $\mathbf{R}$ and Λ respectively.

- Calculate the new data vectors in matrix $\mathbf{V}$ using Equation (9.18). Note that the columns of matrix $\mathbf{V}$ thus obtained are the *principal components*.

---

[7] That this covariance is formally equivalent to that version given earlier is readily demonstrated by writing out the terms explicitly.

The data itself can be expressed in terms of the principal components by noting that the eigenvector matrix is *orthonormal* (i.e. $\mathbf{R}\mathbf{R}^{\mathrm{T}} = \mathbf{R}^{\mathrm{T}}\mathbf{R} = \mathbf{I}$). Multiplying Equation (9.19) from the left by $\mathbf{R}^{\mathrm{T}}$ then gives the original data as $\mathbf{X} = \mathbf{V}\mathbf{R}^{\mathrm{T}}$.

9.10 Principal axes and principal components

There is a subtle difference between the two alternative derivations we have offered. In the first case we considered our variables to define the feature space and derived a set of principal axes. The dimensionality of our feature space was determined by the number of variables and the diagonalizing matrix of eigenvectors $\mathbf{R}$ is used in Equation (9.13) to produce a new set of axes in that 2-D space. Our second formulation leads to essentially the same procedure (i.e. to diagonalize the covariance matrix), but it nonetheless admits an alternative viewpoint. In this instance, we view the observations on each of the variables to define our (higher dimensional) feature vectors. The role of the diagonalizing matrix of eigenvectors $\mathbf{R}$ here is thus to multiply and transform these higher dimensional vectors to produce a new orthogonal (principal) set. Thus, the dimensionality of the space here is determined by *the number of observations* made on each of the variables.

Both points of view are legitimate. We prefer to reserve the term principal components for the vectors which constitute the columns of the matrix $\mathbf{V}$ calculated according to our *second* approach. Adopting this convention, the principal components are in fact formally equivalent to the projections of the original data onto their corresponding principal axes.

9.11 Summary of properties of principal component analysis

It is worth summarizing the key points about the principal axes and principal components.

- The principal axes are *mutually orthogonal*. This follows if we consider how they are calculated. After the first principal axis has been calculated, each successive component is calculated on the residual data (i.e. after subtraction of the projection onto the previously calculated axis). It follows, therefore, that the next component can have no component in the direction of the previous axis and must, therefore, be orthogonal to it.

- The data expressed in the new (principal) axes system is uncorrelated. The new variables do not co-vary and, thus, their covariance matrix has zero entries in the off-diagonal terms.

- The principal axes successively *maximize* the variance in the data with respect to themselves.

- The eigenvalues in the diagonal matrix Λ directly specify the total amount of variance in the data associated with each principal component.

Example 9.5 calculates the principal axes and principal components for the height and weight measurements provided earlier. Figure 9.10 (in part shown in Figure 9.9) summarizes the results.

Example 9.5

Matlab code	What is happening?
W=[65 75 53 54 61 88 70 78 52 95 70 72]'	%1. Form data vector on weight
H=[170 176 154 167 171 184 182 190 166 168 176 175]'	%Form data vector on height
XM=[mean(W).*ones(length(W),1) mean(H).*ones(length(H),1)]	%matrix with mean values replicated
X=[W H]-XM;	%Form mean-subtracted data matrix
Cx=cov(X)	%Calculate covariance on data
[R,LAMBDA,Q]=svd(Cx)	%Get eigenvalues LAMBDA and %eigenvectors R
V=X*R;	%Calculate principal components
subplot(1,2,1), plot(X(:,1),X(:,2),'ko'); grid on;	%2. Display data on original axes
subplot(1,2,2), plot(V(:,1),V(:,2),'ro'); grid on;	%Display PCs as data in rotated space
XR=XM+V*R'	%3. Reconstruct data in terms of PCs
XR-[W H]	%Confirm reconstruction (diff = 0)
V'*V./(length(W)-1)	%4. Confirm covariance terms on %New axes = LAMBDA

Comments

Matlab functions: *cov*, *mean*, *svd*.

This example (1) calculates the principal components of the data, (2) displays the data in the new (principal) axes, (3) confirms that the original data can be 'reconstructed' in terms of the principal components and (4) confirms that the variance of the principal components is equal to the eigenvalues contained in matrix LAMBDA.

Note that we have used the Matlab function *svd* (singular value decomposition) to calculate the eigenvectors and eigenvalues rather than *eig*. The reason for this is that *svd* orders the eigenvalues (and their associated eigenvectors) *strictly according to size* from largest to smallest (*eig* does not). This is a desirable and natural ordering when calculating principal components which are intended to successively describe the maximum possible amount of variance in the data.

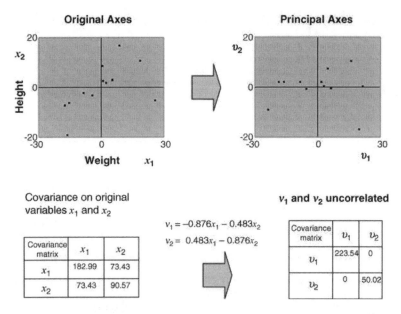

Figure 9.10   The weight–height data referred to the original axes x_1 and x_2 and to the principal axes v_1 and v_2

Example 9.6 shows the basic use of PCA in the calculation of the principal axes of a 2-D shape (the silhouette of an aeroplane). Identification of the principal axes, as shown in Figure 9.11 of a 2-D or 3-D shape, can sometimes assist in applying processing operations such that they are effectively invariant to image rotation.

Example 9.6

Matlab code	**What is happening?**
A=imread('aeroplane_silhouette.png');	%Read in image and convert to grey scale
bw=~im2bw(A,0.5);	%Threshold and invert
subplot(1,2,1), imshow(bw,[]);	%Display image
[y,x]=find(bw>0.5);	%Get coordinates of non-zero pixels
centroid=mean([x y]);	%Get (centroid) of data
hold on; plot(centroid(1),centroid(2),'rd');	%Plot shape centroid
C=cov([x y]);	%Calculate covariance of coordinates
[U,S]=eig(C)	%Find principal axes and eigenvalues
	%Plot the principal axes
m=U(2,1)./U(1,1);	
const=centroid(2)-m.*centroid(1);	
xl=50:450; yl=m.*xl + const	
subplot(1,2,2), imshow(bw,[]); h=line(xl,yl);	%Display image and axes
set(h,'Color',[1 0 0],'LineWidth',2.0)	
m2=U(2,2)./U(1,2);	
const=centroid(2)-m2.*centroid(1);	
x2=50:450; y2=m2.*x2 + const	

h=line(x2,y2); set(h,'Color',[1 0 0],
    'LineWidth',2.0)

Comments
This example calculates the principal axes of the foreground pixels in the binary image of the
    aeroplane silhouette.

Figure 9.11   Calculation of the principal axes of a 2-D object (see Example 9.6 for calculation)

9.12   Dimensionality reduction: the purpose of principal component analysis

Depending on the precise context, there may be several closely related reasons for doing
PCA, but the aim central to most is *dimensionality reduction*. By identifying a dominant
subset of orthogonal directions in our feature space we can effectively discard those
components which contribute little variance to our sample and use only the significant
components to approximate accurately and describe our data in a much more compact way.
To take our 2-D height–weight problem as an example, the basic question we are really
asking is: 'Do we really need two variables to describe a person's basic physique or will just
one variable (what one might call a 'size index') do a reasonable job? We can answer this
question by calculating the principal components. In this particular example, a single index
(the value of the first principal component) accounts for 84 % of the variation in the data.
If we are willing to disregard the remaining 16 % of the variance we would conclude that we
do, indeed, need just one variable. It should be readily apparent that the more closely our
height–weight data approximates to a straight line, the better we can describe our data using
just one variable. Ultimately, if the data lay exactly on a straight line, the first principal axis
then accounts for 100 % of the variance in the data.

   Our height–weight example is, perhaps, a slightly uninspiring example, but it hopefully
makes the point. The *real power* of PCA becomes apparent in multidimensional problems,
where there may be tens, hundreds or even thousands of variables. Such, indeed, is often
the case when we are considering digital images. Here, a collection of primitive image
features, landmarks or even the pixel values themselves are treated as variables which vary

over a sample of correlated images. We will shortly present such an example in Section 9.15. In the next section, we first make an important note on how PCA on image ensembles is carried out. This is important if we are to keep the computational problem to a manageable size.

9.13   Principal components analysis on an ensemble of digital images

We can generally carry out PCA in two senses:

(1)   by taking our statistical average over the ensemble of vectors;

(2)   by taking our statistical average over the elements of the vectors themselves.

This point has rarely been made explicit in existing image-processing textbooks. Consider Figure 9.12, in which we depict a stack of M images each of which contains N pixels. In one possible approach to PCA the covariance is calculated on a *pixel-to-pixel* basis. This is to say that the ijth element of our covariance matrix is calculated by examining the values of the ith and jth pixels in each image, $I(i, j)$, and averaging over the whole sample of images. Unfortunately, such an approach will lead to the calculation of a covariance matrix of quite unmanageable size (consider that even a very modest image size of 256^2 would then require calculation and diagonalization of a covariance matrix of $256^2 \times 256^2 \sim 4295 \times 10^6$ elements). However, the second approach, in which the ensemble averaging takes place over the pixels of the images, generally results in a much more tractable problem – the resulting covariance matrix has a square dimension equal to the number of images in the ensemble and can be diagonalized in a straightforward fashion.[8]

9.14   Representation of out-of-sample examples using principal component analysis

In the context of pattern recognition and classification applications, we speak of *training data* (the data examples used to calculate the principal components) and *test data* (new data examples from the same basic pattern class that we are interested in modelling). So far, we have only explicitly discussed the capacity of PCA to produce a reduced-dimension representation of training data. An even more important aspect of PCA is the ability to model a certain class of object so that new (*out-of-sample*) examples of data can be accurately and efficiently represented by the calculated components. For example, if we calculate the principal components of a sample of human faces, a reasonable hope

---

[8] The derivation and calculation of principal components for image ensembles is covered in full detail on the book's website at http://www.fundipbook.com/materials/.

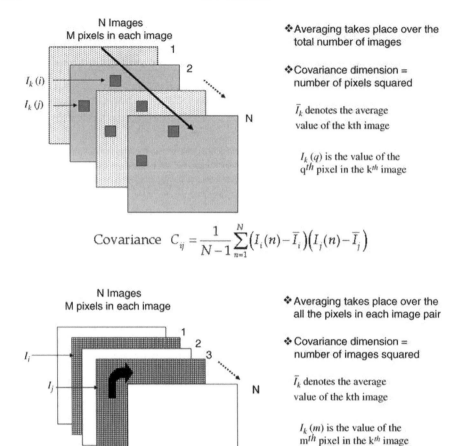

Figure 9.12 Two ways of carrying out PCA on image ensembles. Only the second approach in which the statistical averaging takes place over the image pixels is computationally viable.

is that we might be able to represent *new* unseen faces using these principal components as a basis.

Representing an out-of-sample data vector in terms of a set of precalculated principal components is easily achieved by projection of the data onto the principal component axes. Consider an arbitrary vector I having mean pixel value $\bar{I}$. We wish to express this as a linear combination of the principal components:

$$I - \bar{I} = \sum_{k=1}^{N} a_k P_k \qquad (9.22)$$

To do this we can simply exploit the orthogonality of the $\{P_k\}$. Multiplying from the left by $\boldsymbol{P}_i^{\mathrm{T}}$, since $\vec{\boldsymbol{P}}_i^{\mathrm{T}}\vec{\boldsymbol{P}}_k = \lambda_k\delta_{ki}$, we have

$$\boldsymbol{P}_i^{\mathrm{T}}(\boldsymbol{I}-\bar{\boldsymbol{I}}) = \sum_{k=1}^{N} a_k\boldsymbol{P}_i^{T}\boldsymbol{P}_k = \sum_{k=1}^{N} a_k\lambda_k\delta_{ki} \tag{9.23}$$

And we thus obtain the required expansion coefficients as

$$\hat{a}_i = \frac{\boldsymbol{P}_i^{\mathrm{T}}(\boldsymbol{I}-\bar{\boldsymbol{I}})}{\lambda_i} \tag{9.24}$$

and the vector $\boldsymbol{I}$ is estimated as

$$\hat{\boldsymbol{I}} = \bar{\boldsymbol{I}} + \sum_{k=1}^{N} \hat{a}_k\boldsymbol{P}_k \tag{9.25}$$

Note that, in a really well-trained model, it should be possible to represent new test data just as compactly and accurately as the training examples. This level of perfection is not often achieved. However, a good model will allow a close approximation of all possible examples of the given pattern class to be built using this basis.

9.15 Key example: eigenfaces and the human face

It is self-evident that the basic placement, size and shape of human facial features are similar. Therefore, we can expect that an ensemble of suitably scaled and registered images of the human face will exhibit fairly strong correlation. Figure 9.13 shows six such images from a total sample of 290. Each image consisted of 21 054 grey-scale pixel values.

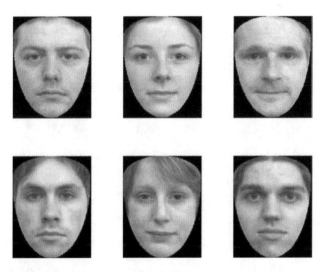

Figure 9.13 A sample of human faces scaled and registered such that each can be described by the same number of pixels (21 054)

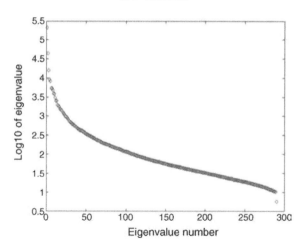

Figure 9.14 The eigenvalue spectrum (on a log scale) after calculation of the principal components of the sample of 290 faces, examples of which are shown in Figure 9.15

When we carry out a PCA on this ensemble of faces, we find that the eigenvalue spectrum dies off quite rapidly. This suggests considerable potential for compression.

The first six facial principal components are displayed in Figure 9.14.

The potential use of PCA for face encoding and recognition has now been studied quite extensively. The essence of the technique is that the face of any given individual may be synthesized by adding or subtracting a multiple of each principal component. In general, a linear combination of as few as 50–100 'eigenfaces' is sufficient for human and machine recognition of the synthesized face. This small code (the weights for the principal components) can be readily stored as a very compact set of parameters occupying ~50 bytes. Figure 9.16 shows the increasing accuracy of the reconstruction as more principal

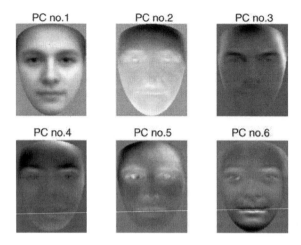

Figure 9.15 The first six facial principal components from a sample of 290 faces. The first principal component is the average face. Note the strong male appearance coded by principal component no. 3

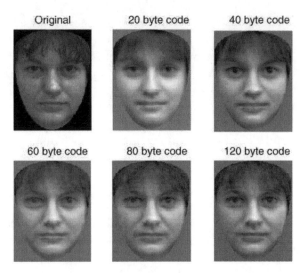

Figure 9.16 Application of PCA to the compact encoding of out-of-sample human faces

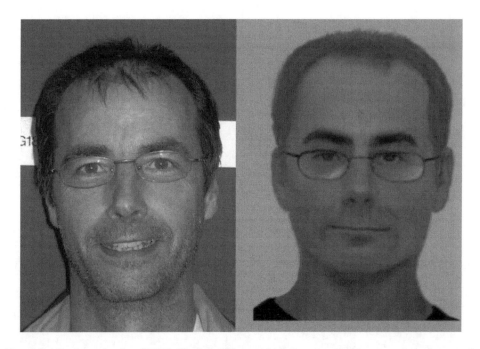

Figure 9.17 Example of an EFIT-V. The facial likeness (right) was created by an evolutionary search to find the dominant principal components in the real face (left) based on an eyewitness's memory. The image detail on the right has been slightly enhanced by use of a computer art package. (Image courtesy of Visionmetric Ltd, www.visionmetric.com)

components are used to approximate the face. Note that the original subject displayed was *not included* in the ensemble used to form the covariance matrix.

The use of PCA modelling of human faces has recently been extended to a new generation of facial composite systems (commonly known as PhotoFIT systems)[9] . These systems are used by police forces in an attempt to create a recognisable likeness of a suspect in a crime based on an eyewitnesss memory. It is hoped that subsequent exposure of the image to the public might then result in positive identification. The basic idea behind these systems is that the face can be synthesized by guiding the witness to evolve an appropriate combination of shape and texture principal components. As the witness is working from memory, this forms a fundamental limitation on the accuracy of the constructed composite. Nonetheless, the global nature of the principal components which carry information about the whole face appears to offer real advantages. Likenesses created in this way can be remarkably good and are often achieved much more quickly than by traditional police methods. Figure 9.17 shows one such example in which a photo of the suspect is compared to the likeness created using a PCA-based, evolutionary system. The image displayed shows the likeness after a small degree of artistic enhancement which is commonplace in this application.

For further examples and exercises see http://www.fundipbook.com

---

[9] See for example the EFIT-V system www.visionmetric.com.

10

Image segmentation

10.1 Image segmentation

Segmentation is the name given to the generic process by which an image is subdivided into its constituent regions or objects. In general, completely autonomous segmentation is one of the most difficult tasks in the design of computer vision systems and remains an active field of image processing and machine vision research. Segmentation occupies a very important role in image processing because it is so often the *vital first step* which must be successfully taken before subsequent tasks such as feature extraction, classification, description, etc. can be sensibly attempted. After all, if you cannot identify the objects in the first place, how can you classify or describe them?

The basic goal of segmentation, then, is to partition the image into mutually exclusive regions to which we can subsequently attach meaningful labels. The segmented objects are often termed the foreground and the rest of the image is the background. Note that, for any given image, we cannot generally speak of a single, 'correct' segmentation. Rather, the correct segmentation of the image depends strongly on the types of object or region we are interested in identifying. *What relationship must a given pixel have with respect to its neighbours and other pixels in the image in order that it be assigned to one region or another?* This really is the central question in image segmentation and is usually approached through one of two basic routes:

- *Edge/boundary methods* This approach is based on the detection of edges as a means to identifying the boundary between regions. As such, it looks for sharp differences between groups of pixels.

- *Region-based methods* This approach assigns pixels to a given region based on their degree of mutual similarity.

10.2 Use of image properties and features in segmentation

In the most basic of segmentation techniques (intensity thresholding), the segmentation is used only on the absolute intensity of the individual pixels. However, more sophisticated properties and features of the image are usually required for successful segmentation. Before

Fundamentals of Digital Image Processing – A Practical Approach with Examples in Matlab
Chris Solomon and Toby Breckon
© 2011 John Wiley & Sons, Ltd

we begin our discussion of explicit techniques, it provides a useful (if somewhat simplified) perspective to recognize that there are three basic properties/qualities in images which we can exploit in our attempts to segment images.

(1) *Colour* is, in certain cases, the simplest and most obvious way of discriminating between objects and background. Objects which are characterized by certain colour properties (i.e. are confined to a certain region of a colour space) may be separated from the background. For example, segmenting an orange from a background comprising a blue tablecloth is a trivial task.

(2) *Texture* is a somewhat loose concept in image processing. It does not have a single definition but, nonetheless, accords reasonably well with our everyday notions of a 'rough' or 'smooth' object. Thus, texture refers to the 'typical' spatial variation in intensity or colour values in the image over a certain spatial scale. A number of texture metrics are based on calculation of the variance or other statistical moments of the intensity over a certain neighbourhood/spatial scale in the image. We use it in a very general sense here.

(3) *Motion* of an object in a sequence of image frames can be a powerful cue. When it takes place against a stationary background, simple frame-by-frame subtraction techniques are often sufficient to yield an accurate outline of the moving object.

In summary, most segmentation procedures will use and combine information on one of more of the properties colour, texture and motion.

Some simple conceptual examples and possible approaches to the segmentation problem are summarized in Table 10.1.

What has led us to suggest the possible approaches in Table 10.1? Briefly: an aeroplane in the sky is (we hope) in motion. Its colour is likely to be distinct from the sky. Texture is unlikely to be a good approach because the aeroplane (being a man-made object) tends to have a fairly smooth texture – so does the sky, at least in most weather conditions. Some form of shape analysis or measurement could certainly be used, but in this particular case would tend to be superfluous. Of course, these comments are highly provisional (and may

Table 10.1   Simple conceptual examples and possible approach to segmentation

Object(s)	Image	Purpose of segmentation	Preferred approach
Aeroplane	Aeroplane in the sky	Tracking	Motion, colour
Human faces	Crowded shopping mall	Face recognition surveillance system	Colour, shape
Man-made structures	Aerial satellite photograph	Intelligence acquisition from aircraft	Texture, colour
Cultivated regions	Uncultivated regions	LandSAT survey	Texture, shape
Granny Smiths apples, pears	Various fruits on a conveyor belt	Automated fruit sorting	Colour, shape

actually be inaccurate in certain specific instances), but they are intended to illustrate the importance of carefully considering the approach which is most likely to yield success.

We will begin our discussion of segmentation techniques with the simplest approach, namely intensity thresholding.

10.3   Intensity thresholding

The basic idea of using intensity thresholding in segmentation is very simple. We choose some threshold value such that pixels possessing values greater than the threshold are assigned to one region whilst those that fall below the threshold are assigned to another (adjoint) region. Thresholding creates a binary image $b(x, y)$ from an intensity image $I(x, y)$ according to the simple criterion

$$b(x, y) = \begin{cases} 1 & \text{if } I(x, y) > T \\ 0 & \text{otherwise} \end{cases} \tag{10.1}$$

where T is the threshold.

In the very simplest of cases, this approach is quite satisfactory. Figure 10.1 (top right) (produced using the Matlab® code in Example 10.1) shows the result of intensity thresholding on an image of several coins lying on a dark background; all the coins are successfully identified. In this case, the appropriate threshold was chosen manually by trial and error. In a limited number of cases, this is an acceptable thing to do; for example, certain inspection tasks may allow a human operator to set an appropriate threshold before automatically processing a sequence of similar images. However, many image processing tasks require full automation, and there is often a need for some criterion for selecting a threshold automatically.

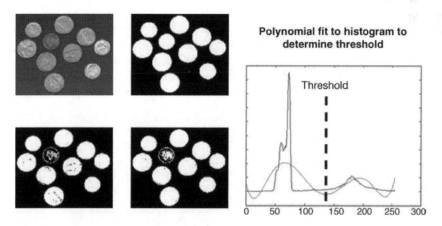

Figure 10.1   Top left: original image. Top right: result after manual selection of threshold. Bottom left: result of automatic threshold selection by polynomial fitting to image histogram. Bottom right: result of automatic threshold selection using Otsu's method. The image on the right shows the histogram and the result of fitting a sixth-order polynomial

Example 10.1

Matlab code	What is happening?
I = imread('coins.png');	%Read in original
subplot(2,2,1), imshow(I);	%Display original
subplot(2,2,2),im2bw(I,0.35);	%Result of manual threshold
[counts,X]=imhist(I);	%Calculate image hIstogram
P = polyfit(X,counts,6); Y=polyval(P,X);	%Fit to histogram and evaluate
[V,ind]=sort(abs(diff(Y))); thresh=ind(3)./255;	%Find minimum of polynomial
subplot(2,2,3), im2bw(I,thresh);	%Result of Polynomial theshold
level = graythresh(I);	%Find threshold
subplot(2,2,4), im2bw(I,level);	%Result of Otsu's method
figure; plot(X,counts); hold on, plot(X,Y,'r-');	%Histogram and polynomial fit

Comments

Matlab functions: *im2bw, imhist, polyfit, polyval, sort, graythresh.*

Automated threshold selection is essentially based on a conceptual or actual consideration of the image histogram. In those situations where thresholding can successfully segment objects from the background, the 1-D histogram of the image will typically exhibit two modes or peaks: one corresponding to the pixels of the objects and one to the pixels of the background. This is, in fact, the case with our chosen example; see Figure 10.1 (image on right). The threshold needs to be chosen so that these two modes are clearly separated from each other.

One simple approach is to calculate the image histogram and fit a polynomial function to it. Provided the order of the polynomial function is chosen judiciously and the polynomial adequately fits the basic shape of the histogram, a suitable threshold can be identified at the minimum turning point of the curve.[1]

A more principled approach to automatic threshold selection is given by Otsu's method. Otsu's method is based on a relatively straightforward analysis which finds that threshold which *minimizes the within-class variance* of the thresholded black and white pixels. In other words, this approach selects the threshold which results in the tightest clustering of the two groups represented by the foreground and background pixels. Figure 10.1 shows the results obtained using both a manually selected threshold and that calculated automatically by Otsu's method.

10.3.1 Problems with global thresholding

There are several serious limitations to simple global thresholding:

- there is no guarantee that the thresholded pixels will be contiguous (thresholding does not consider the spatial relationships between pixels);

---

[1] Two peaks require a minimum of fourth order to model (two maxima, one minimum), but it is usually better to go to fifth or sixth order to allow a bit more flexibility. If there is more than one minimum turning point on the curve, then it is usually straightforward to identify the appropriate one.

- it is sensitive to accidental and uncontrolled variations in the illumination field;

- it is only really applicable to those simple cases in which the entire image is divisible into a foreground of objects of similar intensity and a background of distinct intensity to the objects.

10.4   Region growing and region splitting

Region growing is an approach to segmentation in which pixels are grouped into larger regions based on their similarity according to predefined *similarity criteria*. It should be apparent that specifying similarity criteria *alone* is not an effective basis for segmentation and it is necessary to consider the adjacency spatial relationships between pixels. In region growing, we typically start from a number of seed pixels randomly distributed over the image and append pixels in the neighbourhood to the same region if they satisfy *similarity criteria* relating to their intensity, colour or related statistical properties of their own neighbourhood. Simple examples of similarity criteria might be:

(1)   the absolute intensity difference between a candidate pixel and the seed pixel must lie within a specified range;

(2)   the absolute intensity difference between a candidate pixel and the running average intensity of the growing region must lie within a specified range;

(3)   the difference between the standard deviation in intensity over a specified local neighbourhood of the candidate pixel and that over a local neighbourhood of the candidate pixel must (or must not) exceed a certain threshold – this is a basic roughness/smoothness criterion.

Many other criteria can be specified according to the nature of the problem.

Region splitting essentially employs a similar philosophy, but is the reverse approach to region growing. In this case we begin the segmentation procedure by treating the whole image as a single region which is then successively broken down into smaller and smaller regions until any further subdivision would result in the differences between adjacent regions falling below some chosen threshold. One popular and straightforward approach to this is the *split-and-merge* algorithm.

10.5   Split-and-merge algorithm

The algorithm divides into two successive stages. The aim of the region splitting is to break the image into a set of disjoint regions each of which is regular within itself. The four basic steps are:

- consider the image as a whole to be the initial area of interest;

- look at the area of interest and decide if all pixels contained in the region satisfy some similarity criterion;

- if TRUE, then the area of interest (also called a block) corresponds to a region in the image and is labelled;

- if FALSE, then split the area of interest (usually into four equal sub-areas) and consider each of the sub-areas as the area of interest in turn.

This process continues until no further splitting occurs. In the worst-case scenario, this might happen when some of the areas are just one pixel in size. The splitting procedure is an example of what are sometimes referred to as divide-and-conquer or top-down methods. The process by which each block is split into four equal sub-blocks is known as *quadtree decomposition*.

However, if only splitting is carried out, then the final segmentation would probably contain many neighbouring regions that have identical or similar properties. Thus, a merging process is used after each split which compares adjacent regions and merges them if necessary. When no further splitting or merging occurs, the segmentation is complete. Figure 10.2 illustrates the basic process of splitting via quadtree decomposition and merging.

Figure 10.3 (produced using the Matlab code in Example 10.2) shows the result of a split-and-merge algorithm employing a similarity criterion based on local range (split the block if the difference between the maximum and minimum values exceeds a certain fraction of the image range).

QUADTREE DECOMPOSITION (SPLIT) AND MERGE

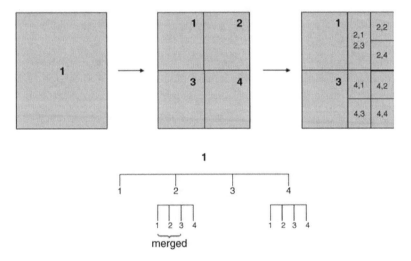

Figure 10.2 The basic split-and-merge procedure. The initial image is split into four regions. In this example, regions 1 and 3 satisfy the similarity criterion and are split no further. Regions 2 and 4 do not satisfy the similarity criterion and are then each split into four sub-regions. Regions (2,1) and (2,3) are then deemed sufficiently similar to merge them into a single region. The similarity criterion is applied to the designated sub-regions until no further splitting or merging occurs

 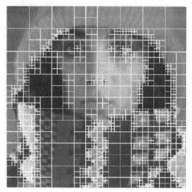

Figure 10.3 The image on the right delineates the regions obtained by carrying out a quadtree decomposition on the image using a similarity criterion based on the range of the intensity values within the blocks

---

Example 10.2

Matlab code	What is happening?
I=imread('trui.png');	%Read in image
S = qtdecomp(I,.17);	%Do quadtree decomposition
blocks = repmat(uint8(0),size(S));	%Create empty blocks
for dim = [512 256 128 64 32 16 8 4 2 1];	%Loop through successively smaller blocks

```
numblocks = length(find(S==dim));
  if (numblocks > 0)
  values = repmat(uint8(1),[dim dim numblocks]);
  values(2:dim,2:dim,:) = 0;
  blocks = qtsetblk(blocks,S,dim,values);
  end
end
blocks(end,1:end) =1;
blocks(1:end,end) = 1;
subplot(1,2,1), imshow(I);
k=find(blocks==1);
A=I; A(k)=255;
subplot(1,2,2), imshow(A);
```

k=find(blocks==1);	%Find border pixels of regions
A=I; A(k)=255;	%Superimpose on original image
subplot(1,2,2), imshow(A);	%Display

Comments

Matlab functions: ***qtdecomp, repmat, length, qsetblk, find.***

The function ***qtdecomp*** performs a quadtree decomposition on the input intensity image and returns a sparse array S. S contains the quadtree structure such that if $S(k, m) = P \neq 0$ then (k, m) is the upper-left corner of a block in the decomposition, and the size of the block is given by P pixels.

10.6   The challenge of edge detection

Edge detection is one of the most important and widely studied aspects of image processing. If we can find the boundary of an object by locating all its edges, then we have effectively segmented it. Superficially, edge detection seems a relatively straightforward affair. After all, edges are simply regions of intensity transition between one object and another. However, despite its conceptual simplicity, edge detection remains an active field of research. Most edge detectors are fundamentally based on the use of gradient differential filters. The most important examples of these (the Prewitt and Sobel kernels) have already been discussed in Chapter 4. However, these filters do not find edges *per se,* but rather only give some indication of where they are most likely to occur. Trying to actually find an edge, several factors may complicate the situation. The first relates to edge strength or, if you prefer, the *context* – how large does the gradient have to be for the point to be designated part of an edge? The second is the effect of noise – differential filters are very sensitive to noise and can return a large response at noisy points which do not actually belong to the edge. Third, where exactly does the edge occur? Most real edges are not discontinuous; they are smooth, in the sense that the gradient gradually increases and then decreases over a finite region. These are the issues and we will attempt to show how these are addressed in the techniques discussed in the following sections.

10.7   The laplacian of Gaussian and difference of Gaussians filters

As we saw earlier in this book, second-order derivatives can be used as basic discontinuity detectors. In Figure 10.4, we depict a (1-D) edge in which the transition from low to high intensity takes place over a finite distance. In the ideal case (depicted on the left), the second-order derivative d^2f/dx^2 depicts the precise onset of the edges (the point at which the gradient suddenly increases giving a positive response and the point at which it suddenly decreases giving a negative response). Thus, the location of edges is, *in principle*, predicted well by using a second-order derivative. However, second-derivative filters do not give the strength of the edge and they also tend to exaggerate the effects of noise (twice as much as first-order derivatives). One approach to getting round the problem of noise amplification is first to smooth the image with a Gaussian filter (reducing the effects of the noise) and then to apply the Laplacian operator. Since the Gaussian and Laplacian filtering operations are linear, and thus interchangeable, this sequence of operations can actually be achieved more efficiently by taking the *Laplacian of a Gaussian* (LoG) and then filtering the image with a suitable discretized kernel. It is easy to show that the Laplacian of a radially symmetric Gaussian is

$$\nabla^2\left\{e^{-r^2/2\sigma^2}\right\} = \frac{r^2-\sigma^2}{\sigma^4}e^{-r^2/2\sigma^2} \qquad (10.2)$$

and a discrete approximation to this function over a local neighbourhood can then easily be achieved with a filter kernel. The use of the LoG filter as a means of identifying edge locations was first proposed by Marr and Hildreth, who introduced the principle of the

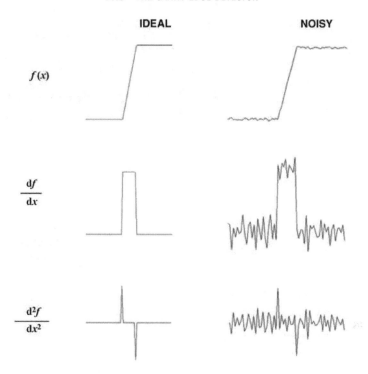

IDEAL   **NOISY**

$f(x)$

$\dfrac{df}{dx}$

$\dfrac{d^2 f}{dx^2}$

Figure 10.4   Demonstrating the sensitivity of differential operators to noise. In the ideal case (depicted on the left), the second-order derivative identifies the precise onset of the edge. However, even modest fluctuations in the signal result in the amplification of the noise in the first- and second-order derivatives (depicted on the right)

zero-crossing method. This relies on the fact that the Laplacian returns a high positive and high negative response at the transition point and the edge location is taken to be the point at which the response goes through zero. Thus, the points at which the LoG-filtered image go to zero indicate the location of the candidate edges.

A closely related filter to the LoG is the *difference of Gaussians* (DoG). The DoG filter is computed as the difference of two Gaussian functions having different standard deviations. It is possible to show formally that, when the two Gaussians have relative standard deviations of 1 and 1.6, the DoG operator closely approximates the LoG operator. Figure 10.5 shows the basic shape of the 2-D LoG filter (colloquially known as the Mexican hat function) together with the results of log filtering of our image and identifying edges at the points of zero crossing.

10.8   The Canny edge detector

Although research into reliable edge-detection algorithms continues, the Canny method is generally acknowledged as the best 'all-round' edge detection method developed to date. Canny aimed to develop an edge detector that satisfied three key criteria:

- A *low error rate*. In other words, it is important that edges occuring in images should not be missed and that there should be no response where edges do not exist.

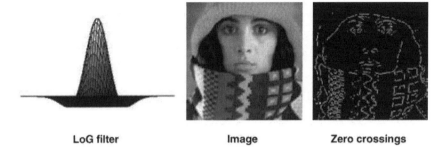

LoG filter   **Image**   **Zero crossings**

Figure 10.5 The basic shape of the LoG filter is shown on the left. The flat surrounding region corresponds to value 0. Filtering of the centre image with a discrete 5 × 5 kernel approximation to the LoG function followed by thresholding to identify the zero crossings results in the image on the right

- The detected edge points should be *well localized*. In other words, the distance between the edge pixels as found by the detector and the actual edge should be a minimum.

- There should be *only one response to a single edge*.

It is outside the scope of this discussion to present the detailed mathematical analysis and reasoning undertaken by Canny, but the basic procedure can be summarized in the following steps:

(1) *The image is first smoothed using a Gaussian kernel*: Gradient operators are sensitive to noise and this preliminary step is taken to reduce the image noise. The greater the width of the kernel, the more smoothing (i.e. noise reduction) is achieved. However, larger kernels result in a greater error in the edge location.

(2) *Find the edge strength*: This is achieved by taking the gradient of the image with the *Sobel operators* in the horizontal and vertical directions and then adding the magnitude of these components as a measure of the 'edge strength'. Thus $E(x, y) = |G_x(x, y)| + |G_y(x, y)|$.

(3) *Calculate the edge direction*: This is easily calculated as

$$\theta = \tan^{-1} \frac{G_y(x, y)}{G_x(x, y)}$$

(4) *Digitize the edge direction*: Once the edge direction is known, we approximate it to an edge direction that can be traced in a digital image. Considering an arbitrary pixel, the direction of an edge through this pixel can take one of only four possible values - 0° (neighbours to east and west), 90° (neighbours to north and south), 45° (neighbours to north-east and south-west) and 135° (neighbours to north-west and south-east). Accordingly, we approximate the calculated θ by whichever of these four angles is closest in value to it.[2]

---

[2] It should be noted that in some formulations of this step in the process, edge orientation is calculated to sub-pixel accuracy using 8 connectivity, SEE Canny, J., A Computational Approach To Edge Detection, IEEE Trans. Pat. Anal. and Mach. Intel., 8(6):679–698, 1986.

(5) *Nonmaximum suppression*: After the edge directions are known, nonmaximum suppression is applied. This works by tracing along the edge in the edge direction and suppressing any pixel value (i.e. set it equal to zero) that is not considered to be an edge. This will give a thin line in the output image.

(6) *Hysteresis*: After the first five steps have been completed, the final step is to track along the remaining pixels that have not been suppressed and threshold the image to identify the edge pixels. Critical to the Canny method, however, is the use of two distinct thresholds – a higher value T_2 and a lower value T_1. The fate of each pixel is then determined according to the following criteria:

- if $|E(x,y)| < T_1$, then the pixel is rejected and is not an edge pixel;

- if $|E(x,y)| > T_2$, then the pixel is accepted and is an edge pixel;

- if $T_1 < |E(x,y)| < T_2$, the pixel is rejected except where a path consisting of edge pixels connects it to an unconditional edge pixel with $|E(x,y)| > T_2$.

Two comments are pertinent to the practical application of the Canny edge detector. First, we are, in general, interested in identifying edges on a particular scale. We can introduce the idea of feature scale by filtering with Gaussian kernels of various widths. Larger kernels introduce greater amounts of image smoothing and the gradient maxima are accordingly reduced. It is normal, therefore, to specify the size of the smoothing kernel in the Canny edge detection algorithm to reflect this aim. Second, we see that the Canny detector identifies weak edges ($T_1 < |E(x,y)| < T_2$) only *if* they are connected to strong edges ($|E(x,y)| > T_2$).

Figure 10.6 demonstrates the use of the LoG and Canny edge detectors on an image subjected to different degrees of Gaussian smoothing. The Matlab code corresponding to Figure 10.6 is given in Example 10.3.

Example 10.3

Matlab code	What is happening?
A=imread('trui.png');	%Read in image
subplot(3,3,1), imshow(A,[]);	%Display original
h1=fspecial('gaussian',[15 15],6);	
h2=fspecial('gaussian',[30 30],12);	
subplot(3,3,4), imshow(imfilter(A,h1),[]);	%Display filtered version sigma=6
subplot(3,3,7), imshow(imfilter(A,h2),[]);	%Display filtered version sigma=12
[bw,thresh]=edge(A,'log');	%Edge detection on original – LoG filter
subplot(3,3,2), imshow(bw,[]);	
[bw,thresh]=edge(A,'canny');	%Canny edge detection on original
subplot(3,3,3), imshow(bw,[]);	%Display
[bw,thresh]=edge(imfilter(A,h1),'log');	%LoG edge detection on sigma=6
subplot(3,3,5), imshow(bw,[]);	

```
[bw,thresh]=edge(imfilter(A,h1),'canny');      %Canny edge detection on sigma=6
subplot(3,3,6), imshow(bw,[]);
[bw,thresh]=edge(imfilter(A,h2),'log');        %LoG edge detection on sigma=12
subplot(3,3,8), imshow(bw,[]);
[bw,thresh]=edge(imfilter(A,h2),'canny');      %Canny edge detection on sigma=12
subplot(3,3,9), imshow(bw,[]);
```

Comments

Matlab functions: *edge*.

The *edge* function can be called according to a variety of syntaxes. In the examples above, appropriate thresholds are estimated *automatically* based on calculations on the input image. See the Matlab documentation doc edge for full details of use.

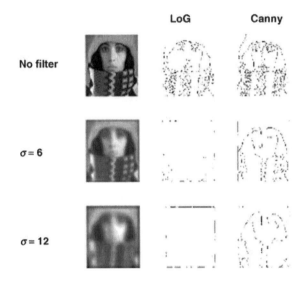

Figure 10.6 Edge detection using the zero-crossing method (LoG filter) and the Canny method for an image at three different degrees of blur. In these examples, the LoG threshold and both the lower and higher threshold for the Canny detector were calculated automatically on the basis of an estimated signal/noise ratio. The superior performance of the Canny filter is evident

10.9 Interest operators

In simple terms, a point is deemed *interesting* if it is distinct from all closely adjacent regions, i.e. all those points in its surrounding neighbourhood. Interest operators are filters which attempt to identify such points of interest automatically. Interest operators do not feature as a 'mainstream' tool for segmentation but are, nonetheless, useful in certain special kinds of problem. They also find application in problems concerned with image matching or registration by geometric transformation, which presupposes the existence of so-called tie points, i.e. known points of correspondence between two images. These are typically points such as spots or corners which are easily distinguishable from their surroundings. The identification of such points of interest is also desirable in other applications, such as

3-D stereographic imaging. The theoretical basis for identification of interesting points is given below.

Consider the change in intensity in an image $I(x, y)$ by displacing from the point (x, y) by some small distance $(\nabla x, \nabla y)$. A first-order Taylor series approximation gives

$$\nabla I = I(x + \nabla x, y + \nabla y) - I(x, y) = \nabla x f_x + \nabla y f_y \qquad (10.3)$$

where the image gradient at point (x, y) is $[f_x, f_y]$.

We note from Equation (10.3) that the change in intensity is *greatest* when the displacement is in the same direction as the gradient vector but is least when the displacement is perpendicular to it. Haralick's criterion deems a point to be interesting when the quantity $(\nabla I)^2$ (squared because negative changes are just as interesting as positive and we do not want them to cancel) summed over a local neighbourhood is large for *any displacement direction*. This condition then satisfies, in the general sense, our constraint that the point be distinct from its surroundings. Using Equation (10.3), we may write a matrix expression for ∇I^2

$$(\nabla I)^2 = (\nabla x f_x + \nabla y f_y)^2 = (\nabla x \quad \nabla y) \begin{pmatrix} \sum_\Omega f_x^2 & \sum_\Omega f_x f_y \\ \sum_\Omega f_x f_y & \sum_\Omega f_y^2 \end{pmatrix} \begin{pmatrix} \nabla x \\ \nabla y \end{pmatrix} = \mathbf{v}^{\mathrm{T}} \mathbf{F} \mathbf{v} \qquad (10.4)$$

where we have denoted the displacement vector

$$\mathbf{v} = \begin{pmatrix} \nabla x \\ \nabla y \end{pmatrix}, \quad \mathbf{F} = \begin{pmatrix} \sum_\Omega f_x^2 & \sum_\Omega f_x f_y \\ \sum_\Omega f_x f_y & \sum_\Omega f_y^2 \end{pmatrix}$$

and the elements inside the matrix $\mathbf{F}$ are summed over a local pixel neighbourhood Ω. Summation over a local neighbourhood is desirable in practice to reduce the effects of noise and measure the characteristics of the image over a small 'patch' with the given pixel at its centre. For convenience (and without imposing any restriction on our final result), we will assume that the displacement vector we consider in Equation (10.4) has unit length, i.e. $\mathbf{v}^{\mathrm{T}} \mathbf{v} = 1$.

Equation (10.4) is a quadratic form in the symmetric matrix $\mathbf{F}$ and it is straightforward to show that the minimum and maximum values of a quadratic form are governed by the eigenvalues of the matrix with the *minimum* value being given by $\|\mathbf{v}\| \lambda_{\min} = \lambda_{\min}$, with $\lambda_{\min}$ the smallest eigenvalue of the matrix $\mathbf{F}$. Equation (10.4) is readily interpetable (in its 2-D form) as the equation of an ellipse, the larger eigenvalue corresponding to the length of the major axis of the ellipse and the smaller eigenvalue to the minor axis. The Haralick method designates a point as interesting if the following two criteria are satisfied:

(1) that $\frac{\lambda_1 + \lambda_2}{2}$ is 'large'. This is roughly interpretable as the 'average radius' of the ellipse or average value of the quadratic over all directions.

(2)   that $\frac{4\det(\mathbf{F})}{\mathrm{Tr}\{\mathbf{F}\}} = 1 - \left(\frac{\lambda_1 - \lambda_2}{\lambda_1 + \lambda_2}\right)^2$ is 'not too small'. This second criterion translates to the requirement that the maximum and minimum values of the quadratic not be too different from each other or, in geometric terms, that the ellipse not be too elongated. It is apparent that a dominant eigenvalue such that $\lambda_1 \gg \lambda_2$ corresponds to a large value in the direction of the first principal axis but a small value in the direction of the second axis.

The question of what is 'large' and 'not too small' in the discussion above is, of course, relative and must be placed in the direct context of the particular image being analysed. 'Large' usually relates to being a member of a selected top percentile, whereas 'not too small' equates to an image-dependent cut-off value which can be systematically adjusted to reduce or increase the number of detected points.

Harris developed a slightly different criterion. He independently defined an interest function given by:

$$R = \lambda_1 \lambda_2 - k(\lambda_1 + \lambda_2)^2 \tag{10.5}$$

which from the basic properties of the trace and determinant of a square matrix can be written as:

$$R = \det[\mathbf{F}] - k(\mathrm{Tr}\{\mathbf{F}\})^2 = \mathbf{F}_{11}\mathbf{F}_{22} - \mathbf{F}_{12}\mathbf{F}_{21} - k(\mathbf{F}_{11} + \mathbf{F}_{22})^2 \tag{10.6}$$

or explicitly as:

$$R(x, y) = (1 - 2k)\sum_{\Omega} f_x^2 \sum_{\Omega} f_y^2 - k\left[\left(\sum_{\Omega} f_x^2\right)^2 + \left(\sum_{\Omega} f_y^2\right)^2\right] - (f_x f_y)^2 \tag{10.7}$$

and the constant k is set to 0.04.

One significant advantage of the Harris function is that it may be evaluated without explicit calculation of the eigenvalues (which is computationally intensive). Note from Equation (10.5), the following limiting forms of the Harris function:

- as $\lambda_2 \to \lambda_1 = \lambda_{\max}$ so $R \to (1 - 2k)\lambda_{\max}^2$ (its maximum possible value)

- as $\lambda_2 \to \lambda_1 \to 0$ so $R \to 0$ (no response – a smooth region)

- if $\lambda_1 \gg \lambda_2$ then $R \to -k\lambda_{\max}^2$.

Thus, a large positive value of R indicates an 'interesting' point, whereas a negative response suggests a dominant eigenvalue (a highly eccentric ellipse) and is 'not interesting'. The basic recipe for detection of interest points is as follows.

Detection of Interest Points

(1)   Smooth the image by convolving the image with a Gaussian filter.

(2)   For each pixel, compute the image gradient.

(3)   For each pixel and its given neighbourhood Ω, compute the 2×2 matrix **F**.

(4)   For each pixel, evaluate the response function $R(x, y)$.

(5)   Choose the interest points as local maxima of the function $R(x, y)$ (e.g. by a nonmaximum suppression algorithm).

For the final step, the non-maximum suppression of local minima, there are several possible approaches. One simple but effective approach (employed in the Example 10.4) is as follows:

- Let R be the interest response function.

- Let S be a filtered version of the interest function in which every pixel within a specified (nonsparse) neighbourhood is replaced by *the maximum value in that neighbourhood.*

- It follows that those pixels whose values are left *unchanged* in S correspond to the local maxima.

We end our discussion with two comments of practical importance. First, image smoothing reduces the size of the image gradients and, consequently, the number of interest points detected (and vice versa). Second, if the size of the neighbourhood Ω (i.e. the *integrative scale*) is too small, then there is a greater probability of rank deficiency in matrix **F**.[3] Larger integrative scale increases the probability of a full-rank matrix. However, it also smoothes the response function $R(x, y)$; thus, too large an integrative scale can suppress the number of local maxima of $R(x, y)$ and, hence, the number of interest points detected.

The example given in Figure 10.7 (produced using the Matlab code in Example 10.4) illustrates the detection of interest points in two images on two different integrative scales using the Harris function. Note in both examples the tendency of smooth regions to have values close to zero and, in particular, of straight edges to return negative values. Straight-edge pixels, by definition, have a significant gradient in one direction (perpendicular to the dark–light boundary) but a zero (or at least very small) gradient response parallel to the edge. This situation corresponds to a dominant eigenvalue in the matrix **F**.

---

[3] Rank deficient matrices have at least one zero eigenvalue.

Response to Harris
interest operator

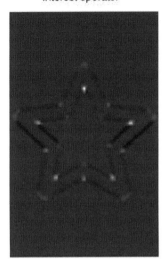

Interesting points
superimposed on original

Figure 10.7 Left: the Harris response function to an image comprising a dark star on a plain background. Right: the significant local maximum points are shown superimposed on the original image

---

Example 10.4

Matlab code	What is happening?
I=imread('zener_star.jpg');	%Read in image and convert to
I=double(rgb2gray(I));	%intensity
wdth=5; sdvn=2;	%Fix smoothing parameters
k=0.04;	%Fix Harris constant
hsmall=fspecial('gaussian',[wdth wdth],sdvn);	%Define Gaussian filter
[Fx,Fy]=gradient(I);	%Calculate gradient
Fx_sq=Fx.^2; Fy_sq=Fy.^2; Fx_Fy=Fx.*Fy;	%Define terms in Harris function
Fx_sq=filter2(hsmall,Fx_sq);	%Perform neighbourhood smoothing
Fy_sq=filter2(hsmall,Fy_sq);	%on each term
Fx_Fy=filter2(hsmall,Fx_Fy);	
R=(1-2.*k).*Fx_sq.*Fy_sq - k.*(Fx_sq.^2 + Fy_sq.^2) - Fx_Fy.^2;	%Calculate Harris function
S=ordfilt2(R,wdth.^2,ones(wdth));	%Maximum filtering over %neighbourhood
[j,k]=find(R>max(R(:))./12 & R==S);	%Get subscript indices of local %maxima
subplot(1,2,1), imagesc(R); axis image; axis off; colormap(gray);	%Display Harris response
subplot(1,2,2), imshow(I,[]);	%Display original image
hold on; plot(k,j,'r*');	%Interest points superimposed
bw=zeros(size(R)); bw([j,k])=1;	%Return logical array of interest
bw=logical(bw);	%locations

Comments
Matlab functions: ***ordfilt2, fspecial, find, logical***.
The interest response is combined with a local maximum filter to ensure that only one local maximum is identified. A subsequent hierarchical thresholding returns the points of maximum interest.

10.10   Watershed segmentation

Watershed segmentation is a relatively recent approach which has come increasingly to the fore in recent years and tends to be favoured in attempts to separate touching objects – one of the more difficult image processing operations. In watershed segmentation, we envisage the 2-D, grey-scale image as a topological surface or 'landscape' in which the location is given by the x,y image coordinates and the *height* at that location corresponds to the image intensity or grey-scale value.

Rain which falls on the landscape will naturally drain downwards, under the action of gravity, to its nearest minimum point. A *catchment basin* defines that connected region or area for which any rainfall drains to the *same* low point or minimum. In terms of a digital image, the catchment basin thus consists of a group of connected pixels. Maintaining the analogy with a physical landscape, we note that there will be points on the landscape (local maxima) at which the rainfall is equally likely to fall into two adjacent catchment basins; this is analogous to walking along the ridge of a mountain. Lines which divide one catchment area from another are called watershed ridges, watershed lines or simply watersheds. An alternative viewpoint is to imagine the landscape being gradually flooded from below with the water entering through the local minima. As the water level increases, we construct dams which prevent the water from the catchment basins spilling or spreading into adjacent catchment basins. When the water level reaches the height of the highest peak, the construction process stops. The dams built in this way are the watersheds which partition the landscape into distinct regions containing a catchment basin. The actual calculation of watersheds in digital images can be performed in several ways, but all fundamentally hinge on iterative morphological operations.

These basic concepts are illustrated for a 1-D landscape in Figure 10.8.

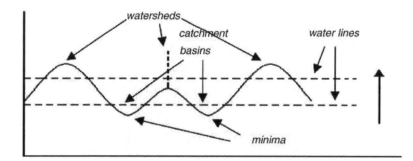

Figure 10.8   Watersheds and catchment basins. Basins are defined as connected regions with one local minimum to which any incident rainfall will flow. The watersheds can be visualized as those ridges from which 'water' would be equally likely to fall towards either of two or more catchment basins

Segmentation using watershed methods has some advantages over the other methods we have discussed in this chapter. A notable advantage is that watershed methods, unlike edge-detection-based methods, generally yield *closed contours* to delineate the boundary of the objects. A number of different approaches can be taken to watershed segmentation, but a central idea is that *we try to transform the initial image (the one we wish to segment) into some other image such that the catchment basins correspond to the objects we are trying to segment.*

10.11 Segmentation functions

A segmentation function is basically a function of the original image (i.e. another image derived from the original) whose properties are such that the catchment basins lie within the objects we wish to segment. The computation of suitable segmentation functions is not always straightforward but underpins successful watershed segmentation. The use of gradient images is often the first preprocessing step in watershed segmentation for the simple reason that the gradient magnitude is usually high along object edges and low elsewhere. In an ideal scenario, the watershed ridges would lie along the object edges.

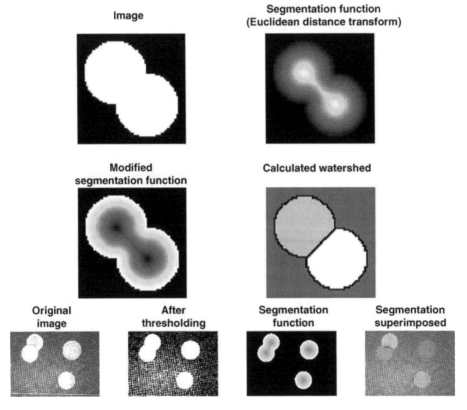

Figure 10.9 In the first, idealized example, the watershed yields a perfect segmentation. In the second image of overlapping coins, morphological opening is required first on the thresholded image prior to calculation of the watershed (*See colour plate section for colour version*)

However, noise and small-scale structures within the objects generally result in many local, small catchment basins (broadly analogous to puddles on the landscape). These spoil things and produce oversegmentation. Figure 10.9 (produced using the Matlab code in Example 10.5) illustrates this.

Example 10.5

Matlab code	**What is happening?**

```
center1 = -10;
```
%Create image comprising two
%perfectly

```
center2 = -center1;
```
%smooth overlapping circles
```
dist = sqrt(2*(2*center1)^2);
radius = dist/2 * 1.4;
lims = [floor(center1-1.2*radius)
eil(center2 + 1.2*radius)];
[x,y] = meshgrid(lims(1):lims(2));
bw1 = sqrt((x-center1).^2
    + (y-center1).^2) <= radius;
bw2 = sqrt((x-center2).^2
    + (y-center2).^2) <= radius;
bw = bw1 | bw2;
```

```
D = bwdist(~bw);
```
%Calculate basic segmentation function
%(Euclidean distance transform
%of ~bw)

```
subplot(2,2,1), imshow(bw,[]);
subplot(2,2,2), imshow(D,[]);
```
%Display image
%Display basic segmentation function
%Modify segmentation function

```
D = -D;
```
%Invert and set background pixels
%*lower*

```
D(~bw) = -inf;
subplot(2,2,3), imshow(D,[]);
L = watershed(D); subplot(2,2,4), Imagesc(L);
```
%than all catchment basin minima
%Display modified segmentation image
%Calculate watershed of segmentation
%function

```
axis image; axis off; colormap(hot); colorbar
```
%Display labelled image – colour
%coded

```
A=imread('overlapping_euros1.png');
bw=im2bw(A,graythresh(A));
se=strel('disk',10); bwo=imopen(bw,se);
D = bwdist(~bwo);
D = -D; D(~bwo) = -255;
```
%Read in image
%Threshold automatically
%Remove background by opening
%Calculate basic segmentation function
%Invert, set background lower than
%catchment basin minima

```
L = watershed(D);
```
%Calculate watershed

```
subplot(1,4,1), imshow(A);
subplot(1,4,2), imshow(bw)
subplot(1,4,3), imshow(D,[]);
```
%Display original
%Thresholded image
%Display basic segmentation function

ind=find(L==0); Ac=A; Ac(ind)=0;   %Identify watersheds and set=0 on
                                                       original

subplot(1,4,4), Imagesc(Ac); hold on   %Segmentation superimposed on
                                                       original

Lrgb = label2rgb(L, 'jet', 'w', 'shuffle');   %Calculate label image
himage = imshow(Lrgb); set(himage,   %Superimpose transparently on original
  'AlphaData', 0.3);

Comments
Matlab functions: ***bwdist, watershed.***
    The Euclidean distance transform operates on a binary image and calculates the distance from each point in the background to its nearest point in the foreground. This operation on the (complement of) the input binary image produces the basic segmentation function. This is modified to ensure the minima are at the bottom of the catchment basins. Computation of the watershed then yields the segmentation. Note the watershed function labels each uniquely segmented region with an integer and the *watersheds are assigned a value of 0.*

In Figure 10.9, the first (artificial) image is smooth and in this case, calculation of a suitable segmentation function is achieved by the Euclidean distance transform.[4] The corresponding image of some real coins in the second example beneath it requires some morphological pre-processing but essentially yields to the same method. However, the segmentation is substantially achieved here (apart from the overlapping coins) by the thresholding. In Figure 10.10 (produced using the Matlab code in Example 10.6) we attempt a gradient-based segmentation function, but this results in the over-segmentation shown. This outcome is typical in all but the most obliging cases.

An approach which can overcome the problem of over-segmentation in certain cases is marker-controlled segmentation, the basic recipe for which is as follows:

(1)   Select and calculate a suitable segmentation function. Recall that this is an image whose catchments basins (dark regions) are the objects you are trying to segment. A typical choice is the modulus of the image gradient.

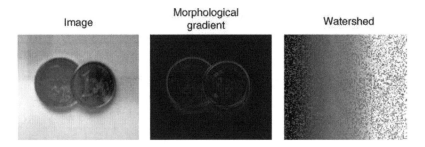

Figure 10.10 Direct calculation of the watershed on gradient images typically results in oversegmentation

---

[4] The Euclidean distance transform codes each foreground pixel as the shortest distance to a background pixel.

Example 10.6

Matlab code	What is happening?
A=imread('overlapping_euros.jpg');	%Read image
Agrad=ordfilt2(A,25,ones(5))-ordfilt2(A,1,ones(5));	%Calculate basic
	%segmentation function
figure, subplot(1,3,1), imshow(A,[]);	%Display image
subplot(1,3,2), imshow(Agrad,[]);	%Display basic segmentation
	%function
L = watershed(Agrad), rgb = label2rgb(L,'hot',[.5 .5 .5]);	%Calculate watershed
subplot(1,3,3), imshow(rgb,'InitialMagnification','fit')	%Display labelled image

Comments

Matlab functions: ***ordfilt2, watershed.***

Watershed on the raw gradient image results in oversegmentation.

(2) Process the original image to calculate so-called foreground markers. Foreground markers are basically connected blobs of pixels lying within each of the objects. This is usually achieved via grey-scale morphological operations of opening, closing and reconstruction

(3) Repeat this process to calculate background markers. These are connected blobs of pixels that are not part of any object.

(4) Process the segmentation function so that it *only possesses minima at the foreground and background marker* locations.

(5) Calculate the watershed transform of the modified segmentation function.

Figure 10.11 shows a (largely) successful marker-controlled segmentation of a relatively complex image.[5] The rationale for this example is explained in Example 10.7.

Marker-controlled segmentation is an effective way of avoiding the problem of over-segmentation. As is evident from the aforementioned example, it generally needs and exploits a priori knowledge of the image and the methods required for marker selection are strongly dependent on the specific nature of the image. Thus, its strength is that it uses the specific context effectively. Conversely, its corresponding weakness is that it does not generalize well and each problem must be treated independently.

---

[5] This example is substantially similar and wholly based on the watershed segmentation example provided by the image processing demos in Matlab version 7.5. The authors offer grateful acknowledgments to The MathWork Inc. 3 Apple Hill Dr. Natick, MA, USA.

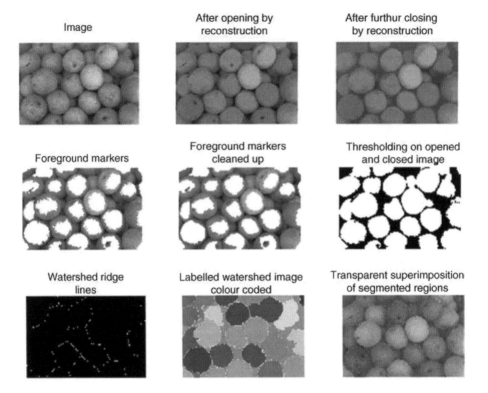

Image | After opening by reconstruction | After furthur closing by reconstruction

Foreground markers | Foreground markers cleaned up | Thresholding on opened and closed image

Watershed ridge lines | Labelled watershed image colour coded | Transparent superimposition of segmented regions

Figure 10.11   Marker-controlled watershed segmentation (*See colour plate section for colour version*)

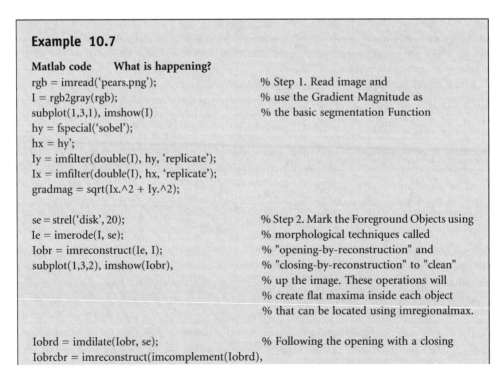

Example 10.7

Matlab code   What is happening?

```
rgb = imread('pears.png');
I = rgb2gray(rgb);
subplot(1,3,1), imshow(I)
hy = fspecial('sobel');
hx = hy';
Iy = imfilter(double(I), hy, 'replicate');
Ix = imfilter(double(I), hx, 'replicate');
gradmag = sqrt(Ix.^2 + Iy.^2);


se = strel('disk', 20);
Ie = imerode(I, se);
Iobr = imreconstruct(Ie, I);
subplot(1,3,2), imshow(Iobr),




Iobrd = imdilate(Iobr, se);
Iobrcbr = imreconstruct(imcomplement(Iobrd),
```

% Step 1. Read image and
% use the Gradient Magnitude as
% the basic segmentation Function

% Step 2. Mark the Foreground Objects using
% morphological techniques called
% "opening-by-reconstruction" and
% "closing-by-reconstruction" to "clean"
% up the image. These operations will
% create flat maxima inside each object
% that can be located using imregionalmax.

% Following the opening with a closing

```
imcomplement(Iobr));
Iobrcbr = imcomplement(Iobrcbr);
subplot(1,3,3); imshow(Iobrcbr);
```
% to remove the dark spots and stem marks.
% Notice you must complement the image
% inputs and output of imreconstruct.

```
fgm = imregionalmax(Iobrcbr);
```
% Calculate the regional maxima of Iobrcbr
% to obtain good foreground markers.

```
I2 = I; I2(fgm) = 255;
figure; subplot(1,3,1); imshow(I2);
```
% To help interpret the result, superimpose
% these foreground marker image on the
% original image.

```
se2 = strel(ones(5,5));
fgm2 = imclose(fgm, se2);
fgm3 = imerode(fgm2, se2);
```
% Some of the mostly-occluded and shadowed
% objects are not marked, which means that
% these objects will not be segmented
% properly in the end result. Also, the
% foreground markers in some objects go
% right up to the objects' edge. That means
% we must clean the edges of the marker
% blobs and then shrink them a bit.

```
fgm4 = bwareaopen(fgm3, 20);
I3 = I; I3(fgm4) = 255;
subplot(1,3,2), imshow(I3);
```
% This procedure leaves some stray isolated
% pixels that must be removed. Do this
% using bwareaopen, which removes all blobs
% that have fewer than a certain number of
% pixels.

```
bw = im2bw(Iobrcbr,
    graythresh(Iobrcbr));
subplot(1,3,3), imshow(bw);
```
%Step 3: Compute Background Markers

% Now we need to mark the background. In
% the cleaned-up image, Iobrcbr, the
% dark pixels belong to the background,
% so you could start with a thresholding
% operation.

```
D = bwdist(bw);
DL = watershed(D);
bgm = DL == 0;
figure; subplot(1,3,1);imshow(bgm);
```
% The background pixels are in black,
% but ideally we don't want the background
% markers to be too close to the edges of
% the objects we are trying to segment.
% We'll "thin" the background by computing
% the "skeleton by influence zones", or SKIZ,
% of the foreground of bw. This can be done
% by computing the watershed transform of the
% distance transform of bw, and
% then looking for the watershed ridge
% lines (DL == 0) of the result.

```
gradmag2 = imimposemin
    (gradmag, bgm | fgm4);
```
% Step 4: Compute the modified

% segmentation function.
% The function imimposemin is used to modify

```
                         % an image so that it has regional minima
                         % only in certain desired locations.
                         % Here you can use imimposemin to modify
                         % the gradient magnitude image so that its
                         % regional minima occur at foreground and
                         % background marker pixels.

L = watershed(gradmag2);    % Step 5: Now compute the Watershed
                            % Transform of modified function
Lrgb = label2 rgb(L, 'jet',  % A useful visualization technique is to display the
    'w', 'shuffle');        % label matrix L as a color
subplot(1,3,2),imshow(Lrgb)  % image using label2rgb. We can then superimpose this
subplot(1,3,3),imshow(I),    %pseudo-color label
    hold on                 % matrix on top of the original intensity image.
himage = imshow(Lrgb);
    set(himage, 'AlphaData', 0.3);
```

10.12 Image segmentation with markov random fields

Except in the most obliging of situations, intensity thresholding alone is a crude approach to segmentation. It rarely works satisfactorily because it does not take into account the spatial relationships between pixels. Segmentation via region growing or splitting enables pixels to be labelled based on their similarity with neighbouring pixels. The use of Markov random field models is a more powerful technique which determines how well a pixel belongs to a certain class by modelling both the variability of grey levels (or other pixel attributes) within each class and its dependency upon the classification of the pixels in its neighbourhood.

Segmentation using Markov random fields is not a simple technique. Our aim here will be strictly limited to discussing the essence of the approach so that an interested reader is then better equipped to go on to examine the literature on this subject. We shall start by stating Bayes' law of conditional probability:

$$p(x|y) = \frac{p(y|x)p(x)}{p(y)} \tag{10.8}$$

In our context of image segmentation, we shall suppose we are given an image containing a range of grey levels denoted by y and a set of possible labels denoted by x. In the simplest segmentation scenarios, there may, for example, be just two labels, referred to as *object* and *background*.

Our basic task is to assign to each pixel the most probable label x given our set of observations (i.e. our image) y. In other words, we wish to maximize $p(x|y)$ which, by Bayes' theorem, is equivalent to maximizing the right-hand side of Equation (10.8). Let us then carefully consider each term in Equation 10.8.

- First, $p(y)$ refers to the probability distribution of pixel grey levels (i.e. the distribution which would be obtained from a grey-level histogram). *For a given image*, this quantity is fixed. Therefore, if $p(y)$ is a constant for a given image, then maximizing $p(x|y)$ reduces to maximising $p(y|x)p(x)$.

- Second, the conditional density $p(y|x)$ refers to the probability distribution of pixel grey levels y for a given class x (e.g. background or object). In the ideal case, a background would have one grey level and an object would have a second distinct grey level. In reality, the grey levels for both classes will exhibit a spread. The densities are often modelled as normal distributions:

$$p(y|x) = \frac{1}{\sqrt{2\pi}\sigma_x} \exp\left[-\frac{(y-\mu_x)^2}{2\sigma_x^2}\right] \tag{10.9}$$

where μ_x and σ_x respectively refer to the mean and standard deviation of the pixel intensity in the class x.

- Third, the term $p(x)$, known as the prior distribution, refers to the level of agreement between the pixel's current label and the label currently assigned to its neighbouring pixels. This requires some elaboration. Intuitively, we would reason that $p(x)$ should return a low value if the pixel's label is *different* from all of its neighbours (since it is very unlikely that an isolated pixel would be surrounded by pixels belonging to a different class). By the same line of reasoning, $p(x)$ should increase in value as the number of identically labelled neighbouring pixels increases.

The whole approach thus hinges on the prior term in Bayes' law, because it is this term which must build in the spatial context (i.e. the correlation relations between a pixel and its neighbours).

To the question 'How exactly is the prior information built into our model?', we cannot give a full explanation here (this would require a long and complex detour into mathematical concepts traditionally associated with statistical physics), but the basic approach is to model the relationship between a pixel and its neighbours as a Markov random field. A Markov random field essentially models the probabilities of certain combination of intensity values occurring at various pixel locations in a neighbourhood region. To specify such probabilities directly is generally a difficult and even overwhelming task because there are a huge number of possible combinations. Fortunately, a powerful mathematical result exists which demonstrates that the right-hand side of Equation (10.8) can be written in the form of an exponential which is known as a Gibbs distribution:

$$p(x|y) = \frac{\exp[-U(x|y)]}{Z_y} \tag{10.10}$$

where Z_y is a normalizing constant which is included to prevent the above equation from returning a probability of greater than one. The exponential part of this equation turns out to be:

$$U(x|y) = \sum_i \left[\frac{1}{2}\ln \sigma_{x_i}^2 + \frac{(y-\mu_{x_i})^2}{2\sigma_{x_i}^2} + \sum_{n=1}^{N} \theta_n J(x_i x_{i+n}) \right] \qquad (10.11)$$

By analogy with arguments from statistical physics, $U(x|y)$ can be loosely interpreted as a 'potential energy' term and it is clear from Equation (10.10) that the maximization of $p(x|y)$ reduces to minimizing the energy term $U(x|y)$.

In Equation (10.11), the outer summation covers all pixels in the image (i.e. the contents are applied to *every* pixel in the image). The first two terms of the outer summation refer to $p(y|x)$ as given in Equation (10.9), which expresses how well the pixel's grey level fits its presumed class label. The third term represents $p(x)$ and expresses the level of agreement between the pixel's label and those of its neighbours. This is itself expressed in the form of a summation over the pixel's immediate neighbours. The function $J(x_i, x_{i+n})$ returns a value of -1 whenever the label for the nth neighbourhood pixel x_{i+n} is the same as the label for the central pixel x_i. The term θ_n is included to adjust the relative weighting between the $p(x)$ and $p(y|x)$ terms.

To complete our formulation of the problem, we need to resolve two questions:

(1) How do we model the class-conditional densities $p(y|x)$ (the parameter estimation problem)?

(2) How do we select the neighbourhood weighting parameters θ_n to properly balance the two contributions to the energy function?

10.12.1 Parameter estimation

Assuming the Gaussian distribution for the class-conditional densities $p(y|x)$ given by Equation (10.9), the two free parameters of the distribution (the mean and variance) can be found by fitting Gaussian curves to a representative (model) image which has been manually segmented into the desired classes. Where such a representative image is not available, the parameters can be estimated by fitting such curves to the actual image histogram. For a two-class segmentation problem (object and background), we would thus fit a Gaussian mixture (sum of two Gaussian functions) to the image histogram $h(y)$, *adjusting the free parameters to achieve the best fit:*

$$h(y) = \sum_{i=1}^{2} \frac{1}{\sqrt{2\pi}\sigma_{x_i}} \exp\left[-\frac{(y-\mu_{x_i})^2}{2\sigma_{x_i}^2}\right]$$

$$= \frac{1}{\sqrt{2\pi}\sigma_{\mathrm{Obj}}} \exp\left[-\frac{(y-\mu_{\mathrm{Obj}})^2}{2\sigma_{\mathrm{Obj}}^2}\right] + \frac{1}{\sqrt{2\pi}\sigma_{\mathrm{Back}}} \exp\left[-\frac{(y-\mu_{\mathrm{Back}})^2}{2\sigma_{\mathrm{Back}}^2}\right]$$

(10.12)

The extension to more than two classes follows naturally.

10.12.2   *Neighbourhood weighting parameter θ_n*

The parameter θ_n represents the weighting which should be given to the local agreement between pixel labels. For example, a pixel which is presently labelled as 'object' should have a higher probability of being correct if, for example, six of its neighbouring pixels were

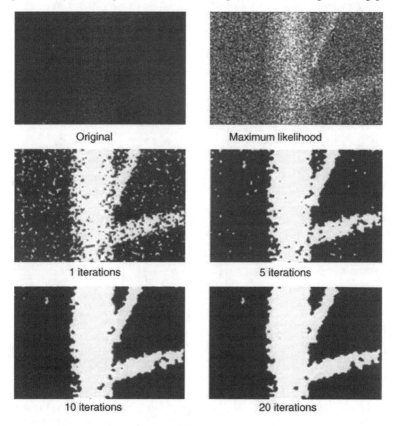

Original   Maximum likelihood

1 iterations   5 iterations

10 iterations   20 iterations

Figure 10.12   Segmentation of a noisy, underwater image of a cylindrical pipe structure. The sequence from left to right and top to bottom shows the original image, the maximum likelihood estimate and the results of segmentation using increasing numbers of iterations of the ICM algorithm. (Image sequence courtesy of Dr Mark Hodgetts, Cambridge Research Systems Ltd.)

identically labelled as 'object' as opposed to three. The analytical approach to determining θ_n would involve initially labelling all pixels according to pixel attribute alone (i.e. in the absence of any contextual information). Every labelled pixel in the image would then be examined and a note would be made of the number of identically labelled neighbouring pixels. Counts would be made for every combination of pixel label and number of agreeing neighbouring pixel labels. These would be used to infer how the probability of the central pixel label being correct varies with the number of consistent neighbouring labels, which would in turn be used to generate estimates for θ_n. In practice, most implementations treat θ_n as a constant θ which is initially estimated at around unity and adjusted until optimal results are obtained. A higher value of θ will tend to emphasize local agreement of pixel labels over the class suitability according to pixel attribute and vice versa.

10.12.3 Minimizing U(x|y): the iterated conditional modes algorithm

Having formulated the problem and estimated the parameters of the class-conditional densities and the neighbourhood weighting parameter, the final step in the segmentation is to actually find a viable means to minimize $U(x|y)$ in Equation (10.11). We briefly describe the iterated conditional modes (ICM) algorithm, as it is conceptually simple and computationally viable. The ICM algorithm is an example of a 'greedy algorithm', as it works on a pixel-by-pixel basis, accepting *only* those changes (i.e. changes from background to foreground or vice versa) which take us successively nearer to our goal of minimizing the potential. This is in contrast to other approaches, such as simulated annealing (Geman and Geman) which allow temporary increases in the potential function to achieve the overall goal of minimization. The ICM algorithm decomposes Equation (10.11) so that it is applied to single pixels in the image

$$U_i(x|y) = \frac{1}{2}\ln\sigma^2_{x_i} + \frac{(y_i-\mu_{x_i})^2}{2\sigma^2_{x_i}} + \sum_{n=1}^{N}\theta_n J(x_i x_{i+n}) \qquad (10.13)$$

- Step 1: label each pixel with its most probable class based on its pixel attribute only, i.e. according to whichever of the class-conditional densities yields the maximum value. This is termed the maximum-likelihood estimate.

- Step 2: for each pixel in the image, update it to the class which minimizes Equation 10.13.

- Repeat Step 2 until stability occurs.

Figure 10.12 shows an example of Markov random field segmentation using the ICM algorithm.

For further examples and exercises see http://www.fundipbook.com

11

Classification

11.1 The purpose of automated classification

The identification of cells in a histological slide as healthy or abnormal, the separation of 'high-quality' fruit from inferior specimens in a fruit-packing plant and the categorization of remote-sensing images are just a few simple examples of *classification* tasks. The first two examples are fairly typical of *binary classification*. In these cases, the purpose of the classification is to assign a given cell or piece of fruit to either of just two possible classes. In the first example above, the two classes might be 'healthy' and 'abnormal'; in the second example, these might be 'top quality' (expensive) or 'market grade' (cheap). The third example typically permits a larger number of classes to be assigned ('forest', 'urban', 'cultivated', 'unknown', etc.).

Autonomous classification is a broad and widely studied field that more properly belongs within the discipline of *pattern recognition* than in image processing. However, these fields are closely related and, indeed, classification is so important in the broader context and applications of image processing that some basic discussion of classification seems essential. In this chapter we will limit ourselves to some essential ideas and a discussion of some of the best-known techniques.

Classification takes place in virtually all aspects of life and its basic role as a necessary first step before selection or other basic forms of decision-making hardly needs elaboration. In the context of image processing, the goal of classification is to identify characteristic features, patterns or structures within an image and use these to assign them (or indeed the image itself) to a particular class. As far as images and visual stimuli go, human observers will often perform certain classification tasks very accurately. Why then should we attempt to build automatic classification systems? The answer is that the specific demands or nature of a classification task and the sheer volume of data that need to be processed often make automated classification the only realistic and cost-effective approach to addressing a problem. Nonetheless, expert human input into the design and training of automated classifiers is invariably essential in two key areas:

(1) *Task specification*   What exactly do we want the classifier to achieve? The designer of a classification system will need to decide what *classes are going to be considered* and *what*

Fundamentals of Digital Image Processing – A Practical Approach with Examples in Matlab
Chris Solomon and Toby Breckon
© 2011 John Wiley & Sons, Ltd

variables or parameters are going to be important in achieving the classification.[1] For example, a simple classifier designed to make a preliminary medical diagnosis based on image analysis of histological slides may only aim to classify cells as 'abnormal' or 'normal'. If the classifier produces an 'abnormal' result, a medical expert is typically called upon to investigate further. On the other hand, it may be that there is sufficient information in the shape, density, size and colour of cells in typical slides to attempt a more ambitious classification system. Such a system might assign the patient to one of a number of categories, such as 'normal', 'anaemic', 'type A viral infection' and so on.

(2) *Class labelling* The process of training an automated classifier can often require 'manual labelling' in the initial stage, a process in which an expert human user assigns examples to specific classes based on selected and salient properties. This forms part of the process in generating so-called *supervised* classifiers.

11.2 Supervised and unsupervised classification

Classification techniques can be grouped into two main types: *supervised* and *unsupervised*. Supervised classification relies on having example pattern or feature vectors which have already been assigned to a defined class. Using a sample of such feature vectors as our training data, we design a classification system with the intention and hope that *new examples* of feature vectors which were not used in the design will subsequently be classified accurately. In supervised classification then, the aim is to use training examples to design a classifier which *generalizes well* to new examples. By contrast, unsupervised classification does not rely on possession of existing examples from a known pattern class. The examples are not labelled and we seek to identify groups directly within the overall body of data and features which enables us to distinguish one group from another. Clustering techniques are an example of unsupervised classification which we will briefly discuss later in the chapter.

11.3 Classification: a simple example

Consider the following illustrative problem. Images are taken at a food processing plant in which three types of object occur: pine-nuts, lentils and pumpkin seeds. We wish to design a classifier that will enable us to identify the three types of object accurately. A typical image frame is shown in Figure 11.1.

Let us assume that we can process these images to obtain reliable measurements on the two quantities of circularity and line-fit error.[2] This step (often the most difficult) is called *feature extraction*. For each training example of a pine-nut, lentil or pumpkin seed we thus obtain a 2-D *feature vector* which we can plot as a point in a 2-D *feature space*. In Figure 11.2,

---

[1] In a small number of specialist applications, it is possible even to attempt to identify identify appropriate patterns and classes automatically; but generally, the expert input of the designer at this stage is absolutely vital.

[2] The details of these measures and how they are obtained from the image do not concern us here.

Figure 11.1 Three types of object exist in this image: pine-nuts, lentils and pumpkin seeds. The image on the right has been processed to extract two: circularity and line-fit error

training examples of pine-nuts, lentils and pumpkin seeds are respectively identified by squares, triangles and circles.

We can see by direct inspection of Figure 11.2 that the three classes form more or less distinct clusters in the feature space. This indicates that these two chosen features are broadly adequate to discriminate between the different classes (i.e. to achieve satisfactory classification). By contrast, note from Figure 11.2 what happens if we consider the use of just one of these features in isolation. In this case, the feature space reduces to a single dimension given by the orthogonal projection of the points either onto the vertical (line-fit error) or horizontal (circularity) axis. In either case, there is considerable overlap or confusion of classes, indicating that misclassification occurs in a certain fraction of cases.

Now, the real aim of any automatic classification system is to *generalize* to new examples, performing accurate classification on objects or structures whose class is *a priori* unknown.

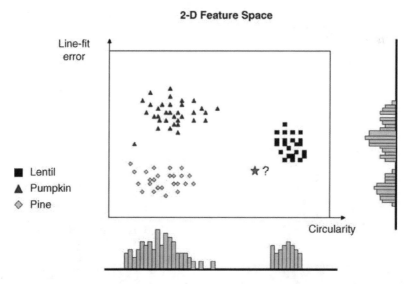

Figure 11.2 A simple 2-D feature space for discriminating between certain objects. The 1-D bar plots in the horizontal *x* and vertical *y* directions show the projection of the data onto the respective axes

Consider, then, a previously unseen object having the feature vector indicated in Figure 11.2 by a star. How should we assign this object to its appropriate class? As this object has features which are closer to the lentils, common sense would suggest that it is more likely to be a lentil, but where should we draw the boundary between classes and how sure can we be about our decision? Most of the classification methods we will explore in the remainder of this chapter are concerned with just such questions and finding rigorous mathematical criteria to answer them. First, however, we move from our simple example to a more general overview of the steps in the design of classifiers.

11.4   Design of classification systems

Figure 11.3 provides a summary flow chart for classifier design. We elaborate below on the main steps indicated in Figure 11.3.

(1)   *Class definition*   Clearly, the definition of the classes is *problem specific*. For example, an automated image-processing system that analysed mammograms might ultimately aim to classify the images into just two categories of interest: normal and abnormal. This would be a *binary classifier* or, as it is sometimes called, a dichotomizer. On the other hand, a more ambitious system might attempt a more detailed diagnosis, classifying the scans into several different classes according to the preliminary diagnosis and the degree of confidence held.

(2)   *Data exploration*   In this step, a designer will explore the data to identify possible attributes which will allow discrimination between the classes. There is no fixed or best way to approach this step, but it will generally rely on a degree of intuition and common sense. The relevant attributes can relate to absolutely any property of an image or image region that might be helpful in discriminating one class from another.

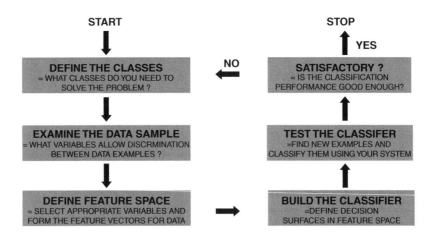

Figure 11.3   Flow diagram representing the main steps in classifier design

Most common measures or attributes will be broadly based on *intensity, colour, shape, texture* or some mixture thereof.

(3) *Feature selection and extraction*   Selection of the discriminating features defines the feature space. This, of course, implicitly assumes that we have established reliable image-processing procedures to perform the necessary feature extraction.[3] In general, it is crucial to select features that possess two key properties. The first is that the feature set be as *compact* as possible and the second that they should possess what, for want of a more exact word, we will call *discriminatory power*. A compact set of features is basically a small set and it is of paramount importance, as larger numbers of selected features require an increasingly large number of training samples to train the classifier effectively. Second, it obviously makes sense to select (i) attributes whose distribution over the defined classes is as widely separated as possible and (ii) that the selected set of attributes should be, as closely as possible, statistically independent of each other. A set of attributes possessing these latter two properties would have maximum discriminatory power.

A simple but concrete illustration of this idea might be the problem of trying to discriminate between images of a lion and a leopard. Defining feature vectors which give some measure of the boundary (e.g. a truncated radial Fourier series) or some gross shape measure like form factor or elongation[4] constitute one possible approach. However, it is clearly not a very good one, since lions and leopards are both big cats of similar physique. A better measure in this case will be one based on *colour*, since the distribution in r–g chromatic space is simple to measure and is quite sufficient to discriminate between them.

(4) *Build the classifier using training data*   The training stage first requires us to find a sample of examples which we can reliably assign to each of our selected classes. Let us assume that we have selected N features which we anticipate will be sufficient to achieve the task of discrimination and, hence, classification. For each training example we thus record the measurements on the N features selected as the elements of an N-dimensional feature vector $\mathbf{x} = [x_1, x_2, \cdots, x_N]$. All the training examples thus provide feature vectors which may be considered to occupy a specific location in an (abstract) N-dimensional feature space (the jth element x_j specifying the length along the jth axis of the space). If the feature selection has been performed judiciously, then the feature vectors for each class of interest will form more or less distinct clusters or groups of points in the feature space.

(5) *Test the classifier*   The classifier performance should be assessed on a *new sample* of feature vectors to see how well it can generalize to new examples. If the performance is unsatisfactory (what constitutes an unsatisfactory performance is obviously application dependent), the designer will return to the drawing board, considering whether

---

[3] Something that is often very challenging but which, in the interest of staying focused, we will gloss over in this chapter.

[4] The formal definitions of these particular measures are given in Chapter 9.

Table 11.1   Some common pattern classification terms with summary descriptions

Term	Description
Feature vector (pattern vector)	An N-dimensional vector, each element of which specifies some measurement on the object
Feature space	The abstract (N-dimensional) mathematical space spanned by the feature vectors used in the classification problem
Training data	A collection of feature vectors used to build a classifier
Test data	A collection of feature vectors used to test the performance of a classifier
Pattern class	A generic term which encapsulates a group of pattern or feature vectors which share a common statistical or conceptual origin
Discriminant function	A function whose value will determine assignment to one class or another; typically, a positive evaluation assigns to one class and negative to another

the selected classes are adequate, whether the feature selection needs to be reconsidered or whether the classification algorithm itself needs to be revised. The entire procedure would then be repeated until a satisfactory performance was achieved.

Table 11.1 summarizes some of the main terms used in the field of pattern classification.

11.5   Simple classifiers: prototypes and minimum distance criteria

A prototype is an instance of something serving as a typical example or standard for other things of the same category. One of the simplest and most common methods of defining a prototype for a particular class is to take a number of training examples for that class and calculate their average. Figure 11.4 shows a set of 8 male faces and 8 female faces and their calculated prototypes. In this particular example, the original images contain many tens of thousands of pixels. By a procedure based on PCA, however, it is possible to represent these faces as lower dimensional, feature vectors $\{x_i\}$ whose components provide a compact description of both the shape and texture of the face and from which the original images can be reconstructed with high accuracy. The prototypes **M** and **F** of the two classes are simply calculated as

$$\mathbf{M} = \frac{1}{N}\sum_{i=1}^{N}\mathbf{x}_i^M \qquad \text{and} \qquad \mathbf{F} = \frac{1}{N}\sum_{i=1}^{N}\mathbf{x}_i^F \tag{11.1}$$

where $\{\mathbf{x}_i^M\}$ correspond to the male examples and $\{\mathbf{x}_i^F\}$ the female examples and there are N examples of each. The reconstructed prototypes shown in Figure 11.4 are given by applying Equation (11.1) and then reconstructing the original images from the feature vectors.

Minimum distance classifiers are conceptually simple and generally easy to calculate. The first step is to calculate the mean feature vector or *class prototype* for each class. When

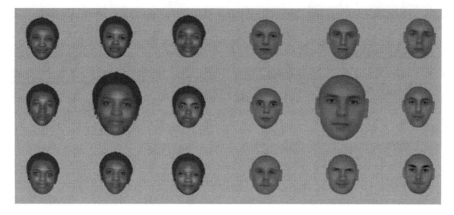

Figure 11.4 Facial prototypes: Left: 8 female faces with the prototype at the centre. Right: 8 male faces with the calculated prototype at the centre. The prototypes are calculated by averaging the shape coordinates of each example and then averaging the RGB colour values of each face after warping to the average shape

presented with a previously unseen feature vector, the *minimum distance classifier will simply assign the new example to that class whose prototype lies closest to it.* We note immediately that the concept of 'distance' in a feature space is a flexible one and admits of a number of other valid definitions apart from the familiar Euclidean measure. However, the principle of a minimum distance classifier is the same irrespective of the precise metric employed. Table 11.2 gives several such metrics which are sometimes used.

11.6 Linear discriminant functions

To apply a minimum distance classifier by calculating the Euclidean distance in the feature space from a given point to the prototype of each class involves a quadratic function of distance and is a straightforward enough procedure. However, it is

Table 11.2 Some common distance measures or *metrics* between two N-dimensional vectors $\mathbf{x}$ and $\mathbf{y}$

Metric	Definition
Euclidean distance (L_2 norm)	$L(\mathbf{x}, \mathbf{y}) = \left[\sum_{i=1}^{N} (x_i - y_i)^2 \right]^{1/2}$
City-block* (L_1 norm)	$L(\mathbf{x}, \mathbf{y}) = \sum_{i=1}^{N} |x_i - y_i|$
Minkowski metric (L_k norm)	$L_k(\mathbf{x}, \mathbf{y}) = \left[\sum_{i=1}^{N} |x_i - y_i|^k \right]^{1/k}$
Mahalanobis distance	$L(\mathbf{x}, \mathbf{y}) = [(\mathbf{x} - \mathbf{y})^{\mathrm{T}} \Sigma^{-1} (\mathbf{x} - \mathbf{y})]^{1/2}$, where $\Sigma = \langle (\mathbf{x} - \vec{\mu}_x)(\mathbf{y} - \vec{\mu}_y)^{\mathrm{T}} \rangle$

*Also known as the Manhattan Distance.

possible to show that application of the minimum distance criterion reduces to a formulation that is of greater utility when we extend our discussion to more complex criteria. It is thus useful to introduce at this point the concept of a (*linear*) *discriminant function*.

For simplicity, consider constructing a line in a 2-D feature space which orthogonally bisects the line connecting a given pair of class prototypes (let us say from the ith prototype possessing coordinates (x_1^i, x_2^i) to the jth prototype (x_1^j, x_2^j)) and passes through the exact midpoint between them. This line divides the feature space into two regions, one of which is closer to the ith prototype, the other to the jth prototype and any point lying on the line is equidistant from the two prototypes. This line is the simplest example of a linear discriminant function. It is easy to show that this function can be expressed in the general form

$$g_{ij}(\mathbf{x}) = g_{ij}(x_1, x_2) = ax_1 + bx_2 + c = 0 \qquad (11.2)$$

If we evaluate $g_{ij}(\mathbf{x})$ at an arbitrary position within the feature space $\mathbf{x} = (x_1, x_2)$, then a positive value indicates that the point lies closer to the jth prototype, whereas a negative value indicates greater proximity to the ith prototype.

In principle, we may calculate a linear discriminant of this kind for every pair of classes. Figure 11.5 depicts the data in the simple three-class problem described in Section 11.3; therefore, we have $\binom{3}{2} = 3$ linear discriminant functions. By calculating linear discriminants for all pairs of prototypes and evaluating each discriminant function, it is possible in principle to classify a feature vector simply by observing the signs of the discriminants and applying basic logic (see table in Figure 11.5).

However, it is normal practice in N-class problems to define discriminant functions (whether linear or otherwise) in the following way:

$$\begin{aligned} &\text{if} \quad\quad g_j(\mathbf{x}) > g_k(\mathbf{x}) \quad\quad \text{for all } j \neq k \\ &\text{then} \quad \text{assign } \mathbf{x} \text{ to class } j \end{aligned} \qquad (11.3)$$

It is evident, then, that the discriminant functions for each class consist of *segments of such linear functions* and combine to define *decision boundaries* which partition the feature space into distinct regions. Figure 11.5 (top right) shows the *decision boundaries* in the 2-D feature space of our example.

Consider male and female faces; these differ in visual appearance, and human observers are generally very adept at distinguishing between them. Could this be achieved through an automatic classifier? In Example 11.1, we use a 2-D feature vector based on the sixth and seventh component values extracted through a PCA on suitably normalized frontal images of the human face. Ten of these subjects are male and ten female and the region of the face subjected to the PCA is shown in Figure 11.6. It is apparent that use of just these two components does a reasonably good job and a simple (Euclidean) minimum distance classifier results in just two misclassifications.

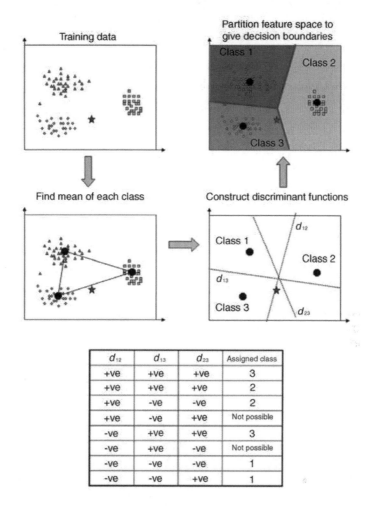

d_{12}	d_{13}	d_{23}	Assigned class
+ve	+ve	+ve	3
+ve	+ve	+ve	2
+ve	-ve	-ve	2
+ve	-ve	+ve	Not possible
-ve	+ve	+ve	3
-ve	+ve	-ve	Not possible
-ve	-ve	-ve	1
-ve	-ve	+ve	1

Figure 11.5 Illustration of the minimum distance classifier, the construction of linear discriminant functions and the resulting decision boundaries

Example 11.1

Matlab Example 11.1	**What is happening?**
load PCA_data.mat;	%Load PCA coefficients for males %and females
f6=female(6,:)./1000; f7=female(7,:)./1000;	%extract female coefficients on %6 and 7
m6=male(6,:)./1000; m7=male(7,:)./1000;	%extract male coefficients on 6 and 7
F_proto=mean([f6' f7']);	
M_proto=mean([m6' m7']);	%calculate Male and female %prototypes
dely=M_proto(2)-F_proto(2);	
delx=M_proto(1)-F_proto(1);	%Difference between prototypes
mid_pt=[F_proto(1)+delx./2 F_proto(2)+dely./2];	%Midpoint between prototypes

```
grad=-1./(dely./delx);                        %Gradient of linear discriminant
a=1./(mid_pt(2)-grad.*mid_pt(1));             %Coefficients of linear discriminant
b=grad./(mid_pt(2)-grad.*mid_pt(1));

figure;
plot(f6,f7,'r*'); hold on; plot(m6,m7,'kd');  %Plot data
x=linspace(-5,5,100); y=(1+b.*x)./a;          %Plot linear discriminant
plot(x,y,'b-');
f_indx=find(a.*f7-b.*f6-1>0)                   %Find index of misclassified females
m_indx=find(a.*m7-b.*m6-1                       %Find index of misclassified males
```

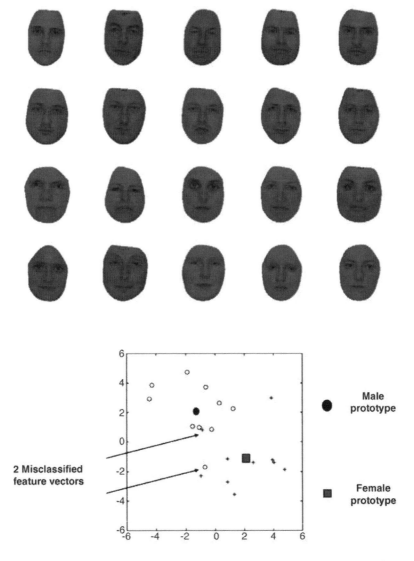

Figure 11.6 Thumbnail images of human faces and their respective positions in a 2-D feature space based on the 6th and 7th principal components. Male faces are plotted as circles and female faces as asterisks. A minimum distance classifier results in two misclassifications

Linear discriminants are important not just for simple minimum-distance classifiers, but find much wider application. For this reason, the next section extends the concept to a general *N*-dimensional feature space and looks at some of their key properties.

11.7 Linear discriminant functions in *N* dimensions

Although a number of interesting classification problems are two-class problems, the majority require the definition of multiple classes. The widespread popularity of linear discriminants in classification problems originates from their simplicity, easy computation and the fact that they extend simply and naturally to higher dimensional feature spaces.

The general form of the *N*-dimensional, linear discriminant $g(\mathbf{x})$ is

$$g(\mathbf{x}) = g(x_1, x_2 \cdots x_N) = \sum_{k=0}^{N} w_k x_k = w_0 + w_1 x_1 + w_2 x_2 + \cdots w_N x_N = 0$$

or expressed in compact, vector form

$$g(\mathbf{x}) = \mathbf{w} \cdot \mathbf{x} + w_0 = 0 \tag{11.4}$$

For a 2-D feature space, the discriminant describes the equation of a straight line and divides a plane into two partitions. For a 3-D feature space, the discriminant describes the equation of *a plane* and divides a volume into two partitions. In the most general *N*-dimensional case, the discriminant describes the equation of *a hyperplane* and divides the *N*-dimensional volume into two partitions.

We note two important properties. Let us consider two arbitrary but distinct vectors $\mathbf{x}_1$ and $\mathbf{x}_2$ both lying on the hyperplane. Since $g(\mathbf{x}) = 0$ for any $\mathbf{x}$ that lies on the hyperplane, it follows that

$$g(\mathbf{x}_1) - g(\mathbf{x}_2) = 0 \quad \Rightarrow \quad \mathbf{w} \cdot (\mathbf{x}_1 - \mathbf{x}_2) = 0 \tag{11.5}$$

Since the arbitrary vector $(\mathbf{x}_1 - \mathbf{x}_2)$ lies *within* the plane,[5] we conclude that the vector $\mathbf{w}$ is perpendicular to any vector lying in the hyperplane.

The discriminant function $g(\mathbf{x})$ also gives an algebraic measure of the shortest (i.e. perpendicular) distance s of the point $\mathbf{x}$ to the dividing hyperplane. It is possible to show that this is:

$$s = \frac{g(\mathbf{x})}{\|\mathbf{w}\|} \tag{11.6}$$

---

[5] More precisely, the vector $(\mathbf{x}_1 - \mathbf{x}_2)$ lies within the subspace defined by the hyperplane.

11.8   Extension of the minimum distance classifier and the Mahalanobis distance

The basic (Euclidean) minimum distance classifier certainly has the virtue of simplicity and is adequate for some applications. However, a little consideration will reveal that it has clear limitations. A classifier that only considers the mean of a distribution of feature vectors (i.e. prototypes) but takes no account of the *statistical distribution of feature vectors about that mean* cannot generally be optimal. Example 11.2 and the resulting Figure 11.7 illustrate this point graphically. In this problem, we simulate two classes of object lying within an arbitrary 2-D feature space. Class 1 (depicted by blue points) and class 2 (depicted by red points) are both normally distributed but have different means and covariances. Clearly, the Euclidean distance of many feature vectors originating from class 1 to the prototype of the second group is less than the distance to the first prototype. They are nonetheless *more likely to have originated from the first distribution*. It is readily apparent that a considerable number of misclassifications result even on the training data (the points additionally circled in green).

Apart from the Euclidean definition, there are a number of possible definitions of distance or *metric* within a feature space. A sensible adjustment we can, then, is to *reconsider our definition of distance in the feature space* to give a probabilistic measure. Referring to Table 11.2, we draw attention here to the *Mahalanobis distance*. For a feature vector $\mathbf{x} = [x_1, x_2 \cdots, x_N]$ belonging to a class with mean vector μ and covariance Σ, this

---

Example 11.2

Matlab Example 11.2	**What is happening?**
clear; NOPTS=200;	
mu1 = [1 -1]; cov1 = [.9 .4; .4 .3];	%Define mean and covariance
d1 = mvnrnd(mu1, cov1, NOPTS);	%Training data for class 1
plot(d1(:,1),d1(:,2),'b.'); hold on;	%Plot class 1 training data
proto1=mean(d1); plot(proto1(1),proto1(2),'ks');	%Calculate & plot prototype 1
[mu2 = [-1 1]; cov2 = [2 0; 0 2];	%Define mean and covariance
d2 = mvnrnd(mu2, cov2, NOPTS);	%Training data for class 1
plot(d2(:,1),d2(:,2),'r.');	%Plot class 2 training data
proto2=mean(d2); plot(proto2(1),	%Calculate & plot prototype 2
proto2(2),'ks');	
axis([-6 4 -5 5]); axis square;	
for i=1:NOPTS	
if norm(d2(i,:)-proto2,2) > norm(d2(i,:)-proto1,2)	%Misclassified points
plot(d2(i,1),d2(i,2),'go');	%Plot in green circle
end	
end	

Comments

Matlab functions ***mvnrnd, norm***

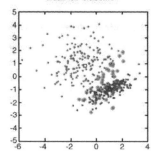

Figure 11.7   Feature vectors which are statistically more likely to have originated from one class may lie closer to the prototype of another class. This is a major weakness of a simple Euclidean minimum distance classifier *(See colour plate section for colour version)*

is defined as

$$D_{\mathrm{mh}} = (\mathbf{x}-\boldsymbol{\mu})^{\mathrm{T}}\boldsymbol{\Sigma}^{-1}(\mathbf{x}-\boldsymbol{\mu}) \qquad (11.7)$$

and the covariance matrix Σ in Equation (11.7) is defined by:

$$\Sigma = \langle(\mathbf{x}-\boldsymbol{\mu})(\mathbf{x}-\boldsymbol{\mu})^{\mathrm{T}}\rangle = \begin{bmatrix} \sigma_1^2 & \sigma_{12} & \cdots & \sigma_{1N} \\ \sigma_{21} & \sigma_2^2 & & \sigma_{2N} \\ \vdots & & \ddots & \vdots \\ \sigma_{N1} & \sigma_{N2} & \cdots & \sigma_N^2 \end{bmatrix} \qquad (11.8)$$

with $\sigma_{ij} = \langle(x_i-\mu_i)(x_j-\mu_j)\rangle$ and $\langle\ \rangle$ denotes the expectation or ensemble average over the sample.

The Mahalanobis distance effectively scales the absolute distances in the feature space by their corresponding standard deviations and, thus, provides an intuitively more 'probabilistic' measure of distance. In the following section, we develop the basic ideas behind classifiers that attempt to make full use of any *information we might possess about the statistical distribution of the feature vectors* within the space. This is the basic idea behind *Bayesian classifiers*.

11.9   Bayesian classification: definitions

The Bayesian approach to optimizing classifier performance is probabilistic in nature. From an operational perspective, a good classifier should obviously maximize the number of correct classifications and minimize the number of incorrect classifications over a definitive number of trials. A little thought, however, will reveal that these are to some degree conflicting requirements[6] and cannot be used as a design principle. Bayesian classifiers are actually designed to assign each feature vector to the *most probable class*. It is probably worth saying at the outset that the Bayesian approach corresponds strictly to a rather idealized situation (in which all relevant probabilities are known) and these conditions are not often met in practice. Despite this, it forms an elegant framework, serving as both an adequate approximation to many problems and as an excellent point of reference for practical classifiers.

We begin by outlining some general definitions and notation. We will assume the following:

- The patterns in whose classification we are interested may be represented as N-dimensional feature vectors $\mathbf{x} = [x_1, x_2, \cdots, x_N]$ whose components $x_1, x_2, \cdots, x_N$ are the measurement parameters which describe the feature space.[7]

- There are a total of C pattern classes, $\omega_1, \omega_2 \cdots \omega_C$ to which a given feature vector $\mathbf{x}$ may be assigned.

- The probability of drawing a feature vector from each of the classes is known and described by the probabilities $p(\omega_1), p(\omega_2), \cdots, p(\omega_C)$. These are known as the *priors*.

- There exists a set of M such examples of feature vectors $\{\mathbf{x}_1, \mathbf{x}_2, \cdots, \mathbf{x}_M\}$ whose corresponding pattern classes are known. These examples form the *design* or *training* set.

11.10   The Bayes decision rule

Let the conditional probability $p(\omega_j|\mathbf{x})$ represent the probability density of belonging to class ω_j given the occurrence of the feature vector $\mathbf{x}$. Note that, in general, $p(\omega_j|\mathbf{x})$ should be interpreted as a probability density (rather than a strict probability), since the feature vector $\mathbf{x}$ is a continuous variable and, thus, $p(\omega_j|\mathbf{x})$ is a function conditional upon $\mathbf{x}$. The Bayes decision rule is an intuitive rule based on probabilities:

$$\text{if} \quad p(\omega_j|\mathbf{x}) > p(\omega_k|\mathbf{x}) \quad \text{for all } k \neq j$$

$$\text{then} \quad \text{assign } \mathbf{x} \text{ to } \omega_j$$

(11.9)

---

[6] For example, if I wish to ensure that patterns from class A are always correctly classified then I can ensure this most easily by *always* assigning to class A whatever the pattern. The consequences for patterns belonging to other classes are obvious.

[7] Naturally, the measurement parameters should correspond directly or indirectly to those specific features thought to be important for classification.

In other words, we simply assign **x** to the most probable class. This decision rule, however, begs the question of how to find or estimate the probabilities $p(\omega_j|\mathbf{x})$ in Equation (11.9). To resolve this, we invoke Bayes' theorem, which for conditional probabilities extends naturally to the notion of probability densities and says

$$p(\omega_j|\mathbf{x}) = \frac{p(\mathbf{x}|\omega_j)p(\omega_j)}{p(\mathbf{x})} \tag{11.10}$$

On substitution into Equation (11.9), our decision rule then becomes

$$
\begin{aligned}
&\text{if} \quad p(\mathbf{x}|\omega_j)p(\omega_j) > p(\mathbf{x}|\omega_k)p(\omega_k) \quad \text{for all } k \neq j \\
&\text{then} \quad \text{assign } \mathbf{x} \text{ to } \omega_j
\end{aligned}
\tag{11.11}
$$

It is important to interpret the terms occurring in Equations (11.10) and (11.11) clearly:

- $p(\mathbf{x}|\omega_j)$ is known as the *class-conditional probability density function*. The term $p(\mathbf{x}|\omega_j)$ may be interpreted as 'the likelihood of the vector **x** occurring when the feature is known to belong to class ω_j.

- The functions $p(\omega_j)$ are known as the *prior probabilities*. Their values are chosen to reflect the fact that not all classes are equally likely to occur. Clearly, the values chosen for the $p(\omega_j)$ should reflect one's knowledge of the situation under which classification is being attempted.

- $p(\mathbf{x})$ is sometimes known as the 'evidence' and can be expressed using the partition theorem or 'law of total probability' as $p(\mathbf{x}) = \sum_j p(\mathbf{x}|\omega_j)p(\omega_j)$. Note that $p(\mathbf{x})$ is independent of the class label ω_j and does not enter into the decision rule described by Equation (11.11).

The key thing to note here is that application of the Bayes decision rule given by Equation (11.11) requires *knowledge of all the class-conditional, probability density functions and the priors.*

It is rare that we will know the functional forms and parameters of these density functions exactly. The approach typically taken in practice is to assume (or estimate) suitable functional forms for the distributions $p(\mathbf{x}|\omega_j)$ and $p(\omega_j)$ for all values of j and estimate their parameters from our training data. We then attempt to apply the rule given by Equation (11.11).

In certain instances, the form of the distribution may actually be known or confidently predicted. By far the most common is the multivariate normal (MVN) distribution. The main reason for this is its analytical tractability and simple properties (other distributions in N dimensions are *much* harder to deal with). However, real-life situations in which the MVN distribution is a fairly good approximation are fortunately quite common.

Before we deal with the specific cases of Bayesian classifiers in which the underlying distributions can be well approximated by the MVN, we will, therefore, dedicate the next section to reviewing its properties.

11.11 The multivariate normal density

The one-dimensional Gaussian or normal density for the random variable x is given by

$$p(x) = \frac{1}{\sqrt{2\pi}\sigma} \exp\left[-\frac{1}{2}\left(\frac{x-\mu}{\sigma}\right)^2\right] \tag{11.12}$$

where μ is the mean of the distribution and σ is the standard deviation. The 1-D normal density is thus defined by just two parameters. For multivariate data $x_1, x_2 \cdots x_N$, represented as a column vector $\mathbf{x} = [x_1, x_2 \cdots x_N]^\mathrm{T}$, the normal distribution has the more general form

$$p(\mathbf{x}) = \frac{1}{(2\pi)N/2|\mathbf{C}|^{1/2}} \exp\left[-\frac{1}{2}(\mathbf{x}-\vec{\mu}_x)^\mathrm{T}\mathbf{C}_x^{-1}(\mathbf{x}-\vec{\mu}_x)\right] \tag{11.13}$$

where $\vec{\mu}_x = E(\mathbf{x})$ is the expectation or average value of $\mathbf{x}$ and $\mathbf{C}_x = E[(\mathbf{x}-\vec{\mu}_x)(\mathbf{x}-\vec{\mu}_x)^\mathrm{T}]$ is the covariance matrix.

The covariance matrix contains the covariance between every pair of variables in the vector $\mathbf{x}$. Thus, the general, *off-diagonal* element is given by

$$\mathbf{C}_x(i,j) = E[(x_i-\mu_i)(x_j-\mu_j)] \qquad i \neq j \tag{11.14}$$

where $\mu_k = E(x_k)$ and the diagonal elements contain the variances

$$\mathbf{C}_x(i,i) = E[(x_i-\mu_i)^2] \qquad i = 1, 2, \cdots, N \tag{11.15}$$

By definition, the covariance matrix is a symmetric matrix and there are $N(N+1)/2$ free parameters. Combining these with the N free parameters for the average vector $\vec{\mu}_x = E(\mathbf{x})$, a total of $N(N+3)/2$ parameters are required to completely specify the MVN density.

We note in passing an important feature of the MVN density. The quadratic form which appears in the exponential is actually the (squared) Mahalanobis distance which we mentioned in Section 11.8:

$$L^2 = (\mathbf{x}-\vec{\mu}_x)^\mathrm{T}\mathbf{C}_x^{-1}(\mathbf{x}-\vec{\mu}_x) \tag{11.16}$$

Setting this term to a constant value, we obtain contours of constant probability density. It is possible to show that this quadratic form defines an hyper-ellipsoid, which reduces to an ellipsoid in three dimensions and to an ellipse in two dimensions and, thus, determines the orientation and spread of the distribution of points within the feature space.

11.12 Bayesian classifiers for multivariate normal distributions

For an arbitrary class ω_j with features $\mathbf{x}$ distributed as an MVN distribution having mean $E(\mathbf{x}) = \vec{\mu}_i$ and covariance matrix $\mathbf{C}_j$, the class-conditional density is given by

$$p(\mathbf{x}|\omega_j) = \frac{1}{(2\pi)N/2|\mathbf{C}|^{1/2}} \exp\left[-\frac{1}{2}(\mathbf{x}-\vec{\mu}_i)^{\mathrm{T}}\mathbf{C}_j^{-1}(\mathbf{x}-\vec{\mu}_i)\right] \qquad (11.17)$$

The Bayes decision rule states that

$$\text{if} \quad p(\mathbf{x}|\omega_j)p(\omega_j) > p(\mathbf{x}|\omega_k)p(\omega_k) \quad \text{for all } k \neq j$$
$$\text{then} \quad \text{assign } \mathbf{x} \text{ to } \omega_j \qquad (11.18)$$

Taking logarithms on both sides of the Bayes decision rule and noting that log is a monotonically increasing function, we obtain the equivalent rule

$$\text{if} \quad g_j(\mathbf{x}) = \log p(\mathbf{x}|\omega_j) + \log p(\omega_j) > g_k(\mathbf{x}) = \log p(\mathbf{x}|\omega_k) + \log p(\omega_k) \quad \text{for all } k \neq j$$
$$\text{then} \quad \text{assign } \mathbf{x} \text{ to } \omega_j$$
$$(11.19)$$

Substituting the normal form for the class-conditional density function, it is easy to show that the discriminant functions take the form

$$g_j(\mathbf{x}) = \log p(\mathbf{x}|\omega_j) + \log p(\omega_j)$$
$$= -\frac{1}{2}(\mathbf{x}-\vec{\mu}_x)^{\mathrm{T}}\mathbf{C}_j^{-1}(\mathbf{x}-\vec{\mu}_x) - \frac{N}{2}\log 2\pi - \frac{1}{2}\log|\mathbf{C}_j| + \log p(\omega_j) \qquad (11.20)$$

These discriminant functions take simpler forms when the covariance matrix $\mathbf{C}_j$ possesses certain properties. First note, however, that the term $(N/2)\log 2\pi$ is the same for all classes. Since our task is to compare discriminants of the form given by Equation (11.20), it can thus be removed from Equation (11.20).

Case I: $\mathbf{C}_i = \sigma^2\mathbf{I}$ If we have a situation in which the features in $\mathbf{x}$ are statistically independent and all have the same variance, then the discriminant for MVN densities reduces to the form

$$g_j(\mathbf{x}) = \log p(\mathbf{x}|\omega_j) + \log p(\omega_j)$$
$$= -\frac{\|\mathbf{x}-\vec{\mu}_x\|^2}{2\sigma^2} + \log p(\omega_j) \qquad (11.21)$$

Expanding the term $\|\mathbf{x}-\vec{\mu}_x\|^2 = (\mathbf{x}-\vec{\mu}_x)^{\mathrm{T}}(\mathbf{x}-\vec{\mu}_x)$ and noting that $\mathbf{x}^{\mathrm{T}}\mathbf{x}$ is the same for all classes and can thus be ignored, we thereby obtain a *linear discriminant* of the standard form

$$g_j(\mathbf{x}) = \mathbf{w}_j^{\mathrm{T}}\mathbf{x} + v_j \qquad (11.22)$$

where

$$\mathbf{w}_j = \frac{\vec{\mu}_j}{\sigma^2} \quad \text{and} \quad v_j = -\frac{1}{2\sigma^2}\vec{\mu}_j^{\mathrm{T}}\vec{\mu}_j + \log p(\omega_j)$$

A classifier that uses linear discriminant functions is called a linear machine. The decision surfaces for a linear machine are thus determined by the intersection of hyperplanes as described by Equation (11.22). Note from Equation (11.22) that, in the case that all classes have the same a priori likelihood of occurring (i.e. $p(\omega_j)$ is the same for all classes), the expression *reduces to the minimum distance criterion* discussed in earlier sections.

Case II: $\mathbf{C}_j = \mathbf{C}$ When the covariance matrices may be assumed to be the same for all classes (i.e. $\mathbf{C}_j = \mathbf{C}$ for all j), we again substitute Equation (11.20) into Equation (11.19), simplify and obtain a linear discriminant:

$$g_j(\mathbf{x}) = \mathbf{w}_j^{\mathrm{T}}\mathbf{x} + v_j \tag{11.23}$$

where this time

$$\mathbf{w}_j = \mathbf{C}^{-1}\vec{\mu}_j \quad \text{and} \quad v_j = -\frac{1}{2}\vec{\mu}_j^{\mathrm{T}}\mathbf{C}\vec{\mu}_j + \log p(\omega_j).$$

In Example 11.3 and the resulting Figure 11.8, we compare the results of these classifiers on some test data for the two-class problem we presented in Section 11.8.

Example 11.3

Matlab Example 11.3	What is happening?
NOPTS=200;	
mu1 = [1 -1]; cov1 = [.9 .4; .4 .3];	%Define mean and covariance
d1 = mvnrnd(mu1, cov1, NOPTS);	%Training data for class 1
subplot(2,2,1), plot(d1(:,1),d1(:,2),'b.'); hold on;	%Plot class 1 training data
proto1=mean(d1); plot(proto1(1),proto1(2),'ks');	%Calculate and plot prototype 1
mu2 = [-1 1]; cov2 = [2 0; 0 2];	%Define mean and covariance
d2 = mvnrnd(mu2, cov2, NOPTS);	%Training data for class 1
plot(d2(:,1),d2(:,2),'r.');	%Plot class 2 training data
proto2=mean(d2); plot(proto2(1),proto2(2),'ks');	%Calculate and plot prototype 2
axis([-6 4 -5 5]); axis square;	
title('TRAINING DATA')	
group=[ones(NOPTS,1); 2.*ones(NOPTS,1)];	%vector to specify actual classes of %training points
%for i=1:NOPTS	
% if norm(d2(i,:)-proto2,2) > norm(d2(i,:)-proto1,2)	%Find misclassified points
% plot(d2(i,1),d2(i,2),'go');	%Plot on top in green circles
% end	
%end	
	%%%%%%%%%%%%%%%%%%%
	%CLASS = CLASSIFY(SAMPLE,
	TRAINING,GROUP,TYPE)

```
                                              %generate TEST DATA

N1=50; N2=100;                                %Number of test points from classes
testgroup=[ones(N1,1); 2.*ones(N2,1)];       %vector to specify actual class of
                                              %test points

sample=[mvnrnd(mu1, cov1, N1);
mvnrnd(mu2, cov2, N2)];                       %generate test data points


subplot(2,2,2),
plot(proto1(1),proto1(2),'ks');
axis([-6 4 -5 5]); axis square; hold on;
plot(proto2(1),proto2(2),'ks');
plot(sample(1:N1,1),sample(1:N1,2),'b*');
plot(sample(N1+1:N1+N2,1),sample(N1+1:N1+N2,2),'r*');
title('TEST DATA');


[class,err]=classify(sample,[d1;d2],group,
    'DiagLinear',[0.9 0.1]);                  %Classify using diagonal
                                              %covariance estimate

subplot(2,2,3),
plot(proto1(1),proto1(2),'ks');
axis([-6 4 -5 5]); axis square; hold on;
plot(proto2(1),proto2(2),'ks');
plot(sample(1:N1,1),sample(1:N1,2),'b*');
plot(sample(N1+1:N1+N2,1),sample(N1+1:N1+N2,2),'r*');


for i=1:size(class,1)
   if class(i)~=testgroup(i)
      plot(sample(i,1),sample(i,2),'go');
   end
end


[class,err]=classify(sample,[d1;d2],group,'DiagLinear',[N1./(N1+N2) N2./(N1+N2)]);
   %Classify using diagonal covariance estimate
subplot(2,2,4),
plot(proto1(1),proto1(2),'ks');
axis([-6 4 -5 5]); axis square; hold on;
plot(proto2(1),proto2(2),'ks');
plot(sample(1:N1,1),sample(1:N1,2),'b*');
plot(sample(N1+1:N1+N2,1),sample(N1+1:N1+N2,2),'r*');


for i=1:size(class,1)
   if class(i)~=testgroup(i)
      plot(sample(i,1),sample(i,2),'go');
   end
end
```

Comments

Matlab functions ***mvnrnd, norm:*** mvnrnd is a Matlab function that generates arrays of normally distributed random variables

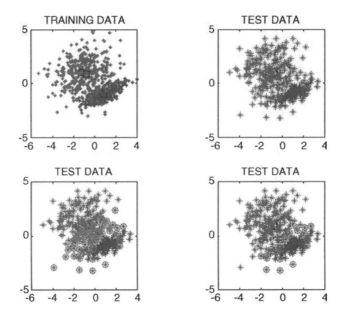

Figure 11.8 A comparison of Bayesian classifiers under different assumptions about the covariance matrix and prior distributions. The class-conditional densities of the training data are both normal. (a) Training data: 200 samples from each class (class 1: blue; class 2: red). (b) Test data: 50 originate from class 1, 100 originate from class 2 (indicated in red). (c) Classification using a Bayesian classifier with a diagonal estimate of the covariance and an erroneous prior distribution of $p(\omega_1) = 0.9$ and $p(\omega_2) = 0.1$. Misclassified test points are circled in green. (d) Classification using a Bayesian classifier with a diagonal estimate of the covariance and correct prior distribution of $p(\omega_1) = 1/3$ and $p(\omega_2) = 2/3$. Misclassified test points are circled in green *(See colour plate section for colour version)*

11.12.1 The Fisher linear discriminant

For simplicity, let us initially consider the case of a two-class problem in a feature space where the training data are described by feature vectors $\{x_i\}$. The basic aim of the Fisher linear discriminant (FLD) is to find that axis from among all the possible axes we could choose within the feature space such that projection of the training data onto it will ensure that the classes are *maximally separated*. The projection of the data $\{x_i\}$ is given by some linear combination $w^T x$ with w to be determined. This basic idea is demonstrated in Figure 11.9 for a 2-D feature space. Note, however, that the concept extends naturally to three dimensions and higher, since we can always project data of an arbitrary dimensionality orthogonally onto a line.

The formal calculation of the FLD relies on maximizing the criterion function:

$$J(w) = \frac{|m_A - m_B|^2}{s_A^2 + s_B^2}$$

where m_A and m_B are the means of the projected data and s_A^2 and s_B^2 are the sample variances of the projected data. Note how this criterion tries to maximize the separation of the projected means but scales them by the total variance of the data.

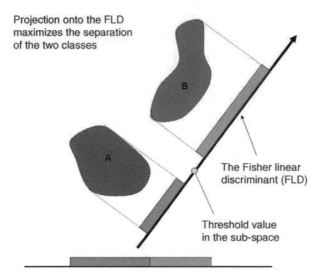

Projection onto the FLD
maximizes the separation
of the two classes

The Fisher linear
discriminant (FLD)

Threshold value
in the sub-space

Figure 11.9   The FLD defines a direction within the feature space that maximizes the separation of the projected feature vectors in each class. This method can achieve a substantial reduction in the dimensionality of the problem (from N dimensions to 1). However, there is, of course, no guarantee that the data will be separable when projected onto a single direction

It is possible to show[8] that the solution is

$$\mathbf{w} = \mathbf{S}_w^{-1}[\mathbf{M}_A^x - \mathbf{M}_B^x]$$

where $\mathbf{M}_A^x$ and $\mathbf{M}_B^x$ are the vectors giving the means of the original (unprojected) data and the total scatter matrix $\mathbf{S}_w$ is given by the sum of the individual scatter matrices $\mathbf{S}_w = \mathbf{S}_A + \mathbf{S}_B$ which, in general, are of the form

$$\mathbf{S}_K = \sum_{\substack{\text{all } \mathbf{X} \\ \text{in class } K}} (\mathbf{x} - \mathbf{M}_K^x)(\mathbf{x} - \mathbf{M}_K^x)^{\mathrm{T}}$$

Once the optimal $\mathbf{w}$ has been found, the classification is then simply a matter of deciding the threshold value along the axis (subspace) defined by the discriminant.

The FLD for the two-class problem reduces the N-dimensional feature space to a single dimension. Whilst such an approach naturally sacrifices the best possible performance for a classifier, this can be acceptable if the advantages of working in just one dimension compensate.

11.12.2   Risk and cost functions

It is self-evident that a basic aim of classification is to assign the maximum number of feature vectors to their correct class whilst making as few misclassifications as possible.

---

[8] See, for example, Duda *et al.* (2001), *Pattern Classification*, John Wiley & Sons, Inc., pp. 117–124.

However, we should recognize that real-life situations are invariably more subtle than this. Depending on the specific situation, *the relative importance of misclassifications can vary significantly*. For instance, a significant amount of research has been devoted to the development of image processing software for automatic breast-screening. Regular screening of the female population was introduced because of the much improved prognosis which generally results if breast cancer can be detected in its early stages. The huge workload involved in manual inspection of these mammograms has motivated the search for reliable, automated methods of diagnosis. The purpose of the automated analysis is essentially to classify to the classes of either 'normal' (in which case no further action is taken) or 'abnormal' (which generally results in further more careful inspection by a radiologist). Abnormal scans are not always the result of abnormal pathology, but because the consequences of classifying what is actually an abnormal mammogram as 'normal' (i.e. missing a pathological condition) can be very serious, the classification tends towards a 'play-safe strategy'. In other words, we associate a *greater cost* or *risk* with a misclassification in one direction. The cost of a *false-negative* event (in which an abnormal scan is assigned to the normal class) is considerably greater than the other kind of misclassification (the *false positive* event, in which a normal scan is assigned to the abnormal class). A reverse philosophy operates in courts of law, hence the old saying 'Better one hundred guilty men go free than one innocent man be convicted'.

By contrast, some financial organizations are beginning to employ sophisticated software to assess the normality of their customers spending patterns, the idea being that unusual and/or excessive spending is a likely indicator of fraudulent use of a card. The prevailing philosophy tends towards the idea that it is better to accept a certain modest amount of fraudulent use than to generate many false-alarms in which the legitimate owner of the card is refused the use of their card.

It is outside the scope of this text to discuss risk or cost function analysis in image classification problems. The theory has been extensively developed and the interested reader can refer to the literature.[9]

11.13   Ensemble classifiers

Ensemble classifiers are classification systems that combine a set of component classifiers in the attempt to achieve an overall performance that is better than any of the individual, components. The principle of consulting several 'experts' before making a final decision is familiar to us in our everyday lives. Before we decide to purchase a particular model of car, undergo a medical procedure or employ a particular job candidate, we typically solicit opinions from a number of sources. Combining the information from our various sources, and weighting them according to our innate level of confidence in each, we generally feel more confident about the final decision we reach.

Automated ensemble classifiers try to achieve a similar goal according to strict mathematical (or at least sound, empirical) principles. A detailed discussion of ensemble classifiers is beyond the scope of this book and we will only attempt here to describe one such ensemble

---

[9]A good starting place is Pattern classification, Duda et al, Wiley 2001

classifier, namely the AdaBoost method, which is rapidly emerging as the most important. Central to this method is the requirement that the component classifiers be as *diverse* as possible. It stands to reason that if we had a single classifier that had perfect generalization to new examples, then we would have no need at all to consider ensemble systems. It is thus implicit that the component classifiers in an ensemble system are imperfect and make misclassifications. Intuitively, the hope would be that each imperfect component mis-classifies different examples and that the combination of the components would reduce the total error. The notion of diversity encapsulates the idea that each classifier be as unique as possible.

11.13.1   *Combining weak classifiers: the AdaBoost method*

A so-called 'weak classifier' is basically a classifier that does not perform well. Indeed, a weak classifier may perform only slightly better than chance.

Consider a set of n feature vectors to constitute our training sample. A subset of these n vectors is first randomly selected to form the training sample for the first component classifier. The probability of random selection for each vector is determined by its associated weight. After the first component classifier has been trained on the data, we increase the weights of those vectors which have been incorrectly classified but decrease the weights of those vectors which have been correctly classified. In this way, we increase the probability that the next component classifier will be trained on more 'troublesome/difficult' feature vectors. The process is then repeated for the second (and subsequent) component classifiers, adjusting the weights associated with each feature vector. Thus, in general, on the kth iteration we train the kth component classifier.

Provided each component classifier performs better than chance (i.e. has an error of less than 0.5 on a two-class problem), it is possible to show that AdaBoost can achieve an arbitrarily small misclassification error. Moreover, there is no restriction placed on the nature of the component classifiers used; thus, the AdaBoost method is generic in nature. This powerful result is not, however, an instant panacea. The choice of component classifiers which satisfy the 'better than chance' requirement cannot be guaranteed beforehand. Even so, AdaBoost performs well in many real problems and has been described as 'the best off-the-shelf classifier in the world'.

11.14   Unsupervised learning: k-means clustering

Unsupervised classifiers do not have the benefit of class-labelled examples to train on. Rather, they simply explore the data and search for naturally occurring patterns or clusters within it. Once these clusters have been found, we may then construct decision boundaries to classify unseen data using broadly similar methods to those used for supervised classifiers. One of the simplest unsupervised learning algorithms is the k-means algorithm.

This procedure, summarized in Figure 11.10 (the figure is generated using the Matlab® code in Example 11.4), outlines a conceptually simple way to partition a data set into a specified number of clusters k. The algorithm aims to iteratively minimize a simple

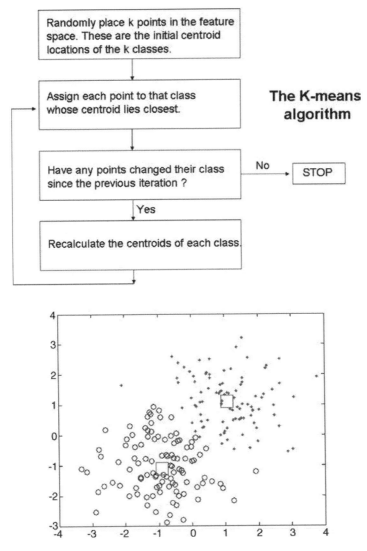

Figure 11.10 The k means algorithm assigns the data to k classes such that the sum of the squared distances from each point to the centroid of the assigned class is as small as possible. This example is for k=2 classes (see example 11.4)

squared error objective function of the form

$$J = \sum_{j=1}^{k} \sum_{\substack{\text{all } i \\ \text{in class } j}} |x_i^j - c_j|^2,$$

where c_k denotes the coordinate vector of the jth cluster and $\{x_i^j\}$ are the points assigned to the jth cluster. Minimizing J equivalently means reaching that configuration at which switching any point to a cluster other than its currently assigned one will only *increase* the

Example 11.4

Matlab Example 11.4	**%What is happening?**
X = [randn(100,2)+ones(100,2); randn(100,2)-ones(100,2)];	%Generate random data array
opts = statset('Display','final');	%Specify statistics options structure
[cidx, ctrs] = kmeans(X, 2, 'Distance','city','Replicates',5, 'Options',opts);	%Run 5 repeats, city-block metric
plot(X(cidx==1,1),X(cidx==1,2),'r.');	%Plot 1st class points ...
hold on; plot(X(cidx==2,1),X(cidx==2,2),'ko');	%Plot 2nd class points ...
plot(ctrs(:,1),ctrs(:,2),'rs','MarkerSize',18);	%plot class centroids as squares

objective function. It is possible to show that the procedure outlined in Figure 11.10 generally achieves this goal, although it is somewhat sensitive to the initial choice of centroid locations. Typically, the k-means algorithm is run multiple times with different starting configurations to reduce this effect.

Arguably, the central weakness of the k-means algorithm is the need to fix the number of clusters 'a priori'. Clearly, increasing the number of clusters will guarantee that the minimum value of the objective function reduces monotonically (ultimately, the objective function will have value zero when there as many clusters as points), so that the algorithm is not inherently sensitive to 'natural clusters' in the data. This last problem is troublesome, since we often have no way of knowing how many clusters exist.

Unfortunately, there is no general theoretical solution to find the optimal number of clusters for any given data set. A simple approach is to compare the results of multiple runs with different values of k and choose the best one according to a given criterion; but we need to exercise care, because increasing k increases the risk of overfitting.

For further examples and exercises see http://www.fundipbook.com.

Further reading

For further more specialist reading on some of the topics covered within this text the authors recommend the following specific texts:

Image processing (general)

- *Digital Image Processing*: Gonzalez, R. & Woods R. (Prentice Hall, 2002)

- *Image Processing & Computer Vision*: Morris, T. (Palgrave Macmillan, 2004)

Image processing (in other programming environments/languages)

- *Learning OpenCV*: Bradski, G. & Kaehler, A. (O'Reilly, 2008) *(C/C++)*

- *Digital Image Processing-An Algorithmic Introduction using Java*: Burger, W & Burge, M. J. (Springer, 2008) *(Java)*

Computer vision

- *Computer Vision – A Modern Approach*: Forsyth, D. & Ponce, J. (Prentice-Hall, 2003)

- *Machine Vision – Theory, Applications & Practicalities*: Davies, E.R. (Morgan-Kaufmann, 2005)

Machine learning and classification

- *Machine Learning*: Mitchell, T. (McGraw-Hill, 1997)

- *Pattern Recognition & Machine Learning*: Bishop, C. (Springer, 2006)

For further examples and exercises see http://www.fundipbook.com

Fundamentals of Digital Image Processing – A Practical Approach with Examples in Matlab
Chris Solomon and Toby Breckon
© 2011 John Wiley & Sons, Ltd

Index

Printed and bound by CPI Group (UK) Ltd, Croydon, CR0 4YY